Timothy Rich
30/9/92

ATLAS OF THE BRYOPHYTES OF BRITAIN AND IRELAND

2. MOSSES
(EXCEPT DIPLOLEPIDEAE)

ATLAS OF THE
BRYOPHYTES
OF
BRITAIN AND IRELAND

VOLUME 2. MOSSES
(EXCEPT DIPLOLEPIDEAE)

EDITED BY

M. O. HILL
Institute of Terrestrial Ecology

C. D. PRESTON
Institute of Terrestrial Ecology

AND

A. J. E. SMITH
University of Wales Bangor

HARLEY
BOOKS

1992

Published by Harley Books
(B. H. & A. Harley Ltd.)
Martins, Great Horkesley,
Colchester, Essex, CO6 4AH, England
for
the British Bryological Society
c/o Department of Botany,
National Museum of Wales,
Cardiff, CF1 3NF

with the support of
The Natural Environment Research Council
and
The Nature Conservancy Council

Designed by Geoff Green
Text set in Ehrhardt by Saxon Printing Ltd, Derby
Printed in Great Britain by St Edmundsbury Press Ltd
Bury St Edmunds, Suffolk

British Library Cataloguing-in-Publication Data
1. Mosses
I. Hill, M. O. (Mark O.) *1945–* II. Preston, C. D. (Christopher David) *1955–*
III. Smith, A. J. E. (Anthony John Edwin) *1935–*
Atlas of the Bryophytes of Britain and Ireland
Volume 2. *Mosses (except Diplolepideae)*
588.0941

ISBN *0 946589 30 5*

CONTENTS

PREFACE

This is the second of a series of three volumes of dot-distribution maps, showing the occurrence of bryophytes in 10-km squares of Britain and Ireland. The first volume (Hill *et al.*, 1991) dealt with the liverworts; the present volume covers about half of the mosses.

The mosses mapped in this volume are those in the isolated classes Sphagnopsida (which includes only one genus, *Sphagnum*) and Andreaeopsida (of which only *Andreaea* occurs in Europe), as well as nine orders of the Bryopsida (the class which includes most of the world's mosses). Most species in the Bryopsida fall into one of two groups: the Haplolepideae and the Diplolepideae. These differ in the details of their peristome development and structure; copiously illustrated explanations are provided by Edwards (1979, 1984). Although the exact nature of the differences is rather technical, members of the Haplolepideae tend to have a single row of peristome teeth, whereas species in the Diplolepideae usually have a robust outer row and a fragile inner row of teeth round the mouth of the capsule. The orders of Bryopsida mapped in this volume are those in the Haplolepideae (Dicranales, Fissidentales, Pottiales and Grimmiales) and those orders, most of them small, which cannot be classified in either group (Tetraphidales, Polytrichales, Buxbaumiales, Archidiales and Encalyptales). The Diplolepideae will be mapped in the final volume.

An account of the British Bryological Society's Mapping Scheme, which was started in 1960, is given in Volume 1. Each volume contains an introductory article on a related subject. In Volume 1, Chris Preston (who is also an editor) wrote about the history of bryophyte recording in the British Isles. In Volume 2, we have pleasure in welcoming Alan Crundwell as a guest to our pages. It is particularly fitting that he should write about British and Irish bryophytes in their European context, for his knowledge of continental bryophytes, and indeed of continental bryologists, is unrivalled.

In the notes on each species, we have attempted to give an up-to-date account not only of the habitat but also of the means by which the species reproduces and disperses. Mosses can regenerate from rhizoid fragments, protonema-gemmae and rhizoidal tubers, as well as from more obvious structures such as bulbils, gemmae, shoots, leaves and spores. New reports of tubers and gemmae occurring in species from which they were previously unknown are published every year. Indeed, for most species, bryologists still have little idea of the relative importance of different dispersal mechanisms. We hope that this atlas will encourage some readers to think about dispersal, as well as about habitat and distribution.

M. O. HILL

Monks Wood, *May, 1992*   C. D. PRESTON

7

ACKNOWLEDGEMENTS

This atlas is based on the fieldwork of members of the British Bryological Society; we thank them for their dedication and enthusiasm. The financial support of the Natural Environment Research Council and the former Nature Conservancy Council has enabled us to turn the results of this work into a published atlas. We thank all those who have helped to prepare the second volume for publication, by data processing, checking draft maps and writing notes on the species. Detailed acknowledgements, including a list of recorders, will be published in the final volume. Publication of the atlas has been assisted by grants from the Linnean Society of London and the Royal Society.

THE BRYOPHYTES OF BRITAIN AND IRELAND IN A EUROPEAN CONTEXT

A. C. CRUNDWELL

The bryophyte flora of the British Isles is probably as well known as that of any area of comparable size in the world. Distribution data have been collected systematically for nearly a century and records have been assembled on the present grid system since 1960. Nevertheless, species are added to the British list nearly every year and much remains to be done. Few grid squares have been mapped exhaustively, and we all make misidentifications from time to time. Anyone expecting maps in this atlas to reveal 'the truth, the whole truth and nothing but the truth' will be disappointed. I guess – and one can only guess – that our present 10-km square records are a little over two-thirds of those we would have were our knowledge complete and that perhaps one in thirty would have to be deleted as erroneous. We can, however, say with confidence that the maps do present a reliable, though necessarily incomplete, picture of the distribution of bryophytes in the British Isles. Only in parts of Ireland is there serious difficulty in telling whether the apparently patchy distribution of a species is real or is the result of inadequate fieldwork.

When studying the maps, the reader unfamiliar with British geography should remember that in England and Wales, and to a lesser extent in Scotland, the mountains are in the western half of the country; in Ireland they are absent from the middle (see Volume 1 of this atlas, p. 336). In England, the south-eastern half of the country is lowland, never reaching 300 m in altitude, and has very little exposure of hard rock, so that rupestral species that rarely grow on brickwork or stonework (e.g. *Cynodontium bruntonii* and many Grimmiaceae) are excluded from large regions which may be climatically quite suitable for them.

If the bryologist from continental Europe bears these limitations in mind, the distribution maps of most species will contain little to surprise him. *Brachythecium rutabulum* is generally distributed and *Ptychomitrium polyphyllum* is western. This article does not deal further with such species but is concerned with the minority of which the British distribution could not be predicted from a knowledge of distribution on the European continent.

SPECIES ABSENT FROM CONTINENTAL EUROPE

The species that occur in the British Isles but not in continental Europe (including those that are only doubtfully recorded on the Continent) fall into three groups. The first of these is Introduced Species (Table 1). Apart from *Atrichum crispum*, recorded

Table 1. Introduced British and Irish bryophytes not definitely known from continental Europe.

Species	Native range	Year of first British collection
Atrichum crispum	N. America	1848
Lophocolea semiteres	Australasia, S. Africa, S. America	1955
Telaranea murphyae	Unknown	1961
Lophocolea bispinosa	Australasia	1962
Hennediella macrophylla	New Zealand	1965
Eriopus apiculatus	Australasia, S. America	1967
Tortula amplexa	Western N. America	1973
Lophozia herzogiana	New Zealand	1986

Table 2. Bryophyte species endemic to Britain and Ireland; those marked with an asterisk are known only from their original localities.

Species	Date of description	Species	Date of description
Weissia mittenii	1850	*Ditrichum plumbicola*	1976
Oxystegus hibernicus	1867	*Anoectangium warburgii*	1978
*Thamnobryum angustifolium**	1886	*Plagiochila britannica*	1979
*Bryum lawersianum**	1899	*Barbula tomaculosa*	1981
B. dixonii	1901	*Brachythecium appleyardiae*	1981
*Tortella limosella**	1907	*Pictus scoticus**	1982
Cephaloziella nicholsonii	1913	*Pohlia scotica*	1982
Fissidens celticus	1965	*Bryoerythrophyllum caledonicum*	1982
Fossombronia fimbriata	1976	*Sphagnum skyense**	1988
Ditrichum cornubicum	1976		

from Belgium and Luxemburg but (according to Arts (1987a)) needing confirmation, all of these species were first found here in the last forty years. Earlier introductions from outside Europe, notably *Orthodontium lineare* and *Campylopus introflexus*, have already spread to the Continent. While it seems unlikely that any of the eight species in the table will become as common or as ecologically important as these, it is worth noting that four of them (*Eriopus*, *Telaranea* and both *Lophocolea* species), and perhaps also a fifth *(Hennediella)*, have each been introduced to the country twice. Some of these introductions have been to places where ornamental plants have been imported, and a search of suitable gardens and estates in southern and western Europe might prove rewarding.

Table 2 gives the nineteen British species at present regarded as endemic. It is a motley list and about the only thing that can be said of it is that in fifty years' time it will certainly look very different. Several species that were termed endemic in Smith's *Moss Flora* (1978) have now been found in other countries or continents (*Barbula mamillosa, B. maxima, Gymnostomum insigne, Weissia multicapsularis*). In fact, most plant species when first described have to be reckoned as endemic to single countries; later on, the vast majority are either found in other countries or reduced to synonymy. Only seven species in this list were described before 1965 and some of these are of doubtful taxonomic status. There are probably very few good species of bryophyte that are truly endemic to the British Isles.

Table 3 lists the native British species that are absent from continental Europe but which do have an extra-European distribution. The first five all belong to the Macaronesian element, with western distributions in the British Isles, presence in the Azores and sometimes also in Madeira, but typically absent from other parts of the world. Most members of the Macaronesian element, e.g. *Radula holtii* and *R. carringtonii*, are known from the Iberian peninsula and it would not be phyto-geographically surprising for any of these five species to be found there.

The last six species of Table 3 are a miscellaneous group of somewhat oceanic species, mainly tropical or American in distribution, all except the *Metzgeria* being rare in the British Isles; but the previous eight species together with the eleven in Table 4 constitute the oceanic-subalpine element, which must be dealt with at some length.

Table 3. Native non-endemic British and Irish bryophytes absent from continental Europe. The species marked with an asterisk occur in the Faeroes.

Species	Extra-European distribution
(a) Macaronesian element	
Acrobolbus wilsonii	Azores, Madeira
Lejeunea flava ssp. *moorei*	Azores, Madeira, Canaries, Cape Verde Islands
L. hibernica	Azores, Madeira, Canaries
Campylopus shawii	Azores, Caribbean Islands
Myurium hochstetteri	Azores, Madeira, Canaries
(b) Oceanic-subalpine element	
Bazzania pearsonii	Western N. America, Japan, China, Thailand, Himalaya, Sri Lanka
Mastigophora woodsii*	Western N. America, China, Himalaya
Plagiochila carringtonii*	Himalaya (ssp. *lobuchensis*)
Andreaea sinuosa	Western N. America
Dicranodontium subporodictyon	China, western N. America
Gymnostomum insigne	Western N. America
Leptodontium recurvifolium	Western N. America
Racomitrium himalayanum	China, Himalaya
(c) Miscellaneous species	
Adelanthus lindenbergianus	Tropical and southern Africa, C. and S. America, Antarctic
Metzgeria leptoneura*	Widespread in tropics and Southern Hemisphere
Radula voluta	Eastern N. America
Barbula maxima	Canada (N.W. Territories)
Bartramidula wilsonii	N., C. and S. America, China, Fernando Po
Didymodon reedii	U.S.A. (Washington and Colorado)

Table 4. Species of the oceanic-subalpine element present in continental Europe.

Species	Extra-European distribution
Anastrophyllum donnianum	Western N. America, China, Himalaya
A. joergensenii	China, Himalaya
Herbertus aduncus ssp. *hutchinsiae*	None
H. borealis	None. The closely related *H. himalayanus* is in western N. America and Himalaya
H. stramineus	Western N. America
Pleurozia purpurea	Western N. America, China, Himalaya
Scapania nimbosa	China, Himalaya
S. ornithopodioides	Western N. America, Hawaii, Japan, Philippines, N. Borneo, China, Himalaya
Sphenolobopsis pearsonii*	Eastern and western N. America, Japan, N. Borneo, China, Himalaya
Andreaea megistospora	Western N. America
Campylopus schwarzii	Western N. America, Japan, Korea, China, Himalaya

* This species has been recorded by Zhukova (1973) from Franz Josef Land and by Konstantinova (1978, 1985, 1987) from the Murmansk region of Arctic Russia. The associated species were – hardly surprisingly – all Arctic rather than oceanic, and the occurrence of this liverwort in a climate so unusual for it should not be accepted without further evidence.

The Oceanic-Subalpine Element

Although eleven of the nineteen species in this element are present on the European continent, their range there, except for the more widely distributed *Campylopus schwarzii*, consists solely of western Norway, where they are local or rare. There is little to suggest from their continental distribution that they are part of a particularly interesting geographical element.

In contrast, in some of the western parts of the British Isles most of the species are locally abundant, often growing together in a mixed hepatic mat, and, particularly when brightly coloured as in *Pleurozia*, *Herbertus* spp. and *Mastigophora*, conspicuous features of the vegetation, catching the eye of the non-bryologist and for the most part so plentiful that no conservationist who has walked over them need worry.

These nineteen species have a number of features in common:

(a) In Scotland, the only country which has all of them, they occupy the middle ranges of altitude. Almost all have their best development between 500 m and 900 m, though they all differ to some extent, some species occurring mainly at the top of this range or even above it, others descending to sea-level. *Andreaea sinuosa* is exceptional as it is confined to altitudes above 900 m. All are terrestrial or rupestral.

(b) They have very disjunct geographical distributions. All are in Scotland, most in Ireland and over half in Norway. Other places where several of these species occur are western North America (the Pacific coast of Alaska and British Columbia, especially the Queen Charlotte Islands), Japan, China (Yunnan and Taiwan) and the Himalayas. To the best of my knowledge there is no other group of organisms – plants or animals – with this pattern of distribution.

(c) All are dioecious with the exception of the autoecious *Pleurozia purpurea*.

(d) Sporophytes are unknown to science or very rare, at least in Europe, except in *Andreaea megistospora*, *A. sinuosa* and *Racomitrium himalayanum*.

(e) All are without gemmae or special methods of vegetative reproduction except *Campylopus schwarzii*, which has fragile leaf tips and shoot tips, and *Scapania nimbosa* and *S. ornithopodioides* in which gemmae, though usually absent, do sometimes occur.

It is tempting to speculate on the origin and history of the species of this phytogeographical element. It is clear that the patterns of distribution are old, for the rarity of spores and gemmae excludes the possibility of long-distance dispersal, and many of the species display a patchiness in their local distribution which indicates the difficulties they may have in getting from one hill to the next. The disjunctions, though striking, can be over-emphasized. There is little or no suitable ground for *Pleurozia purpurea* between the British Isles and the Himalayas, and its distribution is mirrored by those of circumboreal continental species that occur in all the suitable regions of the northern hemisphere, though with thousands of miles of ocean between one part of the range and the next. Presumably much of the range of these oceanic-subalpine species was occupied by them before the onset of continental drift; and initially they must have produced spores frequently. Vegetative reproduction resulting from the branching of shoots accompanied by the decay of older parts must have led to the production of unisexual populations and to a diminution and eventual loss of sexual reproduction, reducing mobility and the capacity of the species to evolve. Monoecious plants in these areas, and dioecious ones that retained the ability to reproduce sexually, were able to evolve and move into other climatic regions or to undergo speciation as, for example, in *Gymnomitrion* sect. *Crenarion* (Grolle, 1966).

This speculation offers no explanation for the absence of similar groups of dioecious, sterile, non-gemmiferous species from other geographical elements (e.g. Arctic-Alpine, Mediterranean-Atlantic), nor for the fact that most of the constituent species of the oceanic-subalpine element, and nearly all the most typical ones, are liverworts not mosses.

Surprising trends and anomalous distributions

Careful study of the maps sometimes reveals distributional features which, although perhaps no great surprise to the European botanist, could not be predicted from a study confined to the Continent. The decline in frequency of *Aulacomnium androgynum* and *Tortula marginata* towards the west is clearly shown. One can see that *Herbertus stramineus*, although its range overlaps with that of *H. aduncus*, is definitely a less western species. Müller (1954) pointed out that the European distributions of *Anastrepta orcadensis* and *Mylia taylorii* were almost identical except for the presence of the *Anastrepta* in the Carpathians, where the *Mylia* is absent. While in broad outline the British distributions of the two species are nearly the same, *Anastrepta* is much the less frequent in Ireland and there are many grid squares in the extreme west of Scotland that have *Mylia* but not *Anastrepta*. While *Mylia* flourishes in very oceanic conditions, *Anastrepta*, which goes farther east, appears to do better in climates that are not quite so extreme. There is a distinction, which is rarely drawn though nevertheless worth making, between wide-ranging oceanic species and suboceanic species that, as uncommon or depauperate plants, sometimes reach the extreme west.

Some of the maps show distributions which are anomalous. *Seligeria oelandica*, known only from the west of Ireland and from Sweden, has a relict distribution. The restriction of *Geocalyx graveolens* in both Scotland and Ireland to localities in the extreme west is a much more puzzling phenomenon, for it is not particularly oceanic in either Norway or North America. *Habrodon perpusillus, Orthotrichum speciosum* and *Tortula princeps* are all more frequent in the south of Europe than in the north but much less rare in Scotland than in southern England. Two such anomalies are worth special attention.

The first concerns *Tritomaria exsecta* and *T. exsectiformis*. In the British Isles the former is much the less common of the two, is decidedly oceanic and has a more northerly geographical distribution. *T. exsectiformis* is not oceanic and is the only species one ever meets with in southern England. Yet Schuster (1969), writing of eastern North America, says that *T. exsectiformis* 'is fundamentally of more northerly overall distribution than *T. exsecta*'. Neither species is oceanic there. There appears to be no consistent geographical distinction between these species on the European continent but the picture may have been confused by some misidentifications. There are presumably some physiological differences between British and American and perhaps continental European populations of one or both species, even though there may be no corresponding morphological differences that would justify the 'splitting' of either species.

A further example of anomalous distribution is given by *Radula lindenbergiana*. Most of the British records are northern, but it also occurs in Cornwall, Devon and the Channel Islands. It has a similar pattern of distribution in continental Europe, combining a boreal-montane range with a Mediterranean one. In England it appears to spurn the good ground in-between, where its near relative *R. complanata* grows. This schizophrenia is unique and demands investigation and explanation. It has been suggested that *R. lindenbergiana* is not specifically distinct from *R. complanata*. When there is taxonomic difficulty with a pair of species a common reaction of the taxonomist is to lump them together as one, or at least to suggest that this may be the solution. The true solution, however, may be to recognize that there are more taxa involved, not fewer. *Radula lindenbergiana* appears to consist in Europe of two geographical races – a southern and a northern. These must have physiological differences, but so far no one has reported morphological differences sufficiently correlated with these or sufficiently clear-cut for taxonomic recognition. An attempt was made last century to distinguish as a third species *R. germana*. This is now regarded as a synonym of *R. lindenbergiana*, but the problem merits re-examination.

ABSENTEES

Negative records may be almost as interesting as positive ones. We have only 21 of the 63 European Marchantiales. Very many of the European species have a Mediterranean distribution pattern, but none a Mediterranean-Atlantic one, though there are a few widespread species present in the Mediterranean region as well as in the British Isles. In spite of the reputation of the British climate, our only endemic *Sphagnum* is the single gathering recently described as *S. skyense*. All our other species occur in Scandinavia, where there are a few more beside. The same applies to some other genera of wet places – *Drepanocladus*, *Hygrohypnum* and *Calliergon*. The ecological point may or may not be significant.

Some distinctive bryophytes that it would be pleasant to have – *Sphagnum pylaisii*, *Splachnum luteum, S. rubrum* – are almost certainly genuinely absent, for otherwise they would have been found by now. Less conspicuous species, however, are still being added now and again to the British flora, and Table 5 is a list of those I consider most likely to be found here. It is a somewhat arbitrary one, but they are all worth looking out for.

Table 5. Some European bryophytes not yet known from the British Isles. Nomenclature follows Grolle (1983) for liverworts and Corley *et al.* (1981) for mosses.

Anastrophyllum assimile	*Cinclidotus riparius*
A. michauxii	*Dicranella howei*
Cephaloziella arctica	*D. humilis*
C. elegans	*Ditrichum pallidum*
Frullania jackii	*Fissidens arnoldii*
Lophocolea minor	*Hypnum pratense*
Lophozia ascendens	*H. recurvatum*
L. laxa	*Lescuraea radicosa*
Metzgeria simplex	*Mnium blyttii*
Nardia insecta	*Pleuridium palustre*
Amblystegium subtile	*Pohlia sphagnicola*
Brachythecium curtum	*Pseudoleskeella tectorum*
B. fendleri	*Rhodobryum spathulatum*
Bryhnia novae-angliae	

The English Channel and the North Sea have been bigger barriers to bryologists than to bryophytes. Nevertheless, several good additions to the British bryophyte flora, including *Brachythecium erythrorhizon*, *Dicranum leioneuron* and *Sphagnum strictum*, have been made by visitors from the Continent. Long may this continue. Continental bryologists who come and look are very welcome, whether or not their findings prove to be of interest to others.

PHYTOGEOGRAPHICAL CONSIDERATIONS

It is a pity that there exists no phytogeographical analysis of the British bryophyte flora as a whole of the type that was done so well for British seed-plants by Matthews (1955). Unfortunately one cannot apply Matthews' work directly to the Bryophyta. The geographical distributions of any group of related species depend upon many factors. These include the place and date of origin of the group; the ratio of sexual to other means of reproduction, and other factors that affect the potential rate of evolution; physiological and morphological characters, including life form, that are especially slow to change in response to evolutionary pressure; and the ways the organisms are dispersed. The Bryophyta and the Spermatophyta differ in all these features and there is no intrinsic reason why the distribution patterns of bryophytes should be more like those of seed-plants than those of, say, spiders. The recognition of patterns of distribution and of geographical elements must come from studies within the group and not from attempts to fit distributions into procrustean beds derived from preconceived notions.

There is a further difficulty that the bryophyte phytogeographer must face. Very few European angiosperms extend beyond Europe and the adjacent parts of Africa and Asia, and they can be classified phytogeographically largely on their distributions within Europe, closely related to their climatic tolerances. Most bryophytes are more widely distributed. European species are often in North America too, sometimes also in the tropics or in the southern hemisphere. Are they to be classified on their distribution within Europe or on their world distribution? Ratcliffe (1968), in his fine study of British Atlantic bryophytes, included in his Northern Atlantic group not only most of the species of the oceanic-subalpine element but also the predominantly Southern Hemisphere hepatic *Adelanthus lindenbergianus*. This is splendid for a primarily ecological treatment, but would be utter nonsense phytogeographically. Yet British species widespread in the Southern Hemisphere, e.g. *Adelanthus lindenbergianus*, *Barbilophozia hatcheri*, *Campylopus pyriformis*, have nothing more than this in common and have quite different distribution patterns in the British Isles and in Europe as a whole. It may be that we need to have two separate classifications, one based upon distributions on a world scale and a second more ecological one based upon distributions within continents.

The problems outlined here, though substantial, are hardly insoluble. An important function of the maps and captions in this atlas is to be of help to those who work toward solutions of them.

MAPS OF MOSS DISTRIBUTIONS

Explanation of maps and accompanying notes

Records are mapped in the 10 × 10-km squares of the Ordnance Survey National Grid in Great Britain and in the Ordnance Survey/Suirbheireacht Ordonais National Grid in Ireland. Records from the Channel Islands are mapped in the 10 × 10-km squares of the Universal Transverse Mercator Grid.

The symbols used are

○   Record made before 1950, or undated
●   Record made in or after 1950.

Where very few symbols appear on a map, the symbols have been encircled to make them more conspicuous. A few outlying symbols that might otherwise be overlooked are marked by arrows.

The numbering and nomenclature of species is basically that of *The Moss Flora of Britain and Ireland* (Smith, 1978). Since 1978, several new species have been added to the flora, and the taxonomy of several groups has been revised. Some of these changes were detailed in the most recent *Census Catalogue* (Corley & Hill, 1981) and in a list of European mosses (Corley *et al.*, 1981). Major revisions, of *Andreaea* (Murray, 1988) and *Racomitrium* (Frisvoll, 1983, 1988) have been published in the last decade.

As a result of this activity, it has been necessary to diverge from Smith's (1978) nomenclature in several cases. In particular, *Oxystegus sinuosus* is placed in *Barbula* (Hill, 1979), *Barbula recurvirostra* and *B. ferruginascens* are transferred to *Bryoerythrophyllum* (Long, 1982), and *Hyophila stanfordensis* is placed in *Hennediella* (Blockeel, 1990). Such changes have been kept to an absolute minimum, and the name in Smith's flora can always be found in the index. All names used in the European list (Corley *et al.*, 1981) are listed in the synonymy, and some extra synonyms from a recent American list (Anderson *et al.*, 1990) have also been included.

Each map is accompanied by notes on the taxon mapped. The first paragraph describes the habitat. The altitudinal range is given (in metres). 'Lowland' denotes altitudes below 300 m. The paragraph ends with a formula indicating the number of 10-km grid squares in which the taxon is mapped. GB 18+11*, IR 6+8* indicates that the plant has been recorded in or after 1950 in 18 grid squares in Great Britain and 6 in Ireland. There are pre-1950 or undated records from an additional 11 squares in Great Britain and 8 in Ireland. Channel Island records are not included in these totals.

The second paragraph describes the sexuality of the taxon, the frequency of sporophytes and indicates whether it has specialized means of vegetative spread, e.g. by gemmae (usually few-celled, more or less undifferentiated structures), bulbils, tubers, branchlets or fragile leaves. The following terms may be used to describe sexuality:

Sterile: antheridia (male sex organs) and archegonia (female sex organs) not produced;

Dioecious: antheridia and archegonia borne on separate plants; and

Monoecious: antheridia and archegonia borne on the same plant.Some authors have distinguished separate categories of monoeciousness:

Autoecious: antheridia and archegonia borne in separate inflorescences;

Paroecious: antheridia naked in the axils of the leaves immediately below the female inflorescence; and

Synoecious: antheridia and archegonia borne in the same inflorescence.

The third paragraph describes the distribution of the taxon outside the British Isles. For many species, information on European distribution is derived from the valuable compilations of Düll (1984, 1985). Information on the extra-European distribution of British and Irish species is readily available for North America, but for other continents is sometimes of doubtful accuracy. Species reported from other continents may not be the same as the European plant given that name, or the European plant may be known by another name elsewhere. Readers should bear in mind this element of uncertainty.

Additional comments may be given in a fourth paragraph.

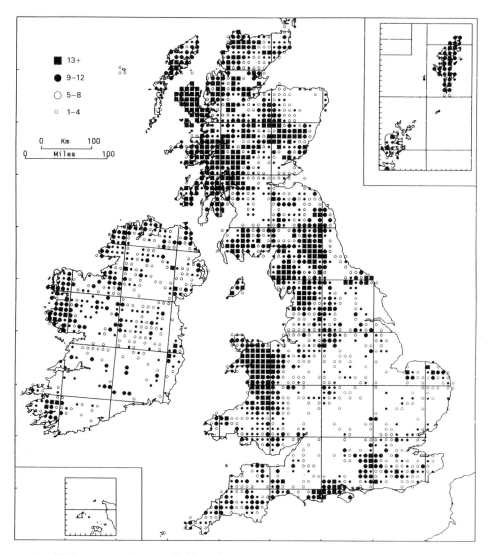

Totals of *Sphagnum* species recorded in 10-km squares

The map shows the number of *Sphagnum* species recorded in each 10-km square. Records have been counted without regard to date, including old records as well as new ones. Infraspecific taxa have not been counted separately. A map showing the total number of all mosses except *Sphagnum* in each 10-km square will appear in Volume 3.

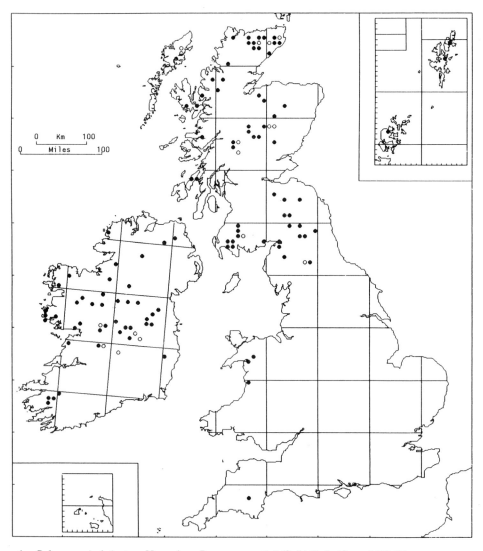

1/1a. **Sphagnum imbricatum** Hornsch. ex Russ. ssp. **austinii** (Sull.) Flatb. (*S. austinii* Sull.)

On ombrotrophic bogs, both raised bogs and blanket bogs, where the peat is deep and wet, frequently forming large hummocks among *Calluna vulgaris*, *Eriophorum vaginatum*, *Trichophorum cespitosum* and other bog sphagna such as *S. capillifolium*, *S. magellanicum* and *S. papillosum*. The tussocks are often abundantly interwoven with bog liverworts, especially *Odontoschisma sphagni*, which is an almost constant associate. 0–500 m (Allt Sowan Hill). GB 64+12*, IR 42+6*.

Dioecious; capsules very rare, found once in Co. Galway.

Frequent in W. Norway, rare round the W. shores of the Baltic, isolated occurrences in Austria (extinct) and Germany (Schwarzwald). Transcaucasia, Atlantic and Pacific coasts of N. America.

Only recently distinguished from ssp. *affine* but probably worthy of specific rank (Hill, 1988a). Peat deposits show that it was formerly abundant on many raised bogs and blanket bogs within its present general range, sometimes in a lax growth form that is now rarely seen. Since the Middle Ages it has suffered widespread decline and is still decreasing.

M. O. HILL

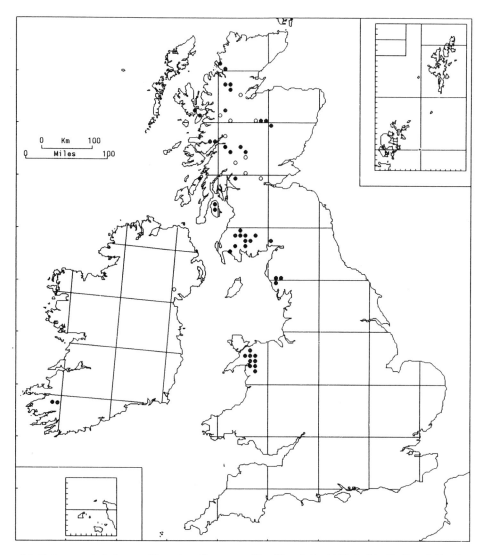

1/1b. **Sphagnum imbricatum** Hornsch. ex Russ. ssp. **affine** (Ren. & Card.) Flatb. (*S. affine* Ren. & Card.)

It grows in a variety of weakly minerotrophic mires, including slightly basic flush-bogs with associates such as *Juncus acutiflorus, J. effusus, Molinia caerulea, Myrica gale, Sphagnum auriculatum, S. recurvum* and *S. papillosum*, as well as on more swampy ground with *Carex rostrata, Menyanthes trifoliata* and *Potentilla palustris*. It has also been found in ditches and on flushed stream-banks. 0–400 m (Migneint). GB 44+9*, IR 2+1*.

Dioecious; capsules rare, known from three localities in Scotland.

Widespread in N. and C. Europe, from Iceland and Norway south to the Alps and from Finland west to France. Eastern N. America.

Not distinguished from ssp. *austinii* until Flatberg's (1984) monograph of the *Sphagnum imbricatum* complex. It is easily overlooked as a depauperate form of *S. papillosum* and has been much under-recorded. It may well prove to be common in W. Scotland.

M. O. HILL

21

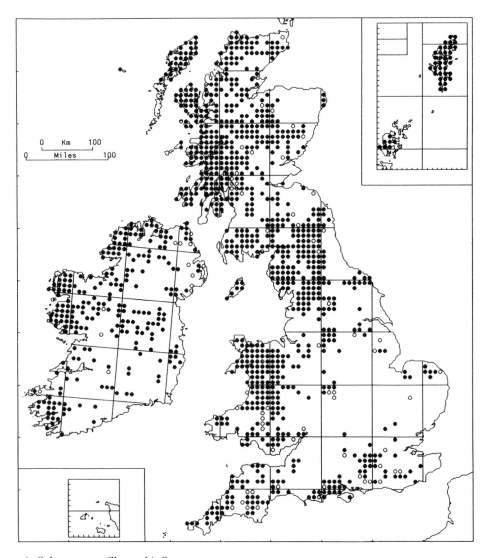

1/2. **Sphagnum papillosum** Lindb.

A characteristic species of open bogs and boggy moorland, growing most abundantly on deep wet peat, where it sometimes forms hummocks with *S. capillifolium* but also occurs in flat carpets. In high-rainfall areas of the north and west, it extends to a wide range of other acid peaty habitats, including boggy grassland, wet heaths, ditches, flushed peaty banks and weakly minerotrophic mires. It is intolerant of shade and rarely occurs in woods except where the canopy is open. It prefers more acid ground than *S. palustre*, and the two rarely grow intermixed. 0–1050 m (Lochnagar). GB 908+64*, IR 241+14*.

Dioecious; capsules occasional, August.

Discontinuously circumboreal, being suboceanic and absent from the continental interior of N. America. All Europe except the south, being most abundant in the boreal zone.

Sensitive to air pollution. Now almost totally absent from the S. Pennines, but peat deposits show that it was abundant there until the Industrial Revolution.

M. O. HILL

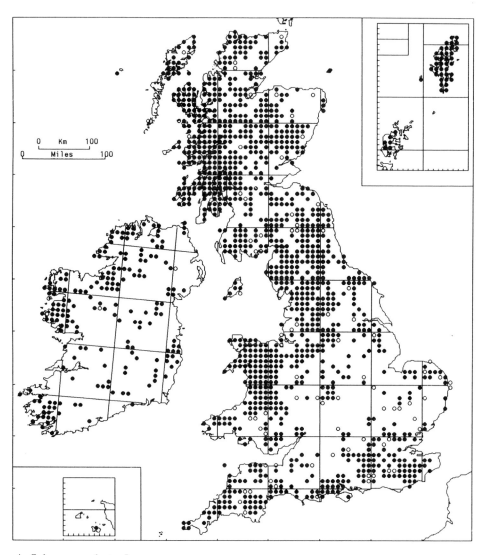

1/3. **Sphagnum palustre** L.

Wet woods, boggy grassland, ditches, flushed peaty banks, marshes and streamsides; tolerant of shade and often abundant. It is one of the less acid-demanding sphagna, often growing with *S. fimbriatum*, *S. squarrosum* and *S. subnitens*. In spite of this, it is common on some oceanic blanket bogs, for example in Shetland, where it occupies small declivities receiving surface flow in wet weather. Often occurring in carr and wet secondary woodland, it is a rapid colonist of new habitats, e.g. Wicken Fen, near Cambridge, where it appeared within a few years of the surface becoming acid. 0–1000 m (Glas Maol). GB 1180+83*, IR 189+4*.

Dioecious; capsules occasional, August.

Circumboreal. Widespread in Europe except the south and far north.

British plants are mostly var. *palustre*, but var. *centrale* (C. Jens.) A. Eddy, which is often treated as a separate species, *S. centrale* C. Jens., has been collected in Scotland (Daniels & Eddy, 1985); intermediates also occur.

M. O. Hill

23

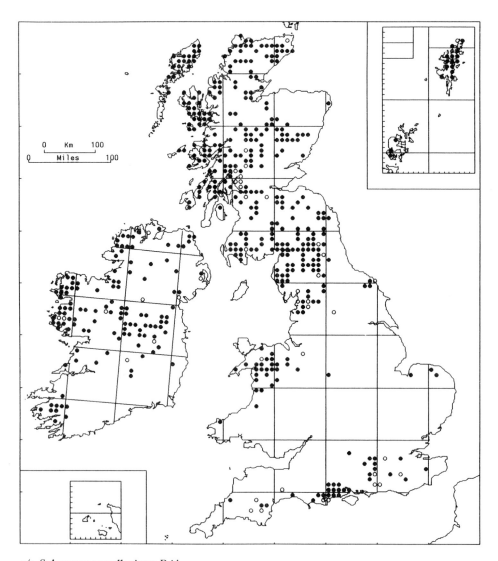

1/4. **Sphagnum magellanicum** Brid.

A characteristic species of well-illuminated bogs on deep uneroded peat, where it occupies a position intermediate between the tops of hummocks and the lower part of lawns. It reaches its greatest abundance on raised bogs and is almost invariably present where these are well developed. It also occurs on blanket bogs and valley bogs where the peat is deep and wet. Associates such as *Erica tetralix, Eriophorum angustifolium, E. vaginatum, Trichophorum cespitosum* and *Sphagnum papillosum* are typical. 0–1030 m (Creag Meagaidh). GB 375+35*, IR 120+10*.

Dioecious; capsules rather rare but regularly produced on some bogs where the species is abundant.

Circumboreal, from the Arctic south to Florida (U.S.A.); also occurring widely in the Southern Hemisphere and scattered in the tropics. Throughout N. Europe; scattered populations in mountains further south.

M. O. HILL

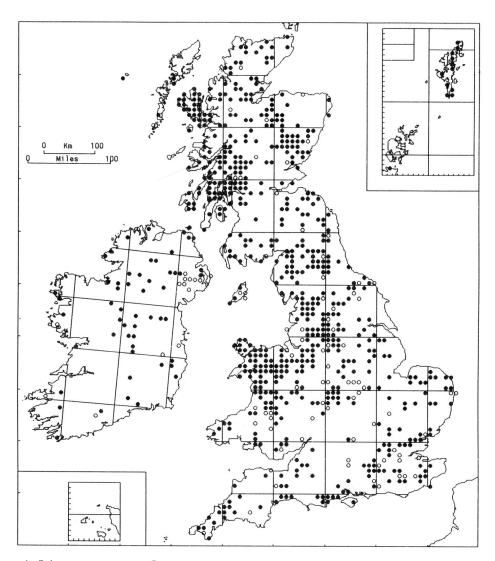

1/5. Sphagnum squarrosum Crome

A plant of swampy mineral-rich ground, growing in wet woods and carr, on rushy stream-banks, in ditches, beside ponds, and in peaty soaks where base-rich water seeps out onto moorland. It normally grows in the shade of higher plants, either in woodland or under *Carex rostrata*, *Juncus* spp. or *Molinia caerulea*. *S. palustre* is a common associate. An effective colonist, *S. squarrosum* occurs frequently in man-made or newly man-modified habitats. Mainly lowland, but occasional in the mountains, to 1030 m in a spring on Creag Meagaidh. GB 610+78*, IR 54+14*.

Monoecious; capsules common, August.

Circumboreal, from the High Arctic to the deciduous forest zone. All Europe except the south.

M. O. HILL

25

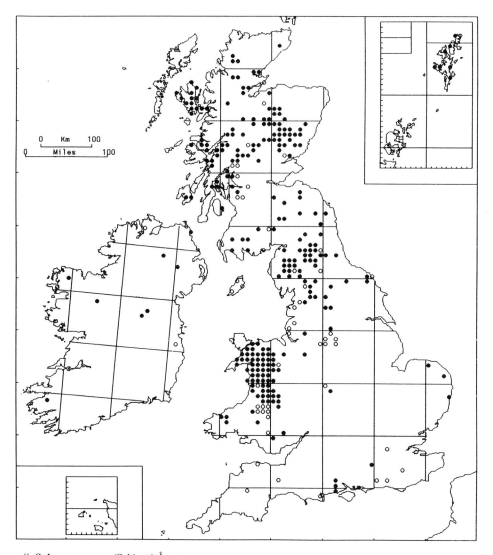

1/6. **Sphagnum teres** (Schimp.) Ångstr.

On peaty, permanently wet ground where there is mineral enrichment of the ground-water, most often among *Carex* spp. and *Juncus* spp. on valley bottoms, but also in moderately basic soaks on hillsides and occasionally in scrub woodland. It is a characteristic indicator of the more base-rich areas on wet moorland, where it grows with a wide range of associates including the base-demanding *Sphagnum contortum*, *S. subsecundum* and *S. warnstorfii*, as well as with commoner sphagna, *Aulacomnium palustre* and several Mniaceae. The habitat distinction from *S. squarrosum* is not sharp, but *S. teres* rarely grows in shade and is characteristic of flushed ground rather than swamps. 0–1050 m (Cairngorms). GB 251+51*, IR 8+3*.

Dioecious; capsules rare, summer.

Circumpolar, chiefly in the boreal zone and the Arctic, but with widespread scattered occurrences further south. Common in N. Europe, frequent in W., C. and E. Europe.

M. O. Hill

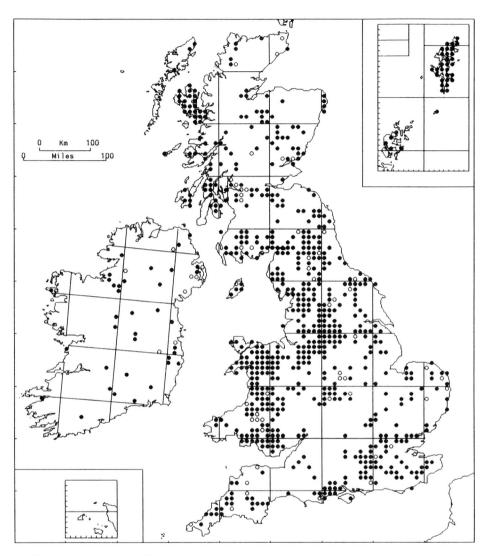

1/7. **Sphagnum fimbriatum** Wils.

Its main habitat is boggy woodland, where it can form extensive carpets on unflushed or weakly flushed ground, often mixed with *Molinia* and *Polytrichum commune*. It is especially characteristic of *Betula* scrub by pools or on disturbed bogland, but also occurs under *Alnus* and *Salix*. In Shetland, where it is common but there are no trees, it occurs on streamsides, in marshes, and on damp banks. It is frequent elsewhere on tussocky flattish moorland, especially where there has been some disturbance such as peat digging, mining or china-clay working. An effective colonist, it rapidly occupies suitable habitats as they become available. 0–350 m (Carn Fadryn), but probably ascends much higher in Scotland. GB 664+52*, IR 36+8*.

Monoecious; capsules abundant, August.

A bipolar species with a wide distribution in the boreal zone and the Arctic, also extending well into the deciduous forest zone. All Europe except the south. Asia, N. America, cool regions of the Southern Hemisphere.

M. O. HILL

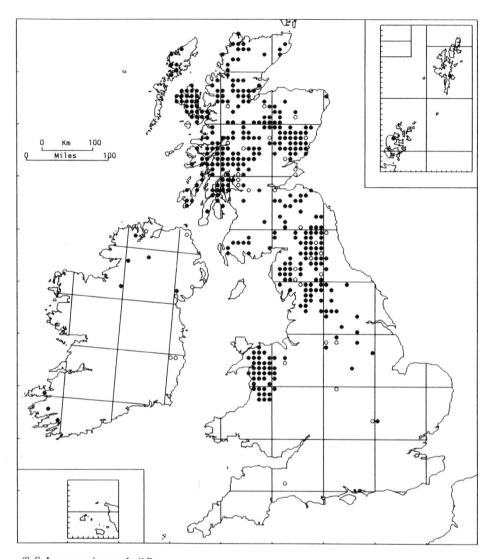

1/8. **Sphagnum girgensohnii** Russ.

This species has a wide ecological range, including moist broad-leaved woodland, damp conifer plantations, flushed grassy and heathery banks, streamsides, and *Juncus effusus* marshes, especially where there is slight base-enrichment of the ground-water. *Polytrichum commune* and *Sphagnum recurvum* are common associates. At lower altitudes it is rather shade-demanding, growing on sheltered banks, in the shade of trees, or under tall-growing *Juncus*; at higher altitudes it occurs also on more open ground, for example in small-sedge communities. Its range of habitats is similar to that of *S. russowii*, with which it frequently grows. 0–1000 m (Glas Maol). GB 373+31*, IR 8+6*.

Dioecious; capsules rare, August.

Circumpolar, chiefly in the boreal zone and the Arctic. N. Europe and mountains further south, with scattered occurrences in the lowlands of C. and W. Europe.

M. O. HILL

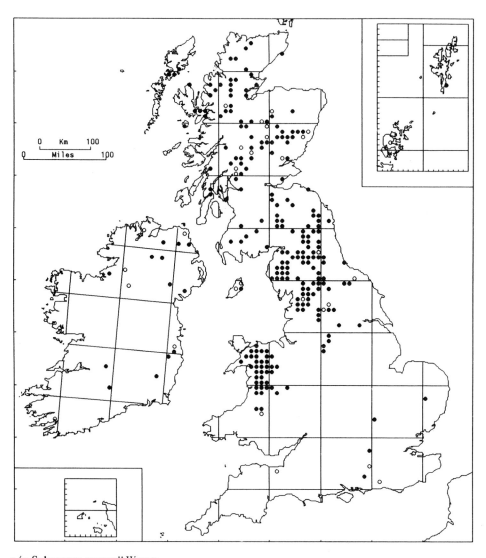

1/9. **Sphagnum russowii** Warnst.

In a wide range of habitats, chiefly on moorlands and mountains. One major habitat is intermittently flushed grassy and rocky banks, both on moors and in upland woods, including conifer plantations. It also grows in slightly basic springs, ditches and soaks, and occasionally on unflushed blanket peat. Lowland occurrences in Berkshire and Bedfordshire are in wet secondary birch woodland, where it is likely to be a recent colonist. 0–1200 m (Cairngorms). GB 217+24*, IR 13+6*.

Dioecious; both sexes are common but capsules are rare.

Circumboreal, north to the Arctic. Widespread in N. Europe and in mountains further south.

Probably under-recorded, as it can easily be mistaken for *S. capillifolium*.

M. O. HILL

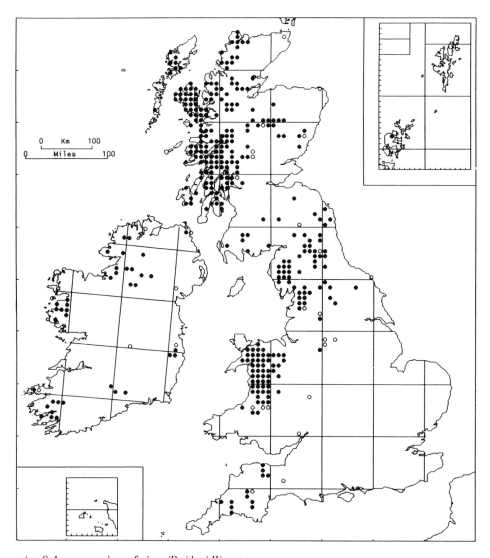

1/10. **Sphagnum quinquefarium** (Braithw.) Warnst.

Unlike most sphagna, it almost always grows on well-drained ground, chiefly in woods, but also on stream-banks, block-strewn mountain-sides and heathery slopes, where shaded or well sheltered. The tufts are often mixed with *Calluna vulgaris* and *Vaccinium myrtillus*; common bryophyte associates are *Scapania gracilis*, *Dicranum scoparium*, *Pleurozium schreberi* and *Sphagnum capillifolium*. 0–550 m (Cader Idris). GB 314+23*, IR 38+8*.

Monoecious; capsules occasional, ripe August and September.

Mountains of C. and W. Europe, and lowlands around the Baltic. Transcaucasia, Japan, eastern and western N. America.

A suboceanic and more southern species than most sphagna.

M. O. HILL

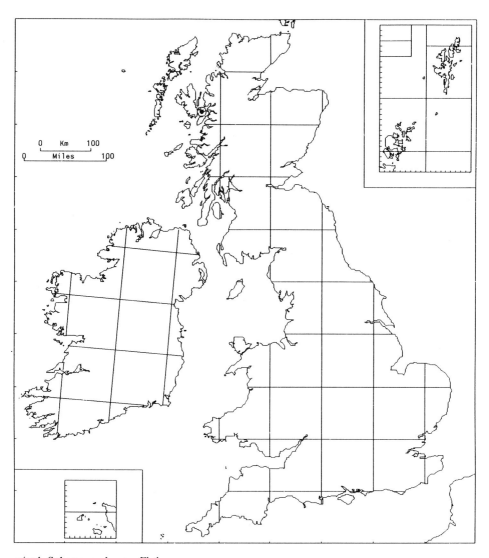

1/10A. **Sphagnum skyense** Flatb.

Among *Calluna vulgaris* and *Erica tetralix* in wet heath, mixed with *Breutelia chrysocoma*, *Hypnum jutlandicum*, *Plagiothecium undulatum*, *Pleurozium schreberi*, *Rhytidiadelphus loreus*, *Sphagnum capillifolium*, *S. quinquefarium* and *Thuidium delicatulum*. Lowland. GB 1.

    Sterile; no data on sexuality.

    Known only from the type collection and possibly endemic.

    Discovered in 1987, this robust species is broadly intermediate between *S. quinquefarium* and *S. subnitens*. In describing it, Flatberg (1988a) suggests that it could either be of hybrid origin, or a Tertiary relict. According to Daniels & Eddy (1990), it is probably conspecific with *S. junghuhnianum* Dozy & Molk., which is mainly Asiatic but also occurs in British Columbia (Queen Charlotte Islands (cf. Crum, 1984)), where the climate is similar to that of Skye.

M. O. HILL

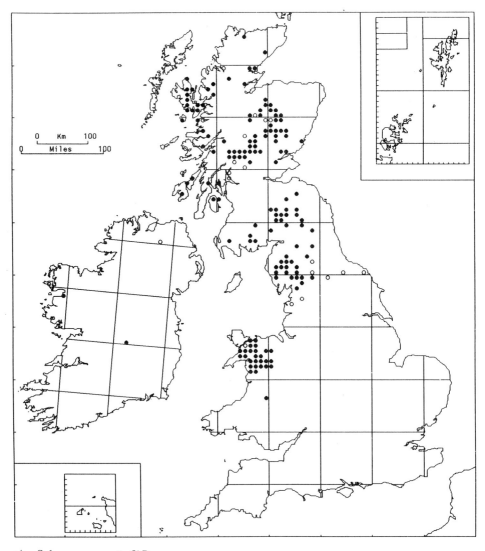

1/11. **Sphagnum warnstorfii** Russ.

One of the most base-demanding sphagna, confined to sites with obvious enrichment of the ground-water. Habitats include flushes on mountains and moorland, wet open scrubby woodland, marshes, stream-banks and flushed valley-bottoms. *Carex panicea, C. rostrata* and *Juncus acutiflorus* are common associates, as are the mesotrophic sphagna *S. contortum, S. subsecundum* and *S. teres*. 0–950 m (Ben Lawers). GB 148+19*, IR 2+2*.

Dioecious; capsules not found in Britain.

Circumpolar, mainly Arctic and boreal. N. Europe and mountains of C. Europe.

M. O. HILL

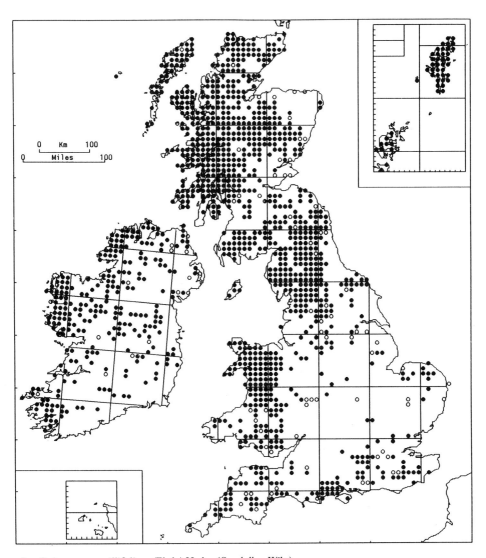

1/12. **Sphagnum capillifolium** (Ehrh.) Hedw. (*S. rubellum* Wils.)

Like its common associate *S. papillosum*, it is most abundant on deep peat-bogs, including raised bogs, valley bogs and blanket bogs, where it forms conspicuous tussocks among Ericaceae and Cyperaceae. In the west and north it also occurs on tussocky moorland, on bouldery slopes, at the edge of flushes, on sheltered banks, and in light woodland, where it is generally commoner than *S. quinquefarium*, even on well-drained banks. Towards the south and east, it is increasingly restricted to saturated open bogs. This is at least partly a pollution effect, as it disappeared from the blanket bogs of the S. Pennines at the time of the Industrial Revolution. 0–1040 m (Aonach Beag). GB 1018+74*, IR 253+15*.

Dioecious in the British Isles; capsules occasional.

Circumboreal. All Europe except the south.

Variation in *S. capillifolium sensu lato* is still the subject of active research (Cronberg, 1989). Many authorities recognize two major subdivisions in Britain and Ireland, corresponding to *S. capillifolium* var. *capillifolium* and *S. capillifolium* var. *tenellum* (Schimp.) Crum (*S. rubellum* Wils.).

M. O. HILL

33

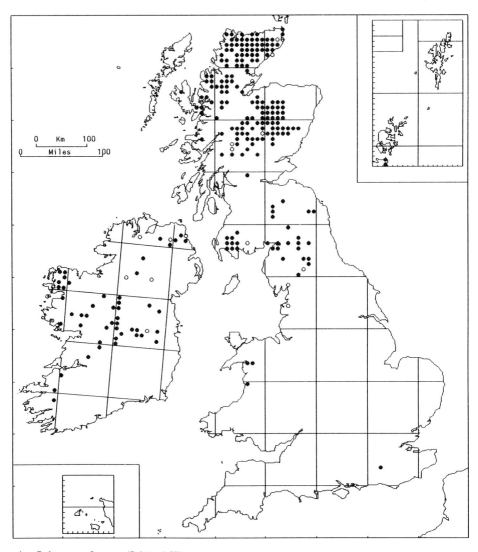

1/13. **Sphagnum fuscum** (Schimp.) Klinggr.

In most of the British Isles, it occurs in raised bogs and other flattish, deep peat-bogs, where it grows among *S. capillifolium*, *S. papillosum* and a wide variety of other bogland species. In mountainous parts of Scotland, it is found mainly on blanket peat above 400 m, forming large hummocks among *Calluna vulgaris*, *Empetrum nigrum*, *Eriophorum vaginatum* and *Rubus chamaemorus*. It sometimes also occurs in flushes. In the E. Highlands, it has been noted as a component of the sphagnum layer beneath *Calluna* on steep N.- and E.-facing slopes. 0–1000 m (Cairngorms). GB 184+12*, IR 48+7*.

Dioecious in Britain; capsules rare.

Circumpolar; an abundant species of much of the boreal zone and tundra, including continental interiors. N. Europe and mountains of C. Europe.

Its lowland habitat is closely similar to that of *S. imbricatum* ssp. *austinii*, with which it frequently grows. *S. fuscum* is also probably decreasing in the lowlands; it has not been seen in S.E. England since 1954.

M. O. Hill

34

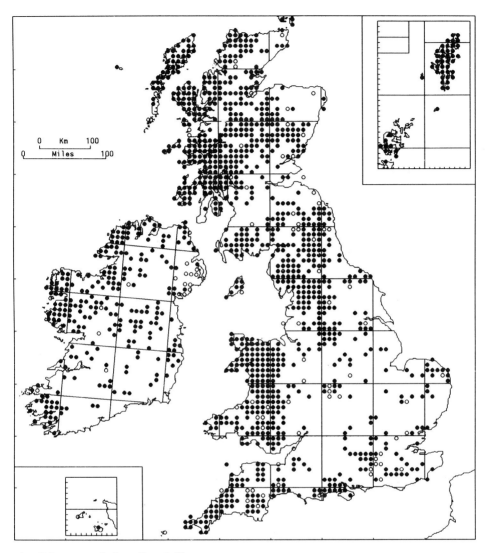

1/14. **Sphagnum subnitens** Russ. & Warnst.

This is one of the less acid-demanding sphagna, occurring in boggy grassland, marshes, fens, runnels, ditches and wet woodland, and also on flushed banks and moist rock-ledges. In the drier parts of the British Isles, it rarely grows with *S. capillifolium*, which prefers more acid ground with less water movement. However, the ecological distinction becomes blurred in the north and west, where the two species often occur together on ombrotrophic bogs, as well as in flushes and on rocky slopes. *S. subnitens* has colonized newly acid ground in Wicken Fen, and presumably spreads rapidly to newly available habitats elsewhere. 0–800 m (Clova). GB 1031+73*, IR 245+18*.

Monoecious; capsules common, August.

Most of Europe but rare in the far north and montane in the south. An oceanic species, occurring in eastern and western N. America but absent from the continental interior.

Irish specimens identified as *Sphagnum subfulvum* Sjörs (Moen & Synnott, 1983) are now attributed to *S. subnitens* ssp. *ferrugineum* Flatb. (Flatberg, 1985). Ssp. *ferrugineum* is not always clearly distinct from ssp. *subnitens*, and is apparently very rare.

M. O. Hill

35

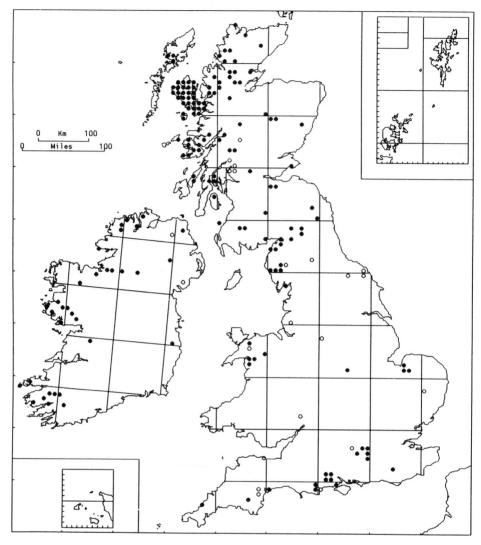

1/15. **Sphagnum molle** Sull.

In areas of high rainfall it grows on damp peaty streamsides, in boggy grassland and in shallow blanket bogs. Elsewhere, it is almost exclusively confined to wet heaths. *Trichophorum cespitosum*, *Sphagnum compactum* and *S. tenellum* are common associates, both in the north-west and on eastern heaths. 0–350 m (Skye). GB 129+20*, IR 33+2*.

Monoecious; capsules frequent, summer.

An oceanic species, occurring along the Atlantic seaboard from Spain to Norway, and round the Baltic. Very rare in central Europe. Atlantic seaboard of N. America.

Under-recorded, probably common in much of W. Scotland; the concentration of records on Skye and Mull is almost certainly due to recorder bias.

M. O. HILL

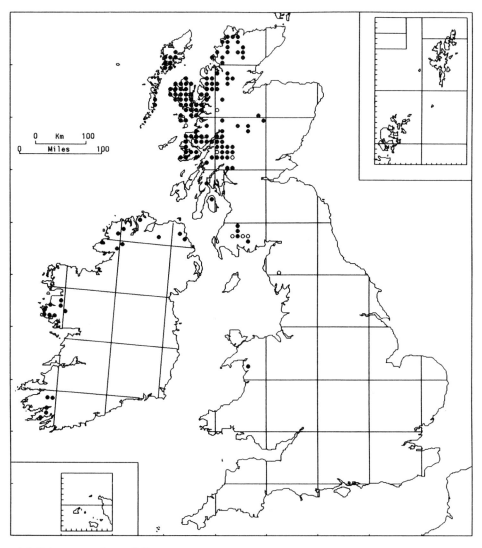

1/16. **Sphagnum strictum** Sull.

On moist peaty banks, in oligotrophic flush bogs, on shallow blanket bogs and in wet heaths, almost invariably among *Molinia caerulea* and often with *Calluna vulgaris*, *Erica tetralix*, *Myrica gale*, *Narthecium ossifragum*, *Trichophorum cespitosum* and *Sphagnum tenellum*. The habitat is rather similar to that of *S. compactum*, and *S. strictum* may also be favoured by disturbance. It normally grows in well-illuminated situations but can persist in shade beside ditches and under trees in young forestry plantations. 0–550 m (Skye). GB 126+9*, IR 22+0*.

Monoecious; capsules frequent, summer.

Switzerland, Germany, Denmark, Iceland, Norway and S. Sweden; very rare except in W. Norway. C. and S. Africa, Madagascar, S.E. Asia, eastern N. America, C. and S. America.

This amphi-Atlantic species was first collected in Scotland in 1885 but was not correctly identified until it was discovered (in three separate localities) by a Norwegian botanist, Johannes Lid, on a visit to Scotland in 1925 (Lid, 1929).

M. O. HILL

37

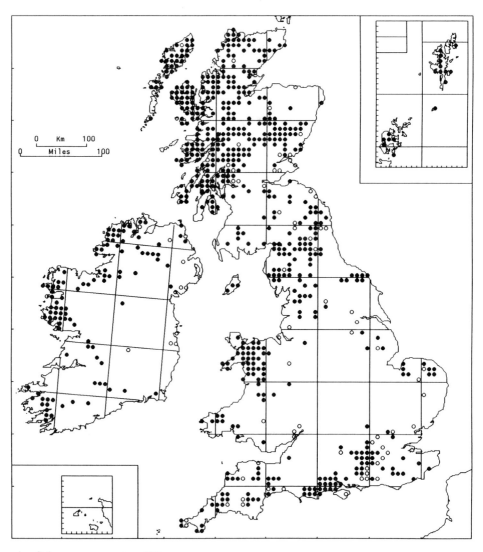

1/17. **Sphagnum compactum** DC.

On open, well-illuminated ground on wet heaths, in damp heathy grassland and on rocky banks, where the peat is shallow and very acid. Typical associates are *Erica tetralix, Trichophorum cespitosum, Sphagnum papillosum* and *S. tenellum*. In Skye, it is often also associated with *Mylia taylorii* and *Pleurozia purpurea*. At higher altitudes in E. Scotland it has been noted as an associate of *Nardus stricta, Narthecium ossifragum* and *Trichophorum cespitosum* in oligotrophic flushes. It regenerates rapidly after fire on wet heaths, and in other habitats also tends to occupy disturbed ground where the associated plants are not too tall. 0–1050 m (Lochnagar). GB 575+70*, IR 102+12*.

Autoecious; capsules frequent, summer.

Circumboreal from the Arctic south to Macaronesia, Turkey, Caucasus, and Louisiana (U.S.A.). W. and N. Europe north to N. Fennoscandia, and in mountains of C. Europe and the Balkans. Disjunct in Hawaii and Colombia.

M. O. HILL

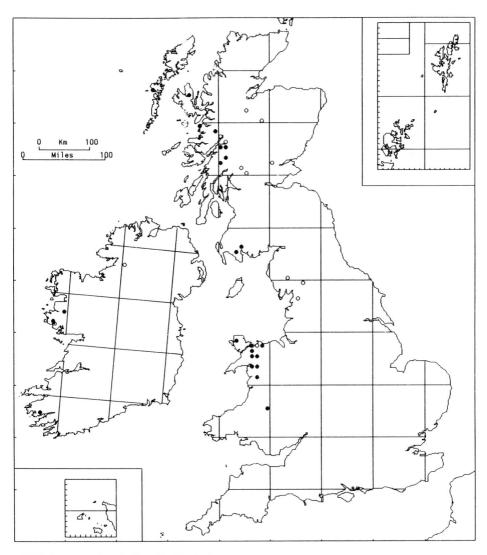

1/18. **Sphagnum platyphyllum** (Braithw.) Warnst.

One of the most base-demanding sphagna, growing chiefly in base-rich flushes in high-rainfall areas among mountains, with associates such as *Drepanocladus revolvens*, *D. vernicosus*, *Sphagnum contortum* and *S. subsecundum*. There are a few lowland records, from basin mires and runnels on wet heath. In many of its localities it must become desiccated during droughts; this may be an ecological factor that favours its occurrence. 0–450 m (Snowdon). GB 20+12*, IR 3+1*.

Dioecious; sex organs and capsules not observed in the British Isles.

Circumboreal, north to the Arctic and present in continental interiors. C. and E. Europe. Disjunct in Costa Rica.

Overlooked by almost all British bryologists, it is probably frequent in W. Scotland. In view of its rather continental world distribution, its western tendency in the British Isles is unexpected.

M. O. HILL

39

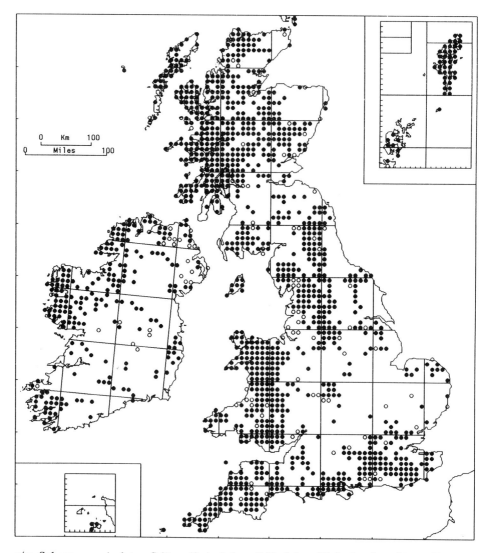

1/19. **Sphagnum auriculatum** Schimp. (*S. denticulatum* Brid., *S. lescurii* Sull., *S. rufescens* (Nees & Hornsch.) Warnst.)

In a wide range of wet, acid habitats, including dripping rocks, oligotrophic springs, wet woodland, bog-pools, ditches, flushes, streamsides and moorland soaks. It sometimes grows submerged down to about 4 m depth in lakes, especially those that have been acidified by atmospheric pollution (Raven, 1986). 0–1100 m (Ben Alder). GB 1093+80*, IR 214+24*.

Dioecious; capsules occasional, summer.

Circumboreal, south to Macaronesia and N. Africa. Widespread in W. and C. Europe, becoming rare towards the south and north and not, or scarcely, reaching the Arctic.

The foregoing remarks apply to var. *auriculatum*. Var. *inundatum* (Russ.) M. Hill (*S. inundatum* Russ., *S. subsecundum* ssp. *inundatum* (Russ.) A. Eddy) occurs in slightly more basic habitats, typically in 'poor fens', on swampy ground by mildly basic pools and ditches, and at the edge of basic flushes on moors. It is less common, but has not been recorded with sufficient consistency to be mapped separately.

M. O. HILL

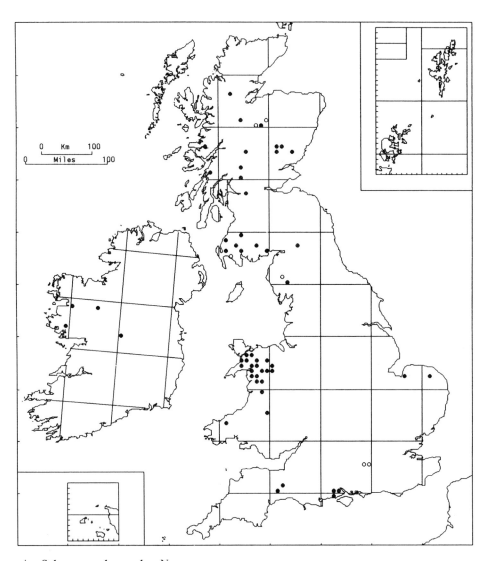

1/20. **Sphagnum subsecundum** Nees

In flushes, soaks, ditches and swamps on heaths and moors, sometimes semi-submerged, always where somewhat basic, often associated with *Drepanocladus exannulatus* and *S. teres*. In swamps it commonly grows among *Carex nigra*, *Menyanthes trifoliata*, *Potentilla palustris* and *S. papillosum*; on firmer ground it is most often found among *Juncus acutiflorus* and *J. effusus*. Its habitat is intermediate between those of *S. contortum* and *S. auriculatum*, and it may occur in mixture with either of them. 0–740 m (Braes of Balquhidder). GB 50+7*, IR 4.

Dioecious; capsules rare, summer.

Circumboreal, from the Arctic south to Himalaya, S.E. Asia and southern U.S.A. (California and N. Carolina). Throughout N., E. and most of C. Europe, becoming rare and scattered towards the south and west.

The map is based on specimens checked by M. O. Hill and a few other recorders, because there was formerly much confusion with *S. auriculatum* var. *inundatum* (Hill, 1975).

M. O. Hill

41

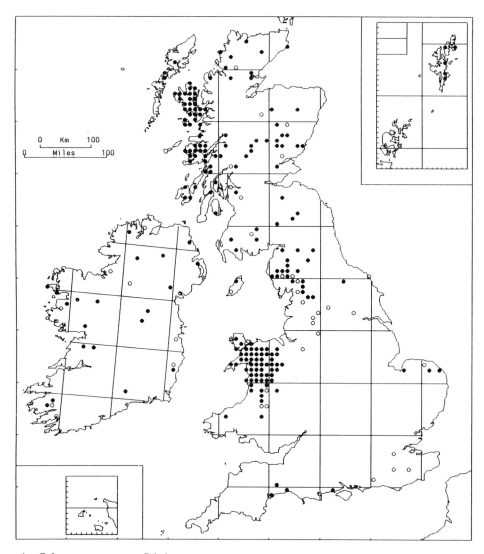

1/21. **Sphagnum contortum** Schultz

The most base-demanding of sphagna, *S. contortum* occurs in permanently wet flushes, in marshes, on runnels near lakes, and, rarely, in open wet woodland. Stands are usually small, being bounded on one side by water that is too basic to support any sphagna at all and on the other by water that is too acid. Common associates are the other two most base-loving sphagna, *S. teres* and *S. warnstorfii*, together with *Aneura pinguis*, *Campylium stellatum*, *Drepanocladus exannulatus*, *D. revolvens* and *Rhizomnium pseudopunctatum*. Associated higher plants are typically small sedges on firmer ground and *Carex rostrata* where the ground is more swampy. 0–800 m (Beinn nan Eachan). GB 186+32*, IR 20+9*.

Dioecious; both sexes present in Britain but capsules not found.

Circumboreal, from the Arctic south to the Caucasus and Colorado. In Europe from the Pyrenees and mountains of C. Europe north to Arctic Fennoscandia (but absent from the extreme north).

M. O. HILL

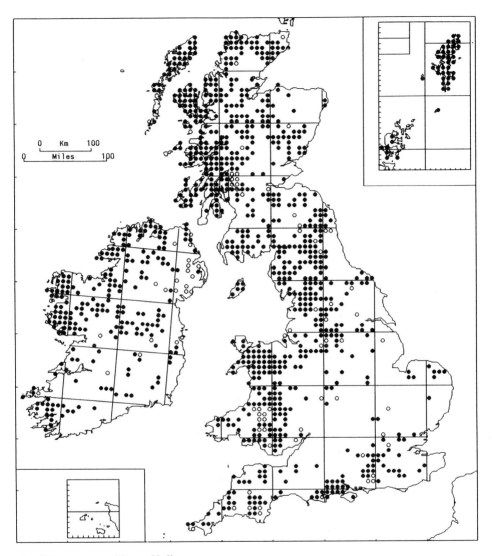

1/22. **Sphagnum cuspidatum** Hoffm.

In runnels, ditches, old peat-diggings, bog-pools and lakes, forming carpets or 'lawns', sometimes semi-submerged or submerged, normally in very acid water. In its less aquatic habitats it is commonly accompanied by *Eriophorum angustifolium*, *S. papillosum*, *S. recurvum* and *S. tenellum*, while in pools it may grow with *Carex rostrata*, *Juncus bulbosus* and *S. auriculatum*. 0–1030 m (Creag Meagaidh). GB 805+61*, IR 216+24*.

Dioecious; capsules occasional.

Discontinuously circumboreal, mainly in oceanic and suboceanic areas. N., W. and C. Europe, commoner in the west and very rare north of the Arctic Circle. Also in S.E. Asia and S. America.

Flatberg (1988b) recognizes two segregates in Britain, namely *S. cuspidatum sensu stricto* and *S. viride* Flatb. Both are apparently common, but *S. viride* is characteristic of the more minerotrophic mires. In N.W. Scotland, *S. viride* occurs in minerotrophic facies of blanket bogs where it forms small carpets in sloping peat depressions. These segregates have not yet been distinguished by British bryologists and are not mapped separately.

M. O. HILL

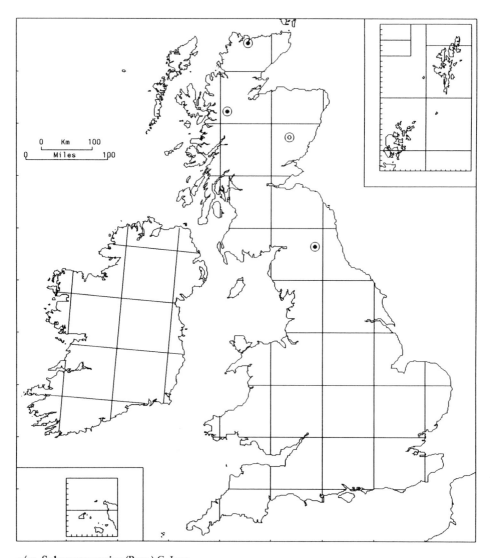

1/23. **Sphagnum majus** (Russ.) C. Jens.

In Northumberland it grows with *S. balticum* and *S. recurvum* in wet depressions in the marginal zone of a large valley bog. In Glen Affric it occurs in an area of Caledonian pine forest. In one place it was associated with *Carex limosa*, *Sphagnum cuspidatum* and *S. recurvum* in a boggy hollow amongst pines; in another it was in a 'lawn' with *S. recurvum* at the edge of a lochan. At its other localities it has been collected only once. In Angus the habitat was boggy ground by a pool; in Sutherland it grew in a slightly mineral-enriched depression in blanket bog. GB 3+1*.

Dioecious; capsules found only in Glen Affric.

A northern species, occurring throughout the boreal zone to the edge of the tundra. Rare and scattered in C. Europe.

First found in Britain in 1947 but not recognized till 1964 (Maass, 1965) and still doubtless overlooked by British bryologists. Flatberg (1987) distinguished two subspecies in Europe. He has examined specimens from all four British localities and identified them as ssp. *norvegicum* Flatb.

M. O. HILL

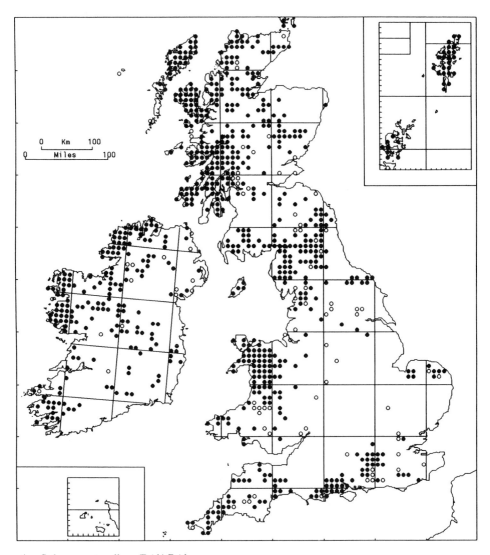

1/24. **Sphagnum tenellum** (Brid.) Brid.

Occurs in a range of unshaded, moist or wet, peaty habitats where the soil is very acid. On heaths it is found on moist ground subject to occasional burning, often as an associate of *S. compactum*. In bogs it occupies a position intermediate between *S. capillifolium* on hummocks and *S. cuspidatum* in hollows, commonly growing with one or other of them and with *S. papillosum* and *S. recurvum*. In areas of high rainfall it also occurs on flushed heathy banks, on mountain crags, and in sheltered crevices among rocks. 0–1090 m (Cairngorms). GB 586+76*, IR 186+15*.

Dioecious; capsules frequent, ripening in summer, often in July, about 3 weeks earlier than those of other *Sphagnum* species.

A suboceanic species, absent from continental interiors and from the Arctic except in Fennoscandia. N., W. and C. Europe. Japan, Kamchatka, western and eastern N. America, S. Greenland, Mexico, Ecuador, Brazil.

M. O. HILL

45

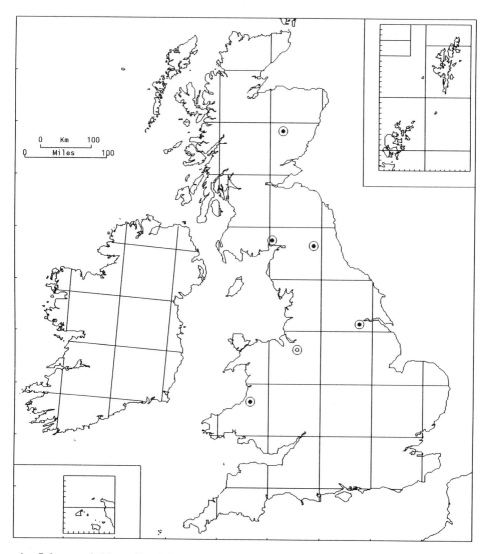

1/25. **Sphagnum balticum** (Russ.) C. Jens.

In wet hollows of bogs, chiefly raised bogs, but also on a blanket bog in the E. Highlands of Scotland. At Muckle Moss in Northumberland it has been found with *S. majus* in the marginal zone of a valley bog (Maass, 1965). At Thorne Waste in Yorkshire it was originally found amongst heather on damp ground, but has recently been seen only in a disused canal on the peat. (The canals were cut to allow removal of peat by boat.) 0–660 m (above Loch Muick). GB 5+1*.

Dioecious; male plants have been found in Northumberland, no other data on sexuality in Britain.

Circumpolar, mainly in the tundra and northern boreal zone, widespread in continental interiors. In Europe, from the Alps and N. European plain north to Svalbard.

M. O. HILL

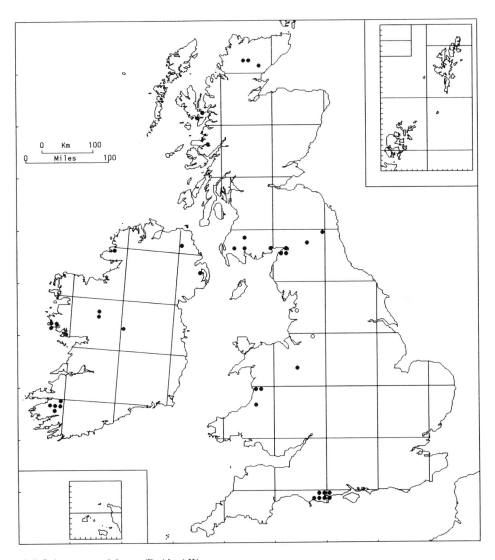

1/26. **Sphagnum pulchrum** (Braithw.) Warnst.

Although rare, this species is locally abundant, forming carpets in depressions (including old peat-diggings) on ombrotrophic bogs and, in S. England, on valley bogs. Its habitat is similar to that of *S. cuspidatum*, with which it sometimes grows. Other common associates are *Drosera rotundifolia*, *Erica tetralix*, *Eriophorum angustifolium*, *Narthecium ossifragum* and *Sphagnum papillosum*. 0–280 m (Butterburn Flow). GB 26+3*, IR 15.

Dioecious; capsules not found in Britain or Ireland.

Suboceanic in Europe, in the Low Countries, Germany, Denmark, S. Fennoscandia and the Baltic States. Japan, Kamchatka, eastern N. America.

Daniels (1982) examined genetic variation in *S. pulchrum* populations from six British sites. He found a high degree of variation, both within and between sites. This could be explained either by independent colonization of different clones or by conditions having latterly become unsuitable for sexual reproduction.

M. O. HILL

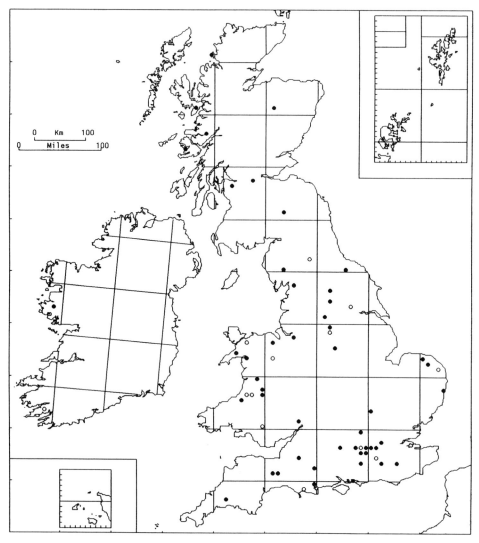

1/27a. **Sphagnum recurvum** P. Beauv. var. **amblyphyllum** (Russ.) Warnst. (*S. flexuosum* Dozy & Molk.)

In weakly minerotrophic marshes, in wet open woodland, and on flushed rock-ledges. In some types of 'poor fen' it can form extensive carpets. Its habitat is very similar to that of var. *mucronatum*, which which it often grows. However, its preference for more minerotrophic sites means that it is scarce in or absent from large tracts of acid country. 0–500 m (Berwyn Mts). GB 47+12*, IR 1+1*.

Dioecious; capsules not recorded but seldom looked for.

Discontinuously circumboreal, mainly in oceanic and suboceanic regions. In Europe from Arctic Fennoscandia south to mountains of Italy and the Balkans.

The varieties of *S. recurvum* are sufficiently distinct to be treated as species by some authorities (e.g. Daniels & Eddy, 1985). Var. *amblyphyllum* is much under-recorded. It is probably frequent or even locally common in many parts of Britain and Ireland.

M. O. HILL

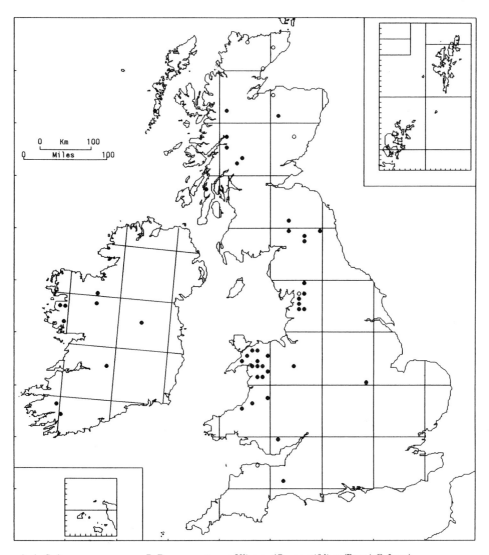

1/27b. **Sphagnum recurvum** P. Beauv. var. **tenue** Klinggr. (*S. angustifolium* (Russ.) C. Jens.)

On flushed ground beside streams, on banks, in marshes, and in open woodland, confined to minerotrophic sites, usually mixed with var. *mucronatum*. It often grows with the more base-demanding sphagna such as *S. subsecundum*, *S. teres* and *S. warnstorfii*, but is not so extreme in its requirements. 0–500 m (Craig y Dulyn). GB 38+8*, IR 9.

Dioecious; capsules known from one locality in Scotland, probably rare.

Circumboreal, commoner in more continental regions and noted for its abundance in the vast bogland of western Siberia. W., N. and C. Europe.

Much under-recorded, it is probably frequent in upland parts of Scotland. Its restriction to minerotrophic flushes in Britain and Ireland contrasts with its occurrence in a wide range of mire types in Fennoscandia.

M. O. HILL

49

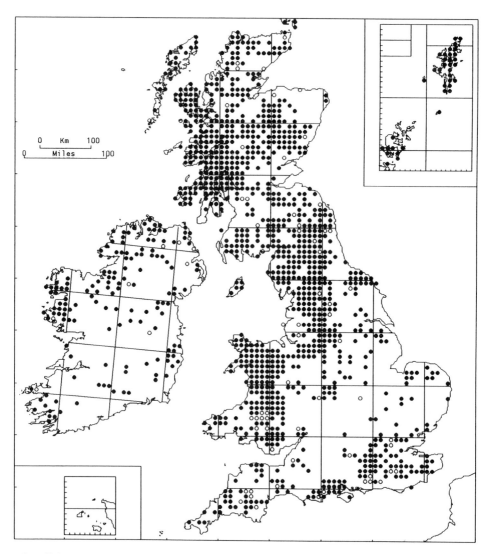

1/27c. **Sphagnum recurvum** P. Beauv. var. **mucronatum** (Russ.) Warnst. (*S. fallax* (Klinggr.) Klinggr., *S. recurvum* var. *brevifolium* (Lindb. ex Braithw.) Warnst.)

Very common and often abundant in a wide range of permanently moist or wet acid habitats. It is particularly plentiful as wide carpets among *Juncus effusus* in wet fields and on valley bottoms, but also occurs in bogs, by streams, in woods, on flushed banks, in ditches, in swamps, and generally on wet moorland. It is characteristic of flushed ground but does not require enrichment with bases; indeed, it is often abundant in pools and runnels on ombrotrophic bogs and on the most acid moorland, surviving even in highly polluted parts of the Pennines. 0–1000 m (Lochnagar). GB 1053+65*, IR 145+9*.

Dioecious; capsules occasional.

Circumboreal, mainly in oceanic and suboceanic regions, scarce or absent in continental interiors. All Europe except the south.

The map is of *S. recurvum sensu lato*, but var. *mucronatum* is so much the commonest variety that its distribution is well indicated. In sphagnum-poor regions of E. Britain, a few records may be of one of the other segregates, but even there, var. *mucronatum* is commonest.

M. O. Hill

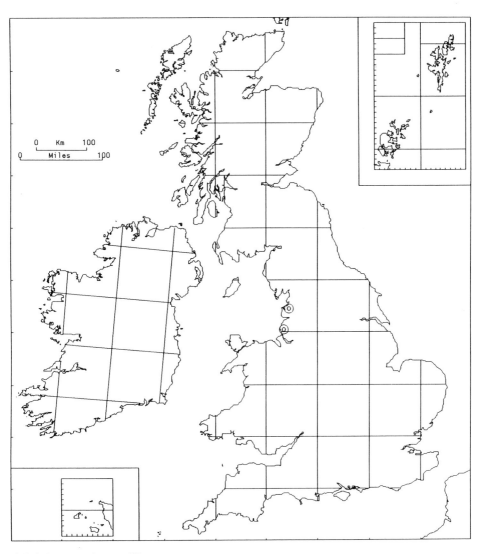

1/28. **Sphagnum obtusum** Warnst.

In peat bogs. At Cockerham Moss, a locality noted also for the occurrence of *S. riparium*, its habitat was described as 'deep bogs amongst birches' (Wheldon & Wilson, 1907). Lowland. GB 2*.

Dioecious; no data on sexuality in Britain.

A circumpolar, boreal and arctic species with a predominantly continental distribution. E. and N. Europe, including Iceland, south to France, Yugoslavia and Hungary.

First found in Britain in 1898 and recorded at intervals up to 1911. Now lost through drainage. In other parts of its range it occurs on rather swampy ground among Cyperaceae at the margins of ponds and streams or in open *Salix* scrub. It is one of the more markedly minerotrophic sphagnum species (Daniels & Eddy, 1985).

M. O. HILL

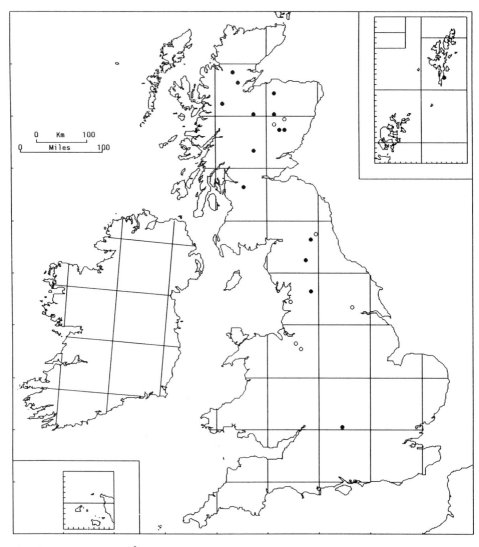

1/29. **Sphagnum riparium** Ångstr.

In a variety of wet or very wet habitats, with weakly or moderately basic water. Near Oxford it occurred in a disused brick-pit under sallow. In Northumberland it is known from the marginal zone of a valley bog; in its former localities near Manchester the habitat was probably similar. In Shetland it has been found in a basic streamside flush with *Carex echinata, C. nigra, Juncus articulatus, J. effusus, Calliergon stramineum* and *Sphagnum squarrosum*. In the Highlands of Scotland it has been noted chiefly from springs and flushes above 600 m, with *Carex curta, C. echinata, Eriophorum angustifolium* and *Sphagnum papillosum*. 0–750 m (Burn of Longshank). GB 15+8*.

Dioecious; capsules very rare, summer.

Circumboreal, mainly in the northern boreal zone and subarctic, reaching 72° N in Greenland. Common in N. Europe, becoming scattered and rare southwards to the Alps and N. European plain.

Its occurrence as a casual near Oxford, where it was recorded in 1961 and 1962, exemplifies the remarkable propensity of bryophytes to colonize new sites from long distances away (Jones, 1986).

M. O. HILL

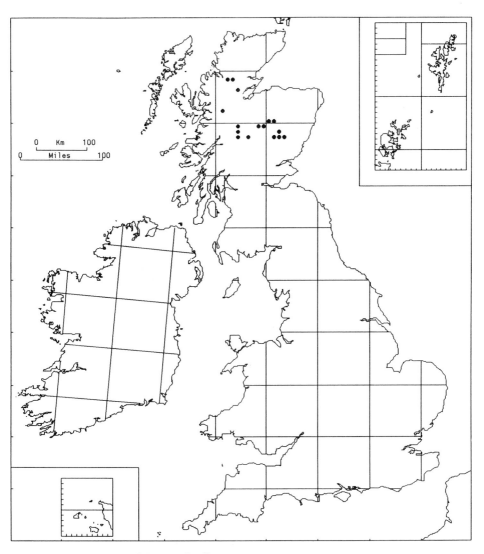

1/30. **Sphagnum lindbergii** Schimp. ex Lindb.

Although it formerly occurred at low altitude in Shetland, all extant localities are in the mountains. Here it grows in oligotrophic flushes and springs, often in drainage-channels and hollows among blanket bog, sometimes near late-snow beds (Ratcliffe, 1958). Its main associates are *Carex echinata*, *Eriophorum angustifolium*, *Polytrichum commune*, *Sphagnum auriculatum*, *S. papillosum* and *S. russowii*. 750 m (Water of Unich) to 1100 m (Beinn a'Bhuird). GB 16+1*.

Autoecious; capsules rare.

Circumpolar, chiefly in the Arctic and the northern boreal zone, south to Austria, Japan, southern Canada (British Columbia) and New England. In Europe it is common in N. Fennoscandia but is scattered and rare further south.

Its absence from low altitudes in Britain is perhaps rather surprising, as it formerly occurred in the Rhineland near the Dutch border and is known from Denmark and Schleswig-Holstein.

M. O. HILL

53

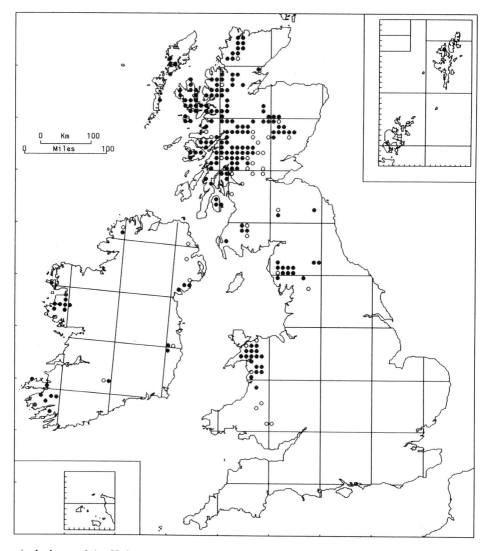

2/1. **Andreaea alpina** Hedw.

On wet to continuously dripping, acid to mildly basic rocks (granite, basalt, schist, mica-schist) in streams and on ledges and cliffs. Also on periodically irrigated outcrops and boulders, and in snow-flushes. Mainly on wet mountains, to over 1300 m on summit of Ben Nevis, occasional at low elevations down to 60 m (Srath Coille na Fearna). GB 176+34*, IR 22+10*.

Monoecious; sporophytes common to occasional, mature in summer.

A bipolar species, known in the Northern Hemisphere only from the N. Atlantic region: Norway, the Faeroes and southernmost Greenland, in addition to the British Isles. Subantarctic and cold temperate Southern Hemisphere islands, New Zealand, S. America north to Colombia.

B. M. MURRAY

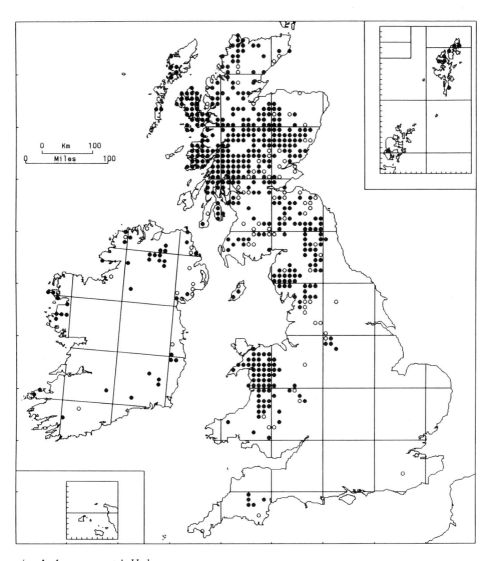

2/3a. Andreaea rupestris Hedw.

Forming cushions on neutral to acid boulders, slabs, cliffs and walls (basalt, mica-schist, sandstone, granite) that are periodically wet but often dry; also in snow-beds. Very rare on lithosols. Generally in drier sites than other *Andreaea* species, and lacking from continuously wet sites that are the habitat of *A. nivalis* and *A. frigida*; often associated with the other members of the genus. Found in lowlands and highlands, 0–1330 m (Ben Nevis). GB 454+77*, IR 40+13*.

Monoecious; sporophytes common, mature in spring and early summer.

Widespread in Europe, common in northern and central parts but rare and sporadic to the south. Widespread elsewhere in the Northern Hemisphere.

Most material is clearly referable to var. *rupestris* but, throughout the British Isles, plants occur with characteristics intermediate to var. *papillosa* (Lindb.) Podp. Although one or two collections could be referred to the latter, no truly unequivocal material of this primarily Arctic variety is known from the British Isles.

B. M. MURRAY

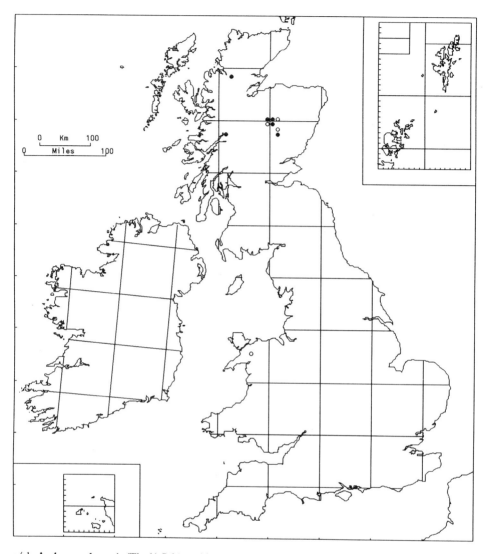

2/3b. **Andreaea alpestris** (Thed.) Schimp. (*A. rupestris* var. *alpestris* (Thed.) Sharp)

Forming small cushions on periodically wet rocks or large mats on lithosols that are almost continuously wet in summer. In alpine sites; usually near snow-beds. 700–1335 m (Ben Nevis). GB 6+4*.

Reportedly autoecious; capsules infrequent, mature in summer.

An arctic-alpine species, found in appropriate sites throughout Europe, commoner in the north but south almost to 40° N in Spain.

The map is based on herbarium specimens checked by B. M. M. The only Welsh collection is scant and its determination equivocal. For further information, refer to Murray (1988).

B. M. MURRAY

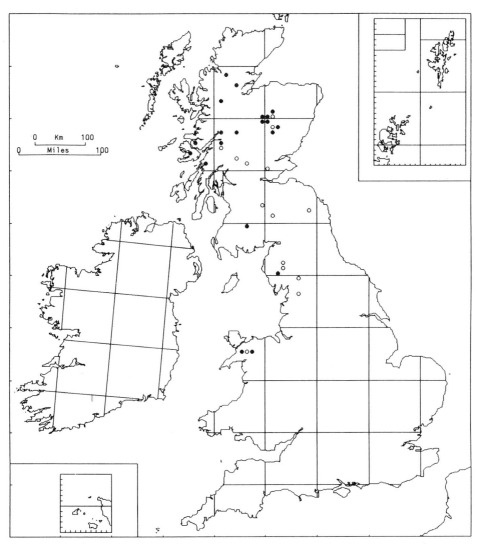

2/3A. **Andreaea mutabilis** Hook. f. & Wils.

On periodically to more continuously wet, more or less exposed, acid boulders, crags or rubble (granite, andesite, basalt, quartzite); rarely on thin soil over rock. Most frequently on high ground from about 600 to over 1300 m (near summit of Ben Nevis), but several collections are from low elevations, 60 m (Pass of Melfort) to 250 m (Glen Etive). GB 20+14*.

Monoecious (and possibly dioecious); sporophytes common, mature in summer.

A bipolar species known elsewhere in the Northern Hemisphere from a few localities in the Faeroes, France and Spain (Pyrenees) and from one N. American collection (British Columbia). Widespread in the Southern Hemisphere.

The map is based on herbarium specimens checked by B.M.M. For further information, refer to Murray (1988).

B. M. Murray

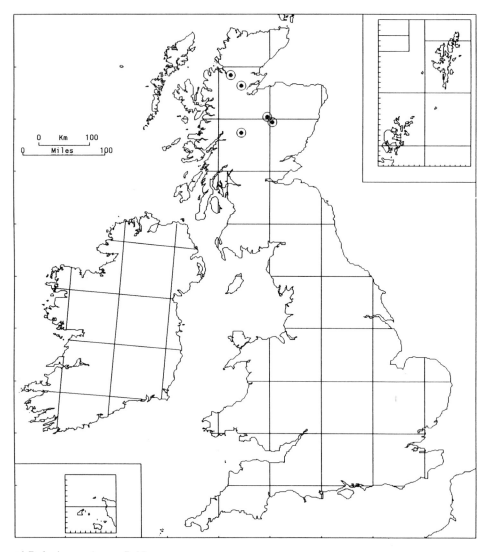

2/3B. **Andreaea sinuosa** B. Murr.

Forming very small patches or occurring as occasional stems among *A. blyttii* on acid rock in extreme snow-beds or deep gullies where snow lies late. 950–1100 m (Ben Macdui). GB 5.

Apparently dioecious but the stems are very short and fragile and it may be monoecious. Sporophytes common, mature in summer.

A very rare disjunct species, known additionally only in oceanic north-western N. America (Aleutian Islands and British Columbia).

The map is based on herbarium specimens checked by B.M.M. The species was described and reported from Scotland by Murray (1986).

B. M. Murray

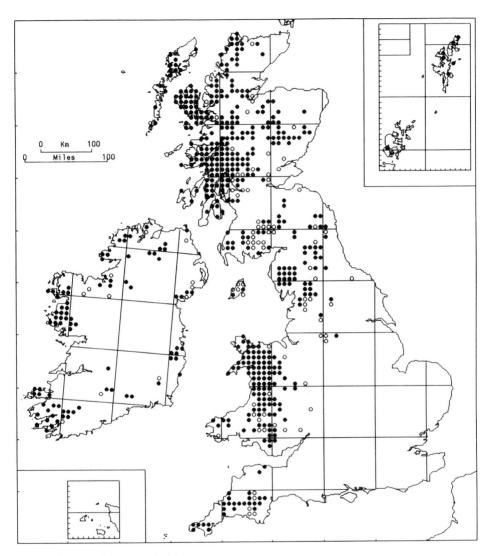

2/4a. **Andreaea rothii** Web. and Mohr

Frequently forming cushions on more or less exposed, periodically wet, acid rock (granite, sandstone, shale, slate). Sometimes forming extensive mats on seeping, gently sloping slabs. 0–900 m (Scafell Pike (checked by B. M. M.)), but recently reported at 1344 m on the summit of Ben Nevis. GB 438+81*, IR 82+13*.

Autoecious; sporophytes common, mature in spring and summer.

Widespread in Europe from about 68°N in Norway south almost to 37°N in the Iberian Peninsula. Eastern and western N. America. Unconfirmed report from Mongolia.

Two subspecies, ssp. *rothii* and ssp. *falcata* (Schimp.) Lindb., are now recognized in Britain, and their distribution has been mapped (Murray, 1988). The present map is based mainly on data collected during the B.B.S. mapping scheme, and the species may be over-recorded in western Scotland, where the newly described *A. megistospora* is frequent. Material named *A. crassinervia* Bruch by British bryologists is actually *A. rothii* ssp. *falcata*. However, a single specimen, from Dubh Loch near Lochnagar, is referable to *A. crassinervia* but its provenance is doubtful.

B. M. MURRAY

59

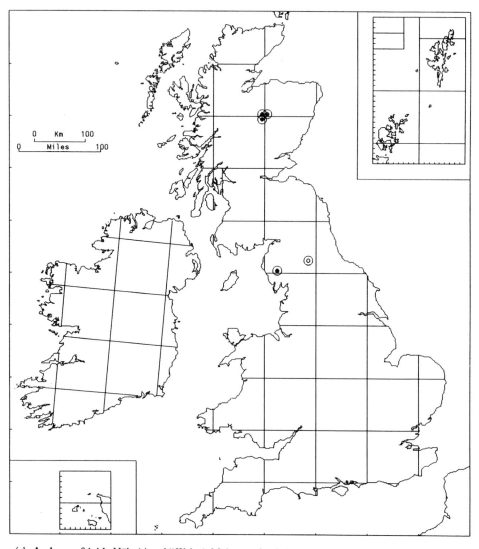

2/4b. **Andreaea frigida** Hüb. (*A. rothii* Web. & Mohr ssp. *frigida* (Hüb.) Schultze-Motel)

Always on wet, acid rock. Locally abundant at Loch Etchachan and on Ben Macdui where it is emergent at the edges of lochs and streams where the water is still or flowing slowly. Very scarce on wet boulders on slopes of Ben Macdui. Mostly 850–1200 m (Ben Macdui) but down to 500 m in Upper Teesdale. GB 4+1*.

Monoecious; sporophytes seen in all collections, mature in summer.

A European endemic species in wet, mountain sites from about 67°N in Norway to 37°N in the Iberian Peninsula, and as far east as Romania.

The map is based on herbarium specimens checked by B. M. M. For further information, refer to Murray (1988).

B. M. MURRAY

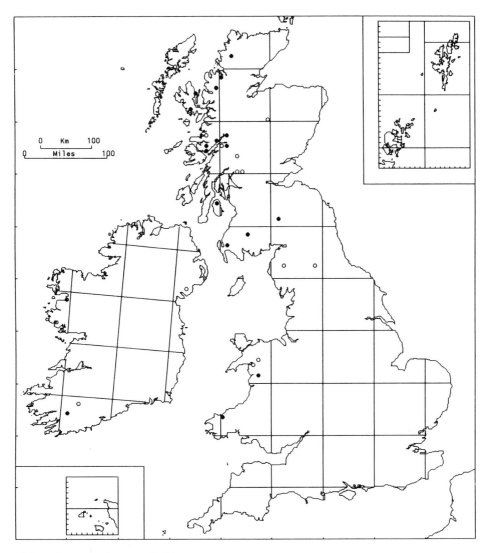

2/5A. **Andreaea megistospora** B. Murr.

On periodically moist, usually exposed rock (quartzite, sandstone, granite). It mostly occurs from near sea-level (Loch Maree) to about 700 m, but there is one collection from 1100 m (Coire an t-Sneachda). GB 16+11*, IR 2+2*.

Autoecious; sporophytes common, mature in spring and summer.

A species of hyperoceanic climates, disjunct between W. Europe and north-western N. America. Very rare elsewhere in Europe, along the coast of Norway as far north as about 63°, and in N.W. Portugal.

The map is based on herbarium specimens checked by B.M.M. European material is ssp. *megistospora*, the type locality being in Scotland (Murray, 1987).

B. M. MURRAY

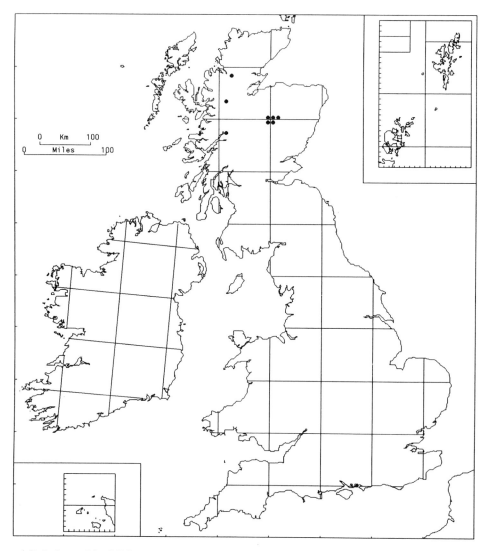

2/5B. **Andreaea blyttii** Schimp.

A rare mountain species found in late-lying snow-beds, where it forms extensive sheets on strongly acid slabs and scree. Wet and often submerged early in the growing season, exposed later. 1000–1200 m (Marquis' Well, Cairn Gorm). GB 8.

Apparently dioecious; capsules occasional, mature in summer.

An arctic-alpine species, known elsewhere in Europe in Svalbard, Fennoscandia, Kola Peninsula and Poland. Arctic Asia, N. America, Greenland.

The map is based on herbarium specimens checked by B. M. M. Specimens from Harris, Outer Hebrides, are correctly identified but their provenance is doubtful. *A. blyttii* was not accepted as a British species until the revision by Murray (1988).

B. M. Murray

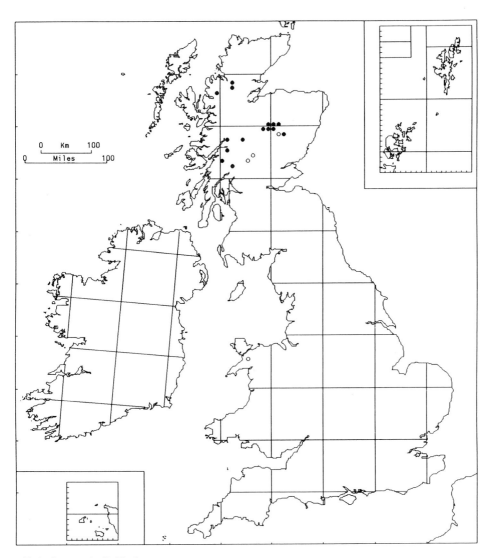

2/6. **Andreaea nivalis** Hook.

Rare but often locally abundant on wet rock (granite, andesite) at edges of streams and flushes that are usually fed by snow-beds; sometimes submerged. Occasionally forming large mats on seeping outcrops and ledges that are flat to gently sloping. Usually near summits, 900–1340 m (Ben Nevis), but down to 700 m in Glen Feshie. GB 15+4*.

Dioecious; sporophytes occasional, mature in early summer.

An alpine species occurring on mountains with an oceanic climate. In Europe it occurs on humid mountains from about 70° N in Norway to about 43°N in Spain. Caucasus, Siberia, Japan, north-western and north-eastern N. America, Greenland.

Herbarium specimens have been checked by Murray (1988). The single Welsh gathering is of doubtful provenance.

B. M. MURRAY

63

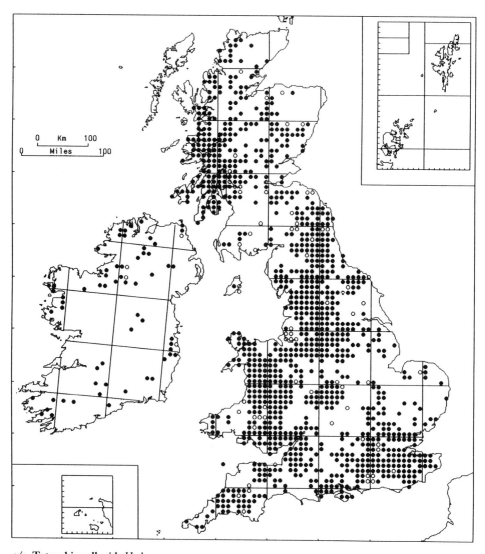

3/1. **Tetraphis pellucida** Hedw.

A calcifuge which colonizes organic matter and sandstone rocks in shaded, humid situations. It is frequent in areas of acid soil on logs and tree-stumps in an advanced state of decay, on sandy and peaty banks in woods, on peaty soil over rocks and between boulders, on decaying grass-tussocks and, more rarely, on acid bark of living trees. It also grows in quantity on base-poor sandstone rocks. In areas of calcareous clay it is much rarer, although it can occasionally be found on slowly rotting tree-stumps and logs; in chalky areas it is almost totally absent. Associates include *Cephalozia* spp., *Diplophyllum albicans*, *Lepidozia reptans*, *Lophozia incisa*, *Scapania umbrosa*, *Dicranum scoparium*, *Hypnum cupressiforme*, *Mnium hornum* and (on sandstone) *Dicranella heteromalla*. 0–380 m (Auchterhouse Hill). GB 1030+71*, IR 61+5*.

Autoecious; sporophytes occasional, maturing throughout the year. Gemmae, borne in terminal cups, are almost always present; they may be splash-dispersed (cf. Brodie, 1951).

Circumboreal. Widespread in Europe, but absent from lowland districts in the Mediterranean region.

Several of the liverworts found with *Tetraphis* (including *Cephalozia catenulata*, *Lepidozia reptans*, *Lophozia incisa* and *Scapania umbrosa*) show the same tendency to grow both on organic matter and on sandstone rocks.

C. D. PRESTON

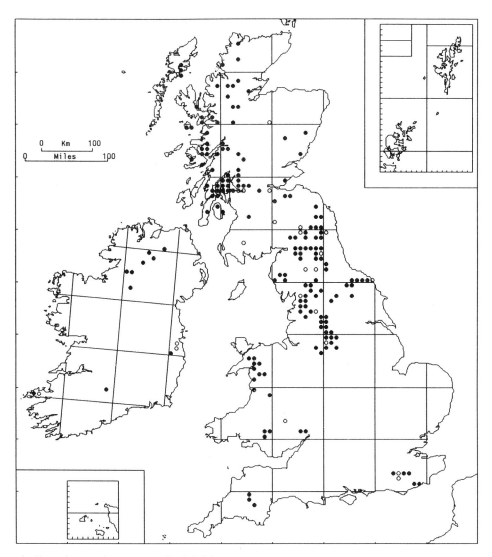

4/1. Tetrodontium brownianum (Dicks.) Schwaegr.

A moss of damp but not wet, siliceous, usually sandstone rocks, which are often slightly basic. It is most often on vertical rock faces, in crevices or under overhangs, usually in shade. It is found mainly in low-lying wooded ravines, but occurs also above the tree-line, reaching 700 m (Beinn Chabhair and Snowdon). GB 161 + 20*, IR 10+4*.

Autoecious; fruits freely in summer. Without special methods of vegetative reproduction.

Westernmost Norway, Pyrenees, C. Europe. Turkey, Caucasus, Himalaya, Japan, N. America. Reported from the Southern Hemisphere in Chile and New Zealand, but an illustration, presumably of the Chilean plant (Mahu, 1985), shows *T. repandum*.

A. C. CRUNDWELL

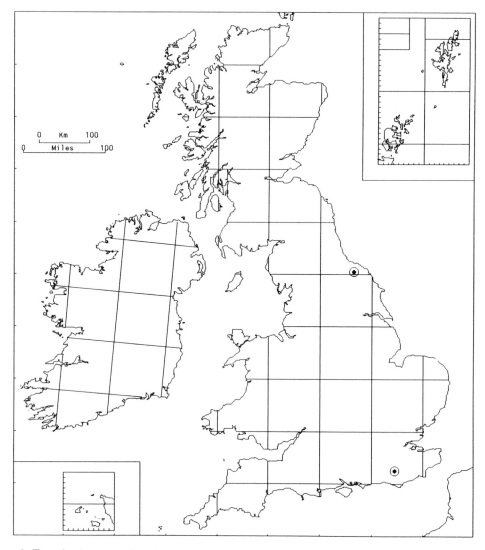

4/2. **Tetrodontium repandum** (Funck) Schwaegr.

On the underside of moist sandstone or gritstone rock. GB 2.

Autoecious; sterile in Britain. Vegetative reproduction by means of flagelliform shoots.

Pyrenees, C. Europe, Fennoscandia, Jan Mayen. Caucasus, China, Japan, western N. America, Newfoundland.

Known in the British Isles with certainty only from Yorkshire (Appleyard, 1956) and Sussex. It has not been refound at either site, and was last seen in 1958. There is a record requiring confirmation from Cheshire (19th century).

A. C. CRUNDWELL

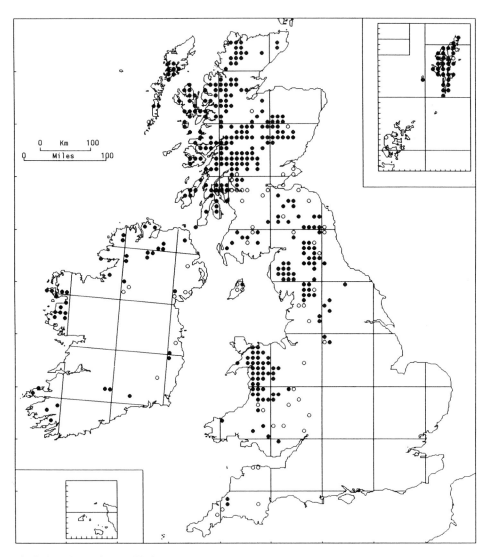

5/1. Polytrichum alpinum Hedw.

A calcifuge pioneer species found on upland heaths, moorlands and most commonly in montane habitats including peaty banks, mossy banks of streams, screes, rocky slopes, cliff-ledges and late-snow areas. Usually commoner on N.- to E.-facing slopes, generally in well-drained and relatively stable habitats. Common associates include a wide range of other calcifuge species, such as *Diplophyllum albicans*, *Dicranum scoparium*, *Hylocomium splendens*, *Mnium hornum*, *Racomitrium lanuginosum* and *Rhytidiadelphus loreus*. Uncommon below 300 m, except in Shetland; frequent up to 1335 m (Ben Nevis). GB 341+53*, IR 39+8*.

Dioecious; capsules common, ripe summer. Fragile leaf laminae, a dispersal mechanism in some Arctic populations, are not recorded in the British Isles.

A bipolar arctic-alpine with a very wide distribution. Common throughout Arctic and boreal Europe; towards the south restricted to mountains.

D. G. Long

67

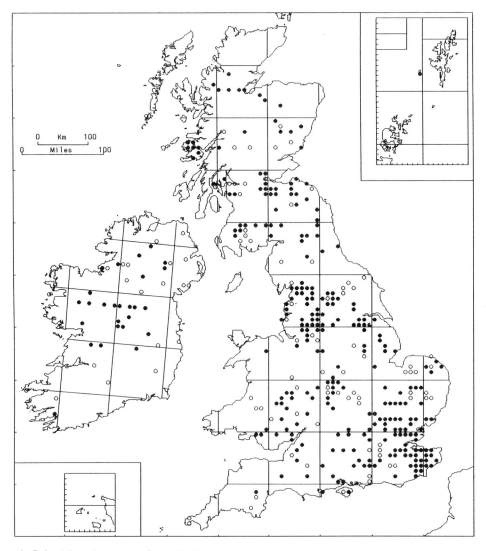

5/2. **Polytrichum longisetum** Sw. ex Brid.

A more weedy species than the other larger members of the genus, often colonizing areas of disturbed peat. Although calcifuge, it grows in a wide range of habitats including lowland heaths, peaty woods (often on logs and stumps), wall-tops, upland moors and mountains on both well-drained and relatively wet substrata. A characteristic habitat is on the peaty sides of drains in young conifer plantations. 0–1030 m (Beinn a'Bhuird). GB 261+78*, IR 26+20*.

Dioecious; capsules common.

Widespread in the Northern Hemisphere. Throughout Europe, but commoner in the north and west where peat deposits are more prevalent. Also occurs in S. America, New Guinea and New Zealand.

D. G. LONG

68

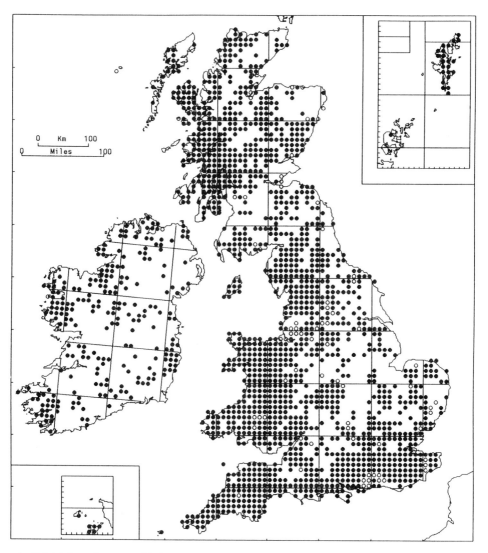

5/3. Polytrichum formosum Hedw.

In Britain this is most typically found in neutral to acid woodland, especially under birch, oak, beech and pine, on poorer soils and peaty substrates. However, it extends less commonly into many other habitats including lowland heaths, moorlands and mountains where it is apparently rare. It sometimes grows with *P. longisetum* but prefers more stable and better-drained situations, often with common pleurocarpous mosses such as *Hypnum jutlandicum*, *Plagiothecium undulatum* and *Thuidium tamariscinum*. It not infrequently grows in quite rocky places such as old walls, block-screes and on rocky slopes. 0–500 m (Snowdon). GB 1527+75*, IR 236+3*.

Dioecious; capsules common, summer.

Mainly circumboreal, but reaching south to New Guinea. Throughout Europe, though rare in the Arctic and on high mountains.

D. G. Long

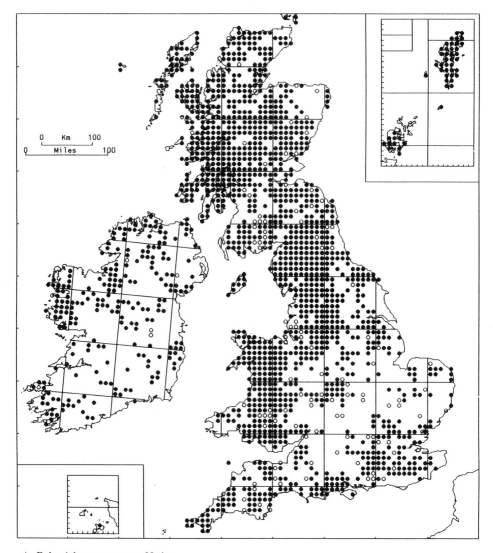

5/4. **Polytrichum commune** Hedw.

In a wide range of acid habitats, including woods, heaths, wet moorlands, ditches, bogs, lake margins, pools and streamsides, usually in damp to wet situations; often abundant in the uplands and growing most luxuriantly in damp woodland, *Salix* scrub, *Betula* carr etc. It is commonly associated with a range of *Sphagnum* species, especially *S. palustre* and *S. recurvum*. It tolerates moderate aerial pollution and will grow in urban habitats such as old railway-banks. 0–1050 m (Lochnagar). GB 1416+112*, IR 227+7*.

Dioecious; sporophytes are frequently produced in abundance, especially in woodland.

Almost cosmopolitan. Throughout Europe; rare in the south and in calcareous areas, commonest in higher-rainfall regions of the north and west.

D. G. LONG

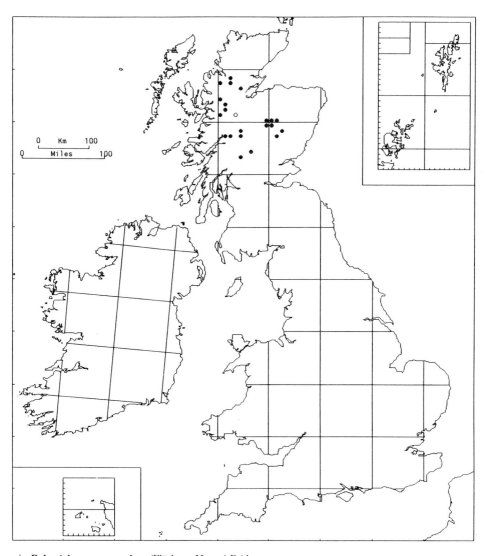

5/5. **Polytrichum sexangulare** (Flörke ex Hoppe) Brid.

Found only on the highest Scottish mountains, where it is restricted to areas of late snow-lie, sheltered gullies, under dripping rock-faces and by pools and springs. It often grows in compact bryophyte-dominated turves with associates such as *Diplophyllum albicans*, *Marsupella* spp., *Moerckia blyttii*, *Dicranum glaciale*, *Kiaeria starkei* and *Pohlia ludwigii*. In such habitats, the plants are often small and rarely produce sporophytes. In late-snow areas it can grow more luxuriantly in damp mossy block-screes, where capsules are not infrequent. 900–1335 m (Ben Nevis). GB 21+1*.

Dioecious; sporophytes occasional.

A disjunctively circumpolar arctic-alpine. Scandinavia, Faeroes, Svalbard and Iceland, south to the higher mountain ranges of S. and C. Europe. Occurs widely in Asia, western N. America and Greenland, but absent from most of the Canadian Arctic.

D. G. LONG

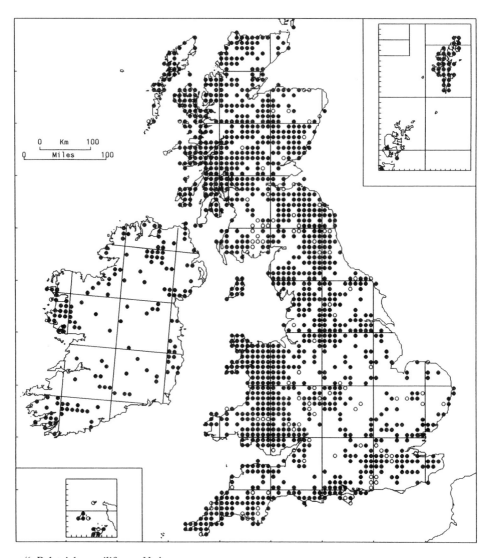

5/6. **Polytrichum piliferum** Hedw.

A calcifuge pioneer species of well-drained sandy or gravelly substrates and dry peaty soils. Lowland habitats include dry soil in woodland clearings, dry heaths, fixed dunes and road-cuttings; in the uplands it grows on shingle banks of streams, stony lake-margins, dry gravelly screes, block-screes, rock outcrops and stony mountain ridge-tops. It often favours wall-tops, large boulders and volcanic rock-outcrops (sometimes quite basic) where a thin layer of leached peaty soil has developed. Typical associates include *Campylopus paradoxus*, *Dicranum scoparium*, *Polytrichum juniperinum*, *Racomitrium lanuginosum* and *Cladonia* species. Common at all altitudes from 0–1170 m (Ben Lawers). GB 1296+101*, IR 151+1*.

Dioecious; capsules common, summer.

Widespread in both Hemispheres, and occurring throughout Europe.

D. G. Long

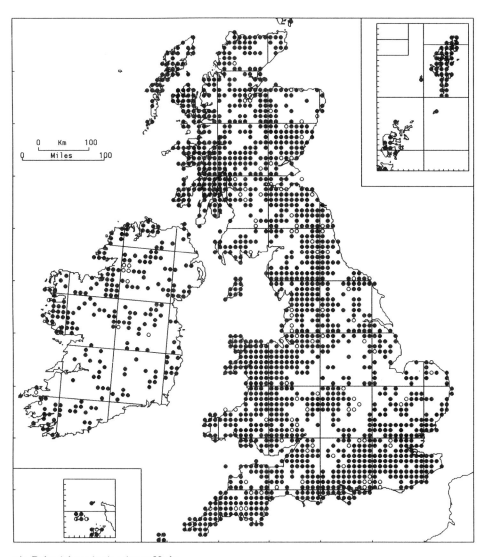

5/7. **Polytrichum juniperinum** Hedw.

A xerophytic species of acid substrates in a wide range of habitats, including coastal fixed dunes, old shingle-beaches, lowland woods, heathland (especially frequent following burning), waste-ground, soil-capped walls and road-cuttings (e.g. in conifer plantations). Typically a pioneer of dry soils, it often grows with *Ceratodon purpureus*, *Dicranum scoparium*, *Hypnum jutlandicum*, *Pleurozium schreberi*, *Polytrichum piliferum* and *Cladonia* species. 0–820 m (Beinn Eighe), but rare above 300 m. GB 1596+101*, IR 238+10*.

Dioecious; capsules are commonly produced.

Widespread in temperate regions of both Hemispheres, and in the Arctic. Throughout Europe; rarer in calcareous regions.

D. G. Long

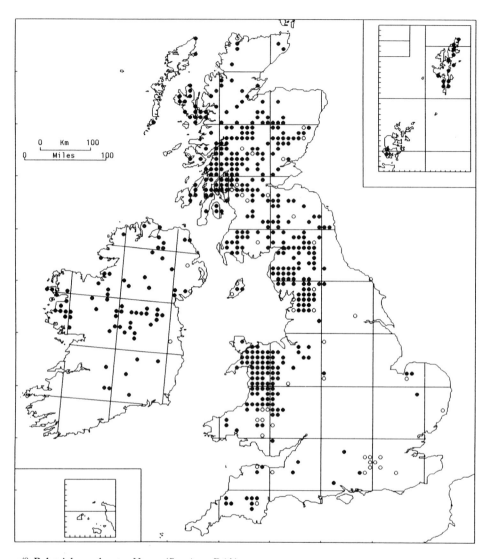

5/8. **Polytrichum alpestre** Hoppe (*P. strictum* Brid.)

Typically grows in dense patches or hummocks, often as almost pure stands or mixed with *Sphagnum*, in raised and blanket bogs and wet heaths, more rarely in damp peaty woodlands, invariably on acid substrates. Associated *Sphagnum* species include *S. capillifolium*, *S. magellanicum* and *S. papillosum*. Suitable ground is mostly in the lower hills and uplands; where peaty sites exist on mountains, it can occur up to 980 m (Glas Maol). GB 409+47*, IR 70+5*.

Dioecious; sporophytes frequent, summer.

Common in the boreal zone and the Arctic. Rare in S. Europe, increasing in frequency towards the north and west where peaty habitats prevail. Occurs also in S. America and Antarctica.

Observations and experiments by Miles & Longton (1990) in N. Wales failed to provide any evidence of spore germination in the field. New shoots were formed by regeneration from shoot fragments, from old shoots and from rhizoid 'wicks'.

D. G. Long

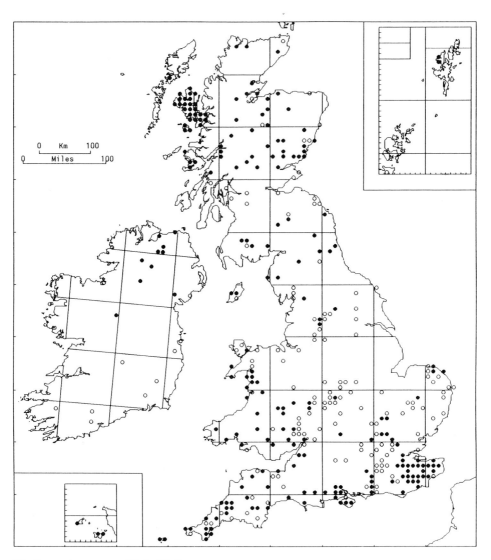

6/1. **Pogonatum nanum** (Hedw.) P. Beauv.

Most occurrences are on acid substrata in the lowlands, where it colonizes relatively bare soil, often in shade, for example along edges of woodland rides, on sandy banks of streams and ditches, or sometimes on bare ground on heaths and moors. It can grow in extensive, almost pure stands or as scattered stems amongst other pioneer species such as *Diplophyllum albicans*, *Jungermannia gracillima* and *Dicranella heteromalla*. 0–550 m (Skye). GB 210+139*, IR 10+12*.

Dioecious; sporophytes, ripening in winter, are common.

Widespread throughout lowland Europe except in calcareous regions. Macaronesia, N. Africa.

It is probably often overlooked, and may be under-recorded in some districts. Some records may be errors due to confusion with *P. aloides*.

D. G. Long

75

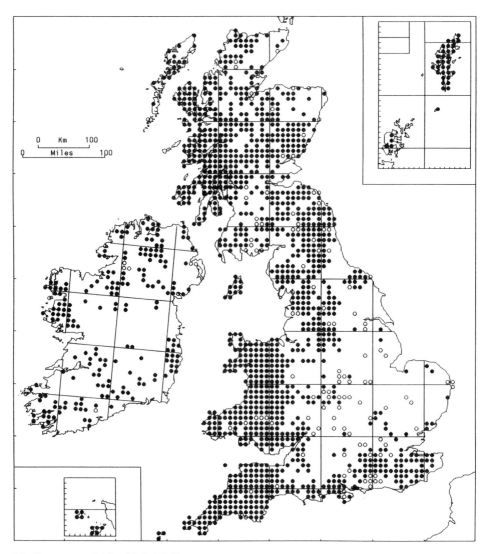

6/2. Pogonatum aloides (Hedw.) P. Beauv.

This species appears to be very similar to *P. nanum* in its ecological requirements, but is generally commoner everywhere, especially in upland areas. It is a typical pioneer species on vertical soil-banks in woodland, shady river-banks, old quarries, and, in upland districts, by ditches and forestry tracks, often forming extensive colonies. A very characteristic habitat is on soil clinging to upturned roots of fallen trees. From sea-level it becomes commoner in hilly districts to 400 m, but is rarer in mountains (600 m, Loch Brandy). GB 1230+117*, IR 196+10*.

Dioecious; capsules common, often abundant.

Common throughout C. and N. Europe, rarer in the south. Macaronesia, N. Africa, Asia.

D. G. LONG

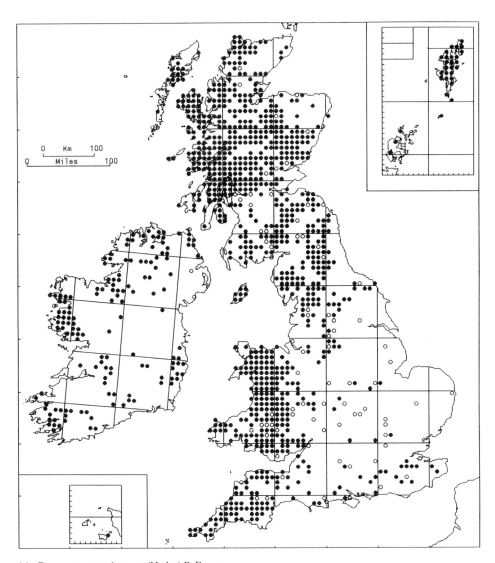

6/3. Pogonatum urnigerum (Hedw.) P. Beauv.

A calcifuge pioneer species often conspicuous on old footpaths, wall-tops, old quarries and gravelly shingle by streams; very common on forestry roads and cuttings, with other pioneers such as *Diplophyllum albicans, D. obtusifolium, Nardia scalaris, Polytrichum juniperinum* and *P. piliferum*. On mountains, it is common in gravelly screes, late-snow areas and on dry eroded slopes. 0–1330 m (Ben Nevis). GB 880+82*, IR 143+4*.

Dioecious; sporophytes common. In the Arctic it propagates by caducous leaf laminae, but this has been observed only once in a British population.

Circumboreal, reaching the High Arctic. Widespread throughout Europe but rarer in the south.

D. G. LONG

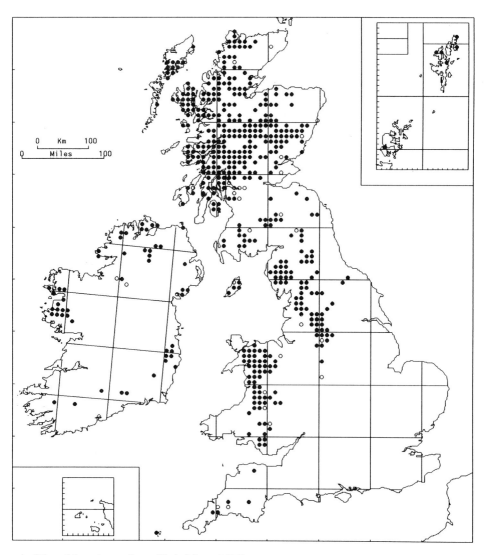

7/1. **Oligotrichum hercynicum** (Hedw.) Lam. & DC.

A calcifuge pioneer of disturbed ground in the mountains, as on shingle by streams, gravelly footpaths, eroding slopes, gullies and stony screes, sometimes in late-snow areas; occasionally on lake margins where it may be intermittently inundated. More local in the lower hills and lowlands. 100–1330 m (Ben Nevis). GB 438+27*, IR 51+4*.

Dioecious; sporophytes are sporadically produced but sometimes locally abundant, late summer.

In Europe, frequent in mountains of central regions and throughout the northern countries; rare elsewhere. Asia, western N. America, Newfoundland, Greenland.

D. G. Long

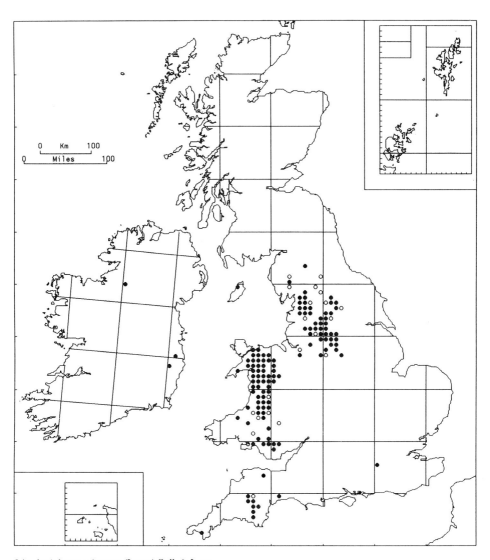

8/1. Atrichum crispum (James) Sull. & Lesq.

A plant of moist semi-open habitats on acid or very acid soils, including wet gravel and bare peat. It is commonest by streams and ditches on moorland, forming large tufts mixed with detritus among grass, rushes, etc. Other habitats include lake-margins, edges of forest roads, and rock crevices by waterfalls. In its typical habitat it has rather few bryophyte associates except for very acid-tolerant species such as *Nardia compressa*, *Scapania undulata* and *Dicranella heteromalla*. 0–450 m (Berwyn Mts). GB 120+21*, IR 3.

Dioecious; British and Irish plants are male. Rhizoidal tubers often abundant (Arts, 1987a).

Eastern N. America from Florida to Newfoundland. Recently reported from Belgium, Luxemburg and Spain, but the records from Belgium and Luxemburg are queried by Arts (1987a).

Presumably introduced from N. America. First found by John Nowell near Rochdale in 1848, it was locally abundant in W. Lancashire at the beginning of this century (Wheldon & Wilson, 1907), by which time it was also known from Devon and Wales. It was first discovered in Ireland in 1957, and in the Isle of Man in 1968.

M. O. Hill

79

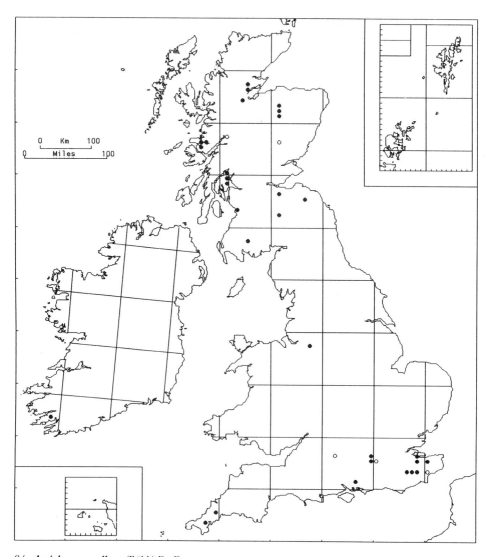

8/2. Atrichum tenellum (Röhl.) Br. Eur.

A sporadic colonist of bare, moist non-calcareous soil, occurring on sand, gravel and clay. Habitats include woodland rides, tracksides, ditches, lake-margins, the bed of a pond, and a disused sand-pit. Most occurrences are in habitats recently disturbed by human activity. In S.E. England, where colonies may be more persistent, it often occurs with *A. angustatum* (Rose, 1975b). Lowland. GB 29+6*, IR 1.

Dioecious; capsules very rare. Tubers are abundant in Belgian specimens, surviving for up to 20 months in herbarium packets (Arts, 1987a); they have not yet been reported in Britain or Ireland.

N., C. and E. Europe. Turkey, Siberia, Japan, western and eastern N. America.

M. O. Hɪʟʟ

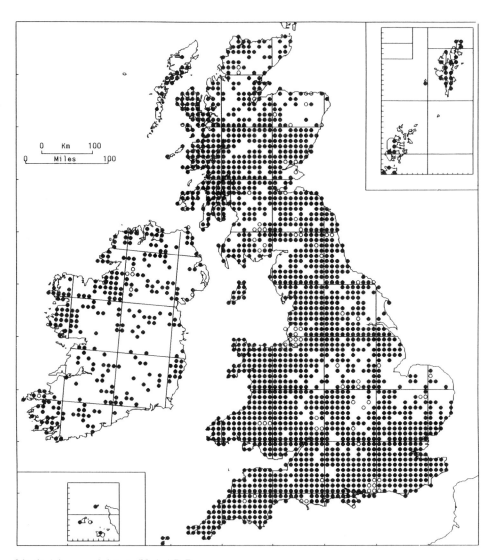

8/3. **Atrichum undulatum** (Hedw.) P. Beauv.

An abundant plant of bare non-calcareous ground that is either sheltered, shaded or moist, especially in woods, on banks, by tracks and by streams. It occurs in a wide range of other habitats, including grassland, moorland, ravines, screes, quarries, gardens and roadsides. 0–500 m (Snowdon). GB 1880+76*, IR 272+7*.

Monoecious; capsules frequent to common, ripe in winter. Lateral spread of the colonies is by an extensive system of rhizoid 'wicks'; new colonies can arise from detached leaves (Longton & Schuster, 1983).

Circumboreal, mainly in the forest zone, becoming rare in the far north; extending south to N. Africa, Mexico and C. America.

At least three cytological races occur in the British Isles, but do not correspond to the recognized varieties. The taxonomic value of the varieties is uncertain and they are not mapped separately.

M. O. Hill

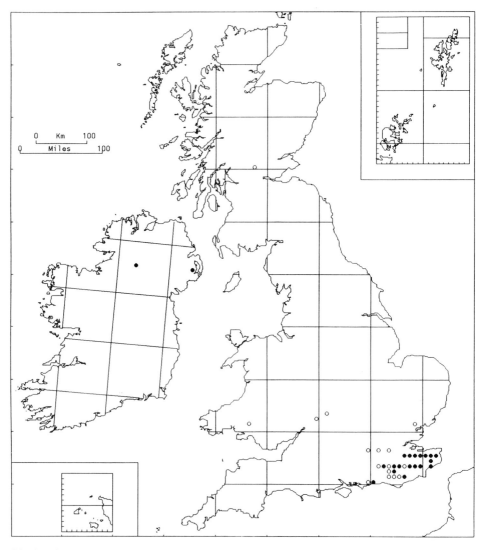

8/4. **Atrichum angustatum** (Brid.) Br. Eur.

In Kent and Sussex this is a plant of damp non-calcareous loamy rides and paths in ancient woodland, where it may be locally abundant with associates such as *Anagallis minima*, *Galium saxatile*, *Radiola linoides*, *Fossombronia wondraczekii*, *Lophozia incisa* and *Nardia scalaris* (Rose, 1975a). Elsewhere its occurrence is sporadic and possibly non-persistent. In Surrey it formerly occurred on heaths and commons; in Co. Tyrone it has been found in a disused sand-pit. Lowland. GB 18+16*, IR 2.

Dioecious; capsules rare.

Most of Europe north to Iceland and S. Sweden and east to European Russia. Azores, Madeira, Turkey, Caucasus, eastern N. America.

M. O. HILL

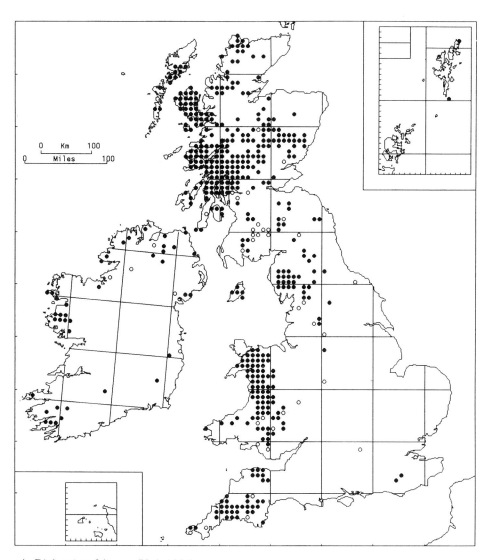

9/1. **Diphyscium foliosum** (Hedw.) Mohr

A frequent plant in the north and west but rarely present anywhere in great abundance. It favours shaded, often overhanging, well-drained bare sandy or earthy banks, generally acid and base-poor, by streams, in wooded valleys, and in low-lying wooded ravines and gullies, where it can tolerate deep shade. It also occurs in shaded soil-filled crevices in rock-walls of ravines, on damp, shaded, earthy or peaty banks amongst boulders in stable block-screes, on shaded earth-banks by tracks and forestry roads, and, more rarely, on soil-covered boulders in moorland, on bare soil on steep moorland slopes, and on soil-capped ledges and crevices of acid crags and montane cliffs. 0–1205 m (Ben Lawers). GB 410+36*, IR 39+6*.

Dioecious; sporophytes frequent.

Iceland, N., W. and C. Europe. Azores, Madeira, Caucasus, Japan, N. and C. America.

H. J. B. Birks

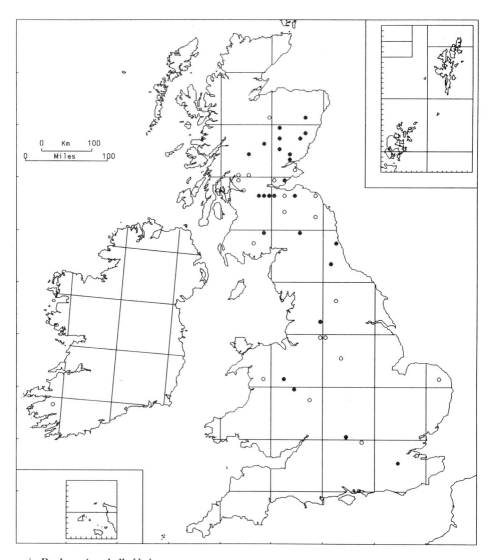

10/1. Buxbaumia aphylla Hedw.

A calcifuge of sporadic and often ephemeral occurrence, normally found scattered on humus-rich sandy soil. More rarely it occurs on rotten wood or litter, especially in planted coniferous woodland. Many recent finds have been from colliery waste, where it prefers N.- and E.-facing slopes of somewhat stabilized shale-debris or raw humus, and may persist for more than 20 years (Steven & Long, 1989). Common associates include *Cephaloziella* spp., *Gymnocolea inflata*, *Lophozia ventricosa*, *Dicranella heteromalla*, *Pohlia nutans*, *Polytrichum* spp. and crustose lichens. Lowland. GB 25+21*, IR 1*.

Dioecious; capsules ripe spring to autumn but long-persistent. No specialized method of vegetative dispersal.

Circumboreal but generally uncommon. Widespread in Fennoscandia including the Arctic, though decreasing to the north and with altitude; less frequent south to the Pyrenees, Alps and Carpathians. Caucasus, N. Asia, N. America; disjunct in New Zealand and Tasmania.

Growth is dependent on the persistent protonemal mat. Uggla (1958) regarded it as a colonist of burnt sites, present only at a certain successional stage.

F. J. Rumsey

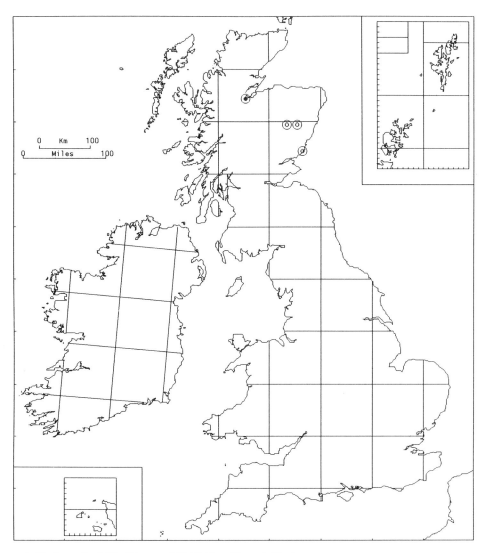

10/2. **Buxbaumia viridis** (Moug. ex DC.) Brid. ex Moug. & Nestl.

A species of sporadic occurrence, restricted to well-decayed wood, especially that of pines in native Caledonian pine forest. Lowland. GB 1+3*.

Dioecious; capsules ripe June to August.

Widely distributed in N. Europe, though not going so far north as *B. aphylla*, and in mountains further south; always in small quantity and uncertain in appearance. Turkey, Caucasus, China, western N. America.

Unlikely to be found unless fruiting, and perhaps under-recorded. Its apparent decrease may reflect the loss of semi-natural conifer woodland with abundant fallen trees.

F. J. RUMSEY

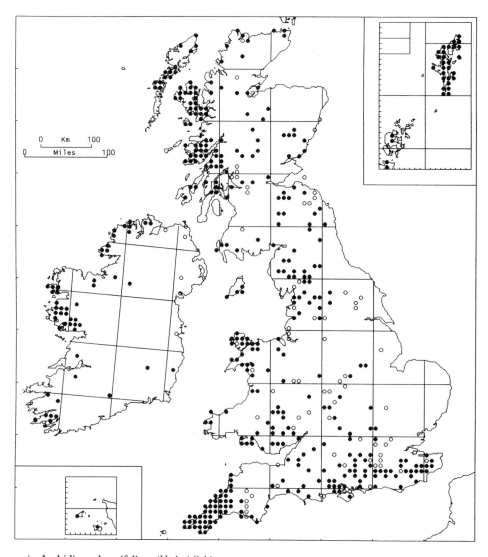

11/1. **Archidium alternifolium** (Hedw.) Schimp.

A plant of moist, non-calcareous open habitats, most often where the ground is kept bare by intermittent submergence, but sometimes also where disturbed by vehicles, trampling or sea-spray. It is commonest and locally abundant by lakes and reservoirs in the north and west, either in nearly pure stands or with associates such as *Fossombronia* spp., *Jungermannia gracillima*, *Scapania irrigua*, *Bryum pallens* and *Pohlia* spp. It also grows in moist quarries, in roadside ditches, and on damp tracks. In coastal habitats it occurs in spray-washed grassland on cliff-tops, on thin soil fringing rock outcrops, and in upper salt-marsh turf. In southern England it is chiefly a plant of damp tracks in woods and on heaths, especially on acid clay. Lowland. GB 378+75*, IR 51+6*.

Monoecious; capsules occasional to frequent.

Europe east to Greece and W. Russia, north to S. Fennoscandia, Faeroes and Iceland. Macaronesia, N. Africa, Turkey, south-eastern U.S.A., Mexico.

M. O. HILL

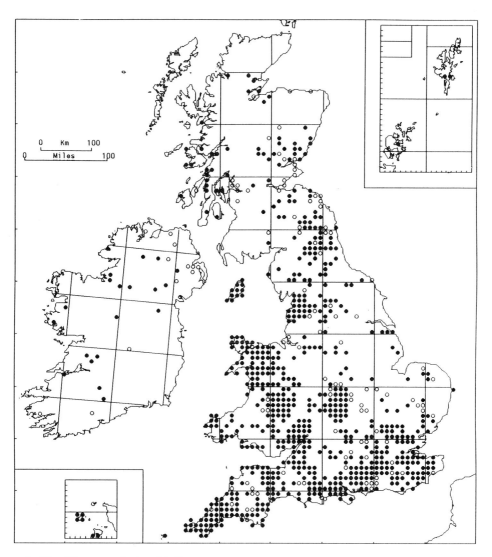

12/1. **Pleuridium acuminatum** Lindb.

This calcifuge plant is a common colonist of bare patches of sandy, clayey or peaty soils of banks, heaths, quarries, woods, pastures and arable land in damp or dry, lightly shaded to exposed places. 0–450 m (Bellavally). GB 659+104*, IR 20+13*.

Paroecious; capsules abundant, spores ripe in spring and summer. Rhizoidal tubers are frequent in specimens from nearby parts of Europe (Arts & Risse, 1988), and doubtless occur in Britain.

Nearly all Europe, from Portugal and the Mediterranean to beyond 60°N in Scandinavia, east to Turkey and W. Russia. Macaronesia, Algeria, E. China, N. America.

P. acuminatum sometimes grows with *P. subulatum*, and intermediates occur which may be hybrids. Some authors, e.g. Crum & Anderson (1981) and Ireland (1982), regard the two taxa as conspecific. The occurrence of tubers in *P. acuminatum* but not *P. subulatum* suggests that, in Europe, they are fairly distinct. In the British Isles, *P. acuminatum* is much more common and widespread than *P. subulatum*, but *P. subulatum* is the commoner species in continental Europe and America, especially in the north.

R. A. FINCH

87

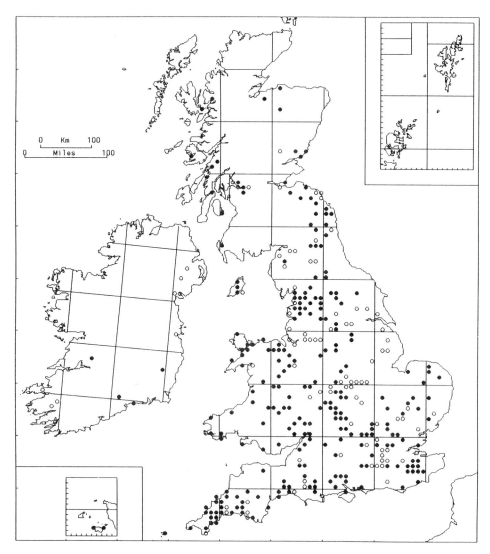

12/2. **Pleuridium subulatum** (Hedw.) Lindb.

P. subulatum colonizes bare patches of clayey, sandy, gravelly or loamy, neutral to acid soils of banks, pits, woods, quarries, pastures and cultivated ground in damp or wet sites, exposed or sometimes in shade. It is less markedly calcifuge than *P. acuminatum*, with which it sometimes grows. 0–380 m (Sunbiggin). GB 246+87*, IR 4+9*.

Autoecious; capsules abundant, spores ripe in spring and summer. No special method of vegetative dispersal, though flagelliform innovations common.

Nearly all Europe from Portugal and the Mediterranean to beyond 63°N near geysers in Iceland, east to Turkey and W. Russia. Macaronesia, N. Africa, Palestine, Caucasus, C. and E. Asia, Japan, N. America, Oceania, New Zealand.

See under *P. acuminatum* for notes on the taxonomy, relative frequencies and extra-European distribution of *P. subulatum* and *P. acuminatum*.

R. A. FINCH

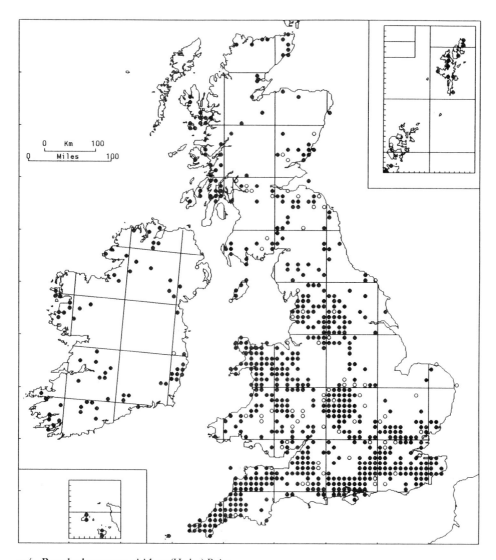

13/1. **Pseudephemerum nitidum** (Hedw.) Reim.

P. nitidum is a common early colonist of bare acid soil in damp places on clay, sand, loam or more rarely peat, occurring in woods, pastures, arable and fallow fields and by streams and pools. It is reported growing among *Puccinellia maritima* in salt-marshes by the Solway Firth (Adam, 1976). It tolerates full exposure and deep shade. Lowland. GB 640+76*, IR 68+3*.

Synoecious; capsules abundant, spores ripe mainly in late summer and autumn, but also throughout the year; stem often innovates below capsule to produce two or three capsules one above another, prolonging fruiting period.

Nearly all Europe from Portugal and the Mediterranean to W. Russia and 62° N in Finland. Macaronesia, Algeria, Morocco, C. Africa, Madagascar, Japan, western N. America (British Columbia), New Zealand.

The species occurs in greatest abundance in ephemeral communities on exposed mud of reservoirs and lakes. Furness & Hall (1981) have recorded a large buried spore-bank in lake-mud in the Pennines, associated with spores of *Physcomitrium sphaericum*. Spores survive in submerged mud for at least 10 years and probably much longer.

R. A. FINCH

89

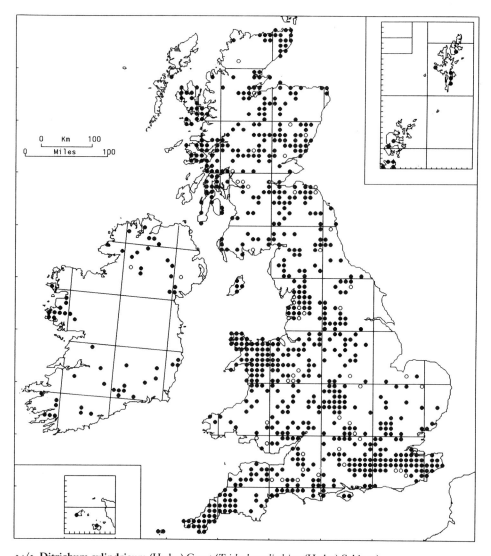

14/1. **Ditrichum cylindricum** (Hedw.) Grout (*Trichodon cylindricus* (Hedw.) Schimp.)

A common plant on non-calcareous disturbed ground such as arable fields. Recorded also from sandy banks, sand-pits and roadsides. *Bryum klinggraeffii, B. sauteri, B. violaceum, Ceratodon purpureus, Dicranella schreberana, D. staphylina, Leptobryum pyriforme, Pottia truncata* and *Pseudephemerum nitidum* are characteristic associates. In the mountains, it has been found on loose soil beneath a cliff, with *Bryum riparium* and *Pogonatum urnigerum*, but it is apparently very rare in this habitat. 0–700 m (Seana Bhraigh). GB 748+51*, IR 56+3*.

Dioecious; most plants are female and sporophytes are very rare, maturing in summer. Rhizoidal tubers are abundant.

Circumboreal, occuring widely in the Arctic as well as the temperate zone. Widespread in Europe.

H. L. K. WHITEHOUSE

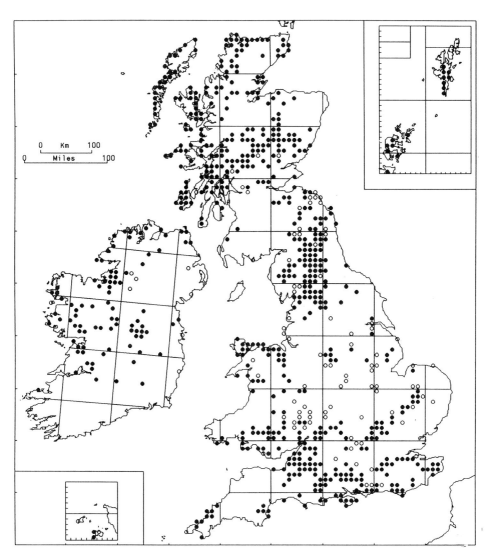

14/2. **Ditrichum flexicaule** (Schimp.) Hampe

A strict calcicole of open, often rocky turf. In chalk grassland on somewhat leached soils it is typically associated with *Campylium chrysophyllum*, *Ctenidium molluscum*, *Eurhynchium swartzii*, *Fissidens cristatus*, *Homalothecium lutescens*, *Tortella tortuosa*, *Trichostomum brachydontium* and *T. crispulum*. On the northern limestones it has many of the same associates as well as *Rhytidium rugosum*. The latter species is also an associate in a distinctive rabbit-grazed Breckland community, together with *Encalypta vulgaris*, *Hypnum cupressiforme* and the lichen *Diploschistes scruposus*. Other habitats include pathsides, sea-cliffs and montane rock-ledges, sometimes with *Dryas octopetala*; also sand-dunes and the machair of N.W. Scotland. 0–1180 m (Ben Lawers). GB 561+88*, IR 88+8*.

Dioecious; male plants rare, sporophytes very rare.

Circumboreal, extending north to the High Arctic, south to Madeira, N. Africa, Himalaya, Guatemala and Colombia; also known from New Zealand and New Guinea.

Frisvoll (1985) recognizes two species in Britain, namely *D. flexicaule sensu stricto* and *D. crispatissimum* (C. Müll.) Par. British bryologists have yet to be convinced of the distinctness of these segregates.

R. D. PORLEY

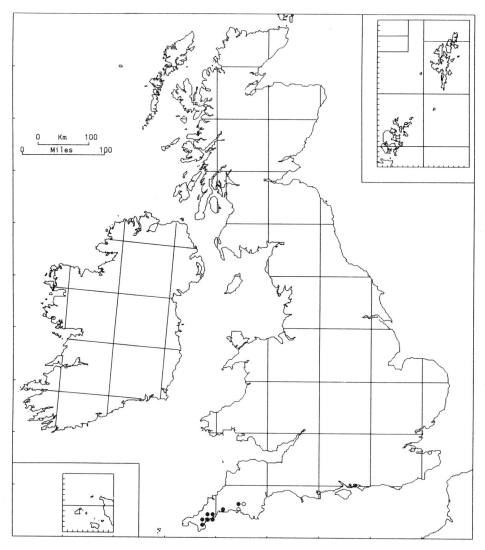

14/3. **Ditrichum subulatum** Hampe

A species of rather open sandy, clayey or loamy soils, most characteristic of crumbling earthy or rocky roadside-banks and earthy rock-crevices, often in light shade and never far from the coast. Associated species are few but it is sometimes found with *Cephaloziella turneri*, *Diplophyllum albicans* and *Dicranella heteromalla*. Lowland. GB 8+3*.

Monoecious; sporophytes common, maturing in winter and spring.

Widespread in the Mediterranean region, extending north along the Atlantic seaboard to W. France and S. England and east to Turkey. Macaronesia.

It is one of a group of species typical of warm, often sunny situations at low altitudes, belonging to the Mediterranean-Atlantic element of the British flora. Others include *Riccia crozalsii*, *R. nigrella*, *Epipterygium tozeri* and *Tortula canescens*.

R. D. PORLEY

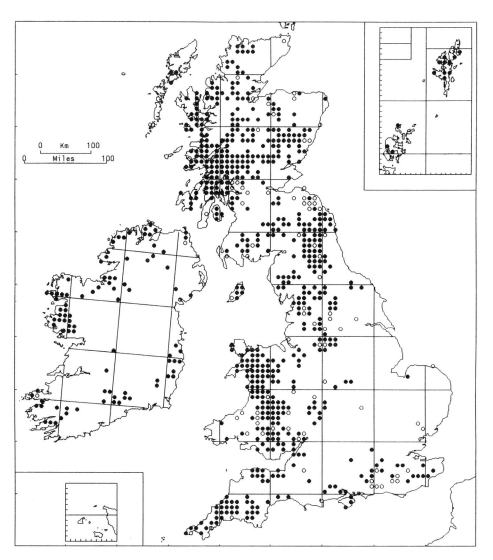

14/4. **Ditrichum heteromallum** (Hedw.) Britt.

A calcifuge growing on gravelly or sandy soils, often disturbed, in a wide variety of habitats. It is most frequent on heathy banks on moorland, in old quarries, around disused mine-workings, on streamsides, in woodland and on tracks and roadsides in hill country. In these situations it is often mixed with *Diplophyllum albicans*, *Jungermannia gracillima*, *Dicranella heteromalla* and *Pogonatum aloides*. Another distinctive habitat is on mountain detritus and amongst scree in the subalpine zone, where it is typically associated with *Nardia scalaris*, *Oligotrichum hercynicum* and *Pogonatum urnigerum*. 0–820 m (Ben More, Mull). GB 587+83*, IR 90+4*.

Dioecious; sporophytes common, maturing in spring. Rhizoidal tubers are reported from Germany, Norway and Poland but are unknown in British plants; protonemal gemmae have been produced by a tuber-bearing plant in culture (Risse, 1985b).

Throughout N. Europe including the Arctic, reaching Iceland and Jan Mayen, becoming rare and montane in the south; absent from Macaronesia. N. Africa, N., W., and E. Asia, Himalaya, western N. America.

R. D. PORLEY

93

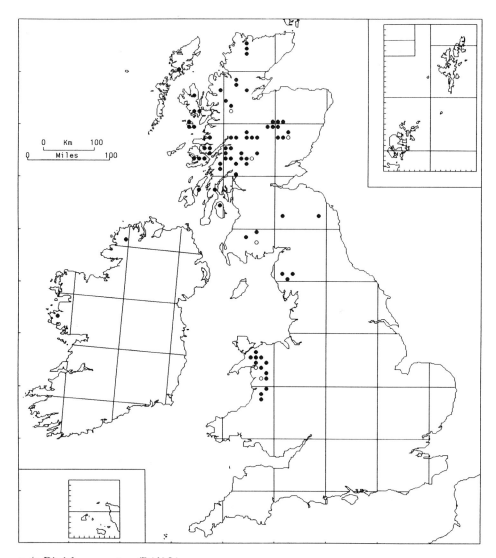

14/5. **Ditrichum zonatum** (Brid.) Limpr.

A species of acidic skeletal soils and rocks, occurring in crevices and amongst block-scree in corries and gullies, as well as on exposed ridges. It is highly characteristic of the hepatic-mat communities of late-snow beds and flushes in relatively sheltered situations with N. and E. aspects. In this habitat it is associated with many typical chionophilous species, including *Anthelia* spp., *Gymnomitrion* spp., *Lophozia* spp., *Marsupella* spp., *Moerckia blyttii*, *Nardia breidleri*, *Pleurocladula albescens*, and *Conostomum tetragonum*. When growing amongst block-scree and in rock crevices it is often associated with *Diplophyllum taxifolium*, *Marsupella* spp., *Andreaea* spp., *Kiaeria* spp., *Oligotrichum hercynicum*, *Polytrichum* spp. and *Racomitrium* spp. 800–1335 m (Ben Nevis), rarely down to 400 m. GB 74+6*, IR 2+1*.

Dioecious; sporophytes very rare.

Fennoscandia, Pyrenees, mountains of C. Europe. Western N. America.

Two varieties are recognized, although doubtfully distinct. Var. *zonatum* is constant in snow-bed communities whereas var. *scabrifolium* Dix. is more characteristic of periodically irrigated skeletal soils.

R. D. PORLEY

94

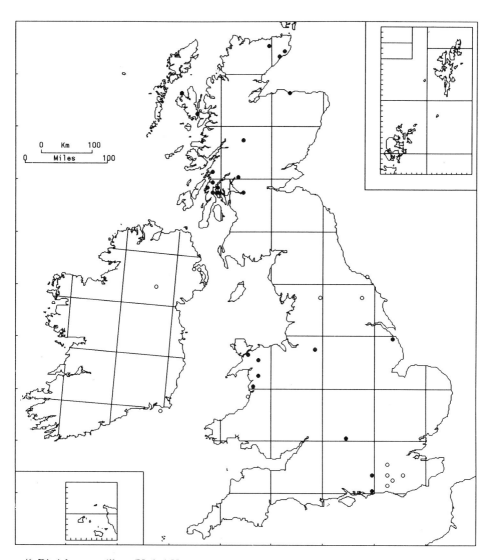

14/6. **Ditrichum pusillum** (Hedw.) Hampe

A rare species of sandy banks and quarries, also recorded from a number of non-calcareous arable fields. In these fields its associates have been *Riccia sorocarpa*, *Bryum rubens*, *Ceratodon purpureus*, *Dicranella staphylina*, *Ditrichum cylindricum*, *Pohlia annotina*, *Pottia truncata* and *Pseudephemerum nitidum*. In the mountains, it has been found on thin gravelly soil overlying limestone rocks. 0–900 m (Coire Cheap, Ben Alder range). GB 24+10*, IR 4*.

Dioecious; sporophytes frequent in undisturbed habitats, maturing in winter. The arable-field plants have all been female where the sex is known and have lacked sporophytes. Rhizoidal tubers are frequent in non-fruiting plants. Protonemal gemmae have been observed in culture.

Circumboreal, south to Florida (U.S.A.). Widespread in Europe.

Vegetative plants are very inconspicuous, and the species is almost certainly more frequent in Scotland than the map suggests.

H. L. K. Whitehouse

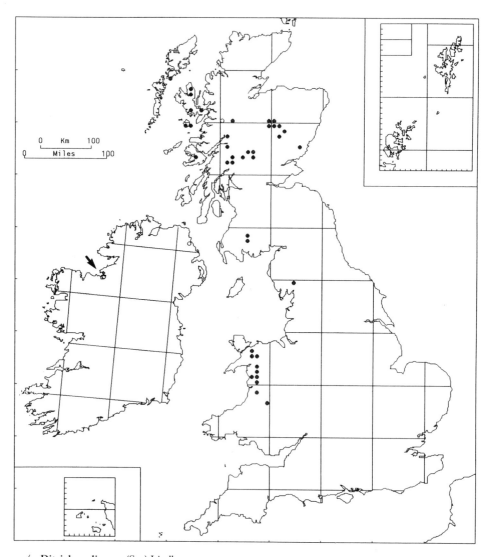

14/7. **Ditrichum lineare** (Sw.) Lindb.

It occurs in patches or as scattered shoots on bare, often disturbed, acidic mineral soils. At high altitudes, on exposed ridges and by late-snow beds, it is typically associated with *Anthelia juratzkana*, *Diplophyllum albicans*, *Gymnomitrion* spp., *Lophozia* spp., *Marsupella* spp., *Moerckia blyttii*, *Nardia breidleri*, *Ditrichum zonatum*, *Kiaeria* spp., *Oligotrichum hercynicum* and *Polytrichum* spp. At lower altitudes it is most frequently found on friable sandy banks, often by forest roads, associated with *Calypogeia fissa*, *Cephalozia bicuspidata*, *Diplophyllum* spp., *Jungermannia gracillima*, *Lophozia* spp., *Nardia scalaris*, *Dicranella heteromalla*, *Ditrichum heteromallum* and *Pogonatum aloides*. It occasionally grows by streams and lane-banks, and has been found, presumably as a casual, on a soil-capped derelict wall. The only English locality for this plant is unusual in being on clay overlying Carboniferous limestone. 100–1040 m (Beinn Heasgarnich). GB 39+1*, IR 2*.

Dioecious; only female plants known in Britain and Ireland. Rhizoidal tubers have been reported from Japan but are unknown in Britain.

N. Europe and mountains of C. Europe. Japan, eastern N. America.

In north-east N. America, where sporophytes are frequent, it is almost ubiquitous on acid roadside banks.

R. D. PORLEY

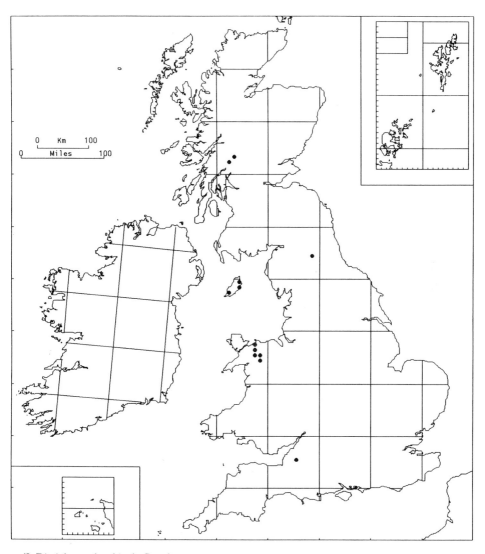

14/8. **Ditrichum plumbicola** Crundw.

This species is restricted to lead-mine spoil that is largely devoid of vascular plants. It occurs in small pure patches or as scattered plants, often with *Weissia controversa* var. *densifolia*, which is also characteristic of soils with a high lead content. Also frequently associated with it are *Cephaloziella* spp., *Diplophyllum albicans*, *Jungermannia gracillima*, *Nardia scalaris*, *Barbula fallax*, *B. rigidula*, *Bryum pallens*, *B. pseudotriquetrum*, *Ceratodon purpureus*, *Dicranella varia*, *Polytrichum aloides* and *P. piliferum*. In Wales it has been found on soil with pH in the range 4.5 to 6.5 (Hill, 1988b). Mostly at low altitudes, but one locality is at 460 m (Allenheads). GB 12.

Gametangia and sporophytes unknown. Gemmae also unknown; the main method of dispersal is presumably by stem fragmentation.

Endemic.

Unique among British bryophytes in having a distribution centred on the Isle of Man, where it was first found in 1914, although not described till many years later (Crundwell, 1976b).

R. D. PORLEY

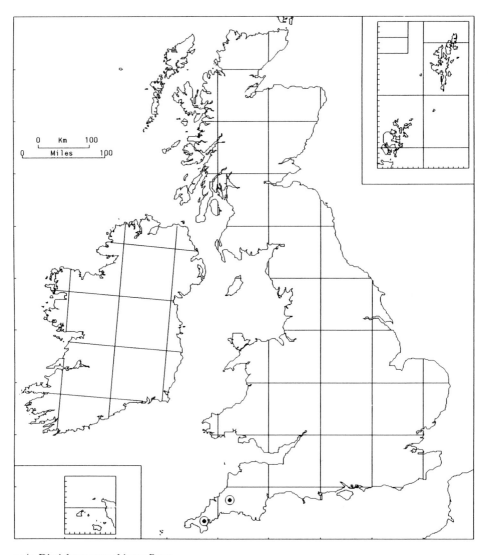

14/9. **Ditrichum cornubicum** Paton

Known only from copper-mine waste in two granite areas of Cornwall. In both areas it was associated with *Marsupella funckii*, *Ceratodon purpureus* and *Pohlia annotina*. In the eastern locality it also grew with *Cephaloziella* sp., *Jungermannia gracillima*, *Scapania irrigua*, *Dicranella rufescens*, *Oligotrichum hercynicum*, *Racomitrium aciculare*, *R. canescens* agg. and *Rhytidiadelphus squarrosus*. Associates noted only in the western locality were *Diplophyllum albicans*, *Gymnocolea inflata*, *Scapania compacta*, *Bryum rubens*, *B. ruderale*, *B. sauteri*, *Dicranella staphylina*, *Ditrichum cylindricum* and *Polytrichum juniperinum*. Lowland. GB 2.

Only male plants are known. Rhizoidal tubers are abundant.

Endemic.

First discovered in 1963 and described, new to science, by Paton (1976).

H. L. K. WHITEHOUSE

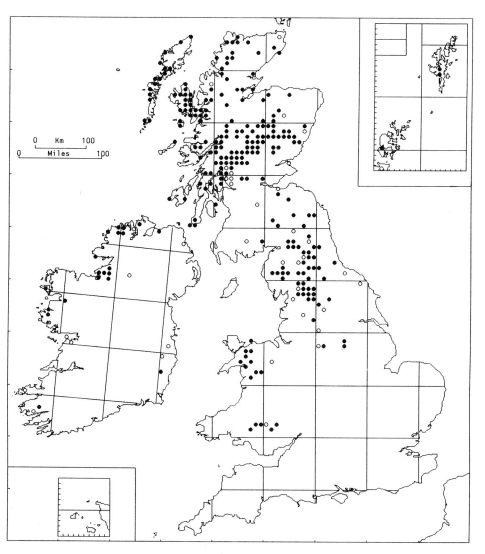

15/1. **Distichium capillaceum** (Hedw.) Br. Eur.

A frequent plant of damp basic rocks in the uplands of Britain, often abundant, especially on the schists and basalts of the Scottish Highlands. It prefers shaded fissures and overhangs where it will not dry out, but it usually avoids rocks over which water flows. On more acidic rocks, these fissures often follow dykes or intrusions which provide a sufficiently basic substrate. It occurs in similar situations in ravines. *D. capillaceum* is less frequent in the lowlands; it grows in calcareous turf, by rivers and streams, and very occasionally in dune-slacks, where it can look confusingly like *D. inclinatum*. 0–1205 m (Ben Lawers). GB 230+30*, IR 20+9*.

Monoecious; sporophytes common and often plentiful.

In cold places throughout the world, except for S.E. Asia, extending north to the High Arctic. Widespread in Europe, with a boreal-montane distribution.

P. H. PITKIN

99

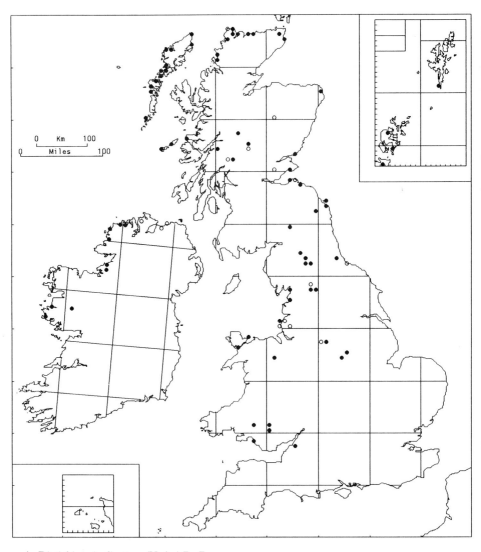

15/2. **Distichium inclinatum** (Hedw.) Br. Eur.

On the coast, *D. inclinatum* is a plant of calcareous dune-slacks, where it grows in closely grazed turf that is rich in herbs and bryophytes. It usually forms dense and hard, pure tufts, and is particularly frequent in the dune systems of the N. coast of Scotland and the Outer Hebrides. Inland, it occurs in basic grassland, on calcareous rocky soil, and on base-rich rock-ledges and rock debris, sometimes colonizing man-made habitats such as quarries and a railway cutting. In dune-slacks it often grows with *Amblyodon dealbatus*, *Catoscopium nigritum* and *Meesia uliginosa*, which are similar in their adaptation to both coastal and upland habitats. 0–950 m (Carn Gorm, Glen Lyon). GB 62+13*, IR 9+3*.

Monoecious; sporophytes often abundant.

Throughout the cooler parts of the Northern Hemisphere, from the High Arctic south to the mountains of N. Africa, Himalaya and Colorado (U.S.A.). In Europe generally a mountain plant rather than a coastal one.

It has been lost from several sites in Lancashire, where dune-slacks have become overgrown with thick grass and *Salix repens* scrub.

P. H. Pitkin

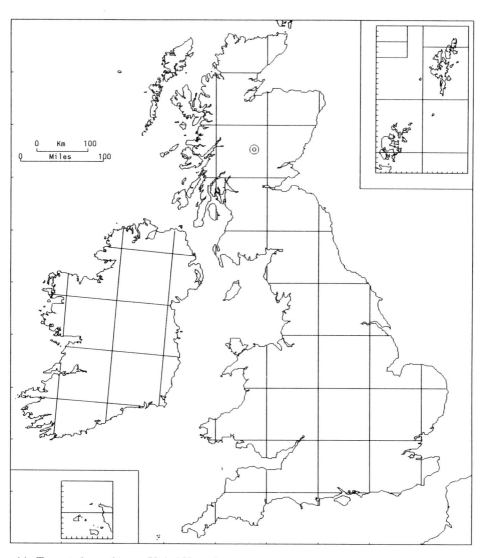

16/1. **Trematodon ambiguus** (Hedw.) Hornsch.

On damp, somewhat disturbed ground rich in organic material. The single British find was from a moorland track where it was growing amongst *Bryum pallens* at about 350 m. GB 1*.

Autoecious; fruiting late summer. The only British material consisted of a tuft with nine sporophytes.

Scattered throughout Fennoscandia and N. Europe, including Iceland; rare in C. Europe. Himalaya, Japan, Kamchatka, Alaska, British Columbia, eastern N. America from Labrador to New York.

Gathered on 4 August 1883 on the moorland between Tummel Bridge and the northern face of Schiehallion. Never refound but is easily overlooked when not fruiting.

F. J. RUMSEY

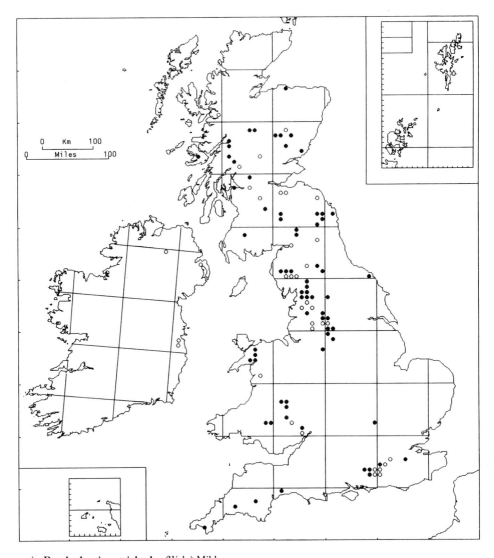

17/1. Brachydontium trichodes (Web.) Milde

In hill districts this species is found locally on vertical or slightly overhanging faces of ravines, mountain cliffs and boulders, on the sides of small boulders in stony flushes, and on stones in stable screes and old quarries. Unshaded sites probably always have a N. to E. aspect. The substrate, which may be moderately calcareous to strongly acid, is most commonly sandstone, but can be any of a wide range of soft siliceous rocks, including the crumbling weathered surfaces of igneous rocks such as granite and basalt. Associated species include *Fissidens pusillus* and *Tetrodontium brownianum*. In lowland districts it almost invariably grows on shaded sandstone. 0–1000 m (Snowdon). GB 70+31*, IR 5*.

Autoecious; sporophytes abundant, ripening in autumn.

A subatlantic species, found mainly in mountain districts from Portugal and northern Spain east to the Carpathians, north to Poland and central Norway. Caucasus, U.S.A. (Appalachian Mts, Washington State).

M. F. V. CORLEY

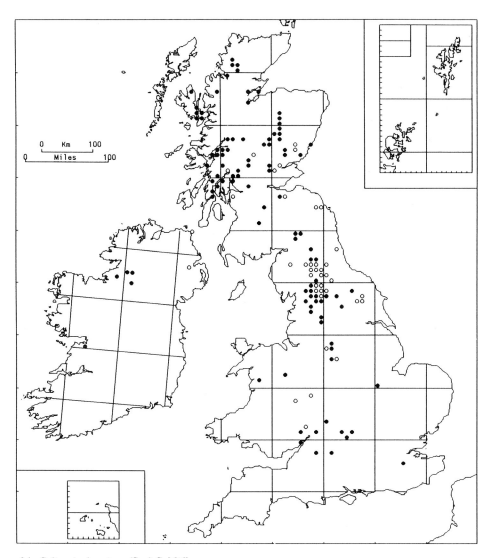

18/1. **Seligeria donniana** (Sm.) C. Müll.

A species of shaded or well-sheltered calcareous rocks, most often on cliffs, but also in crevices, under overhangs, in pot-hole entrances, and on small limestone fragments on the ground in woodland. The commonest substrate is limestone of some type, but not including chalk; locally it is found on calcareous sandstone and schist. Among the most frequent associates are *Jungermannia atrovirens* and *Fissidens pusillus*. 0–730 m (Knock Fell). GB 97+37*, IR 5+2*.

Autoecious; fruit is rather frequent, maturing in summer.

Boreal-montane, ranging in Europe from the Pyrenees and Alps to Fennoscandia. Widespread in the Northern Hemisphere but local or rare through most of its range.

M. F. V. CORLEY

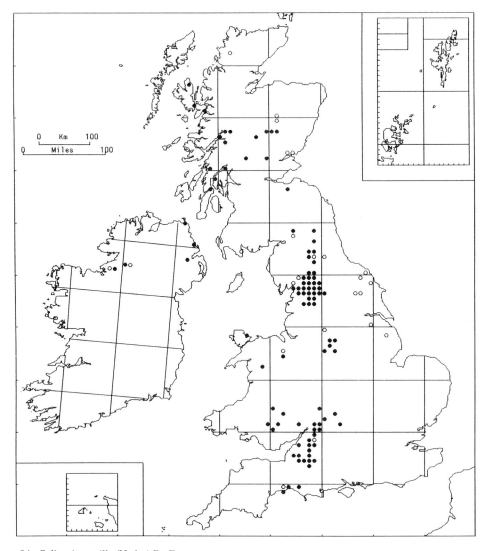

18/2. **Seligeria pusilla** (Hedw.) Br. Eur.

A minute plant of shaded basic rocks, especially limestone, but also basalt and calcareous sandstone, rarely on chalk. It occurs on cliffs and ravine walls, where it is found on vertical and overhanging surfaces. 0–470 m (Wasset Fell). GB 90+23*, IR 5+2*.

Autoecious; sporophytes common, maturing in summer.

Scattered through Europe north to Sweden. Scattered also in Asia and eastern N. America, with disjunctive occurrences in Arctic Alaska and at 81°N in Arctic Canada (Ellesmere Island).

M. F. V. Corley

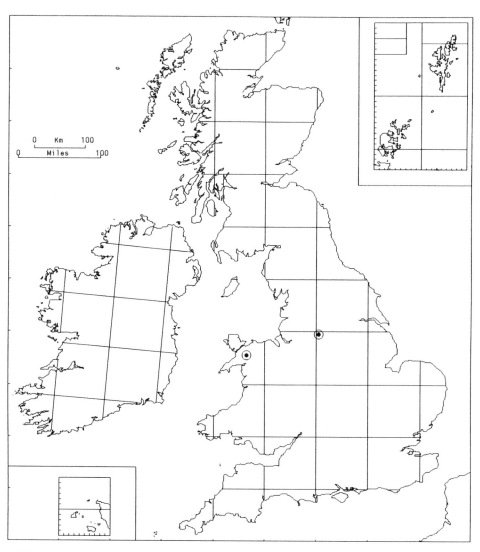

18/2A. **Seligeria brevifolia** (Lindb.) Lindb.

In moist, sheltered situations under overhanging rock, sometimes on the upper surface growing vertically downwards. In Wales, the substratum is pumice tuff, a porous base-rich volcanic rock; associates include *Conocephalum conicum*, *Anoectangium aestivum* and *Pohlia cruda*. In Derbyshire the substratum is Millstone Grit, where it is known only in a single small patch, associated with *Blindia acuta* and *Tetrodontium brownianum*. 400 m (Derbyshire) and 700 m (Snowdon). GB 2.

Autoecious; capsules abundant, ripe September.

N. Europe and mountains of C. Europe, rare and scattered. There are a few scattered localities also in Siberia, the Caucasus and eastern N. America.

First found in Britain in 1978 (Hill, 1980). Owing to its small size, it is easily overlooked.

M. O. HILL

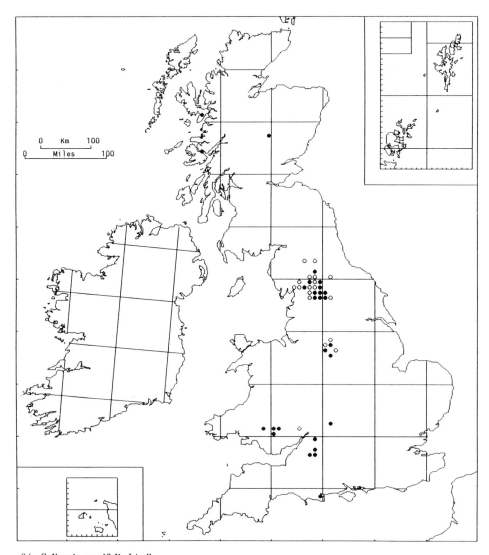

18/3. **Seligeria acutifolia** Lindb.

On moist, shaded limestone rocks, where it occurs in crevices and under overhanging rocks. In some western localities it occurs on metamorphosed chalk, where it may become impregnated with tufa. Associated species include *Jungermannia atrovirens* and *Orthothecium intricatum*. 0–450 m (Churn Milk Hole, Pen-y-Ghent). GB 27+18*.

Autoecious; sporophytes are usually present, ripening in summer.

Scattered through Europe, from France, Italy and Crete north to the British Isles and Fennoscandia. Caucasus, Turkey, Japan, N. America (Vancouver Island).

M. F. V. CORLEY

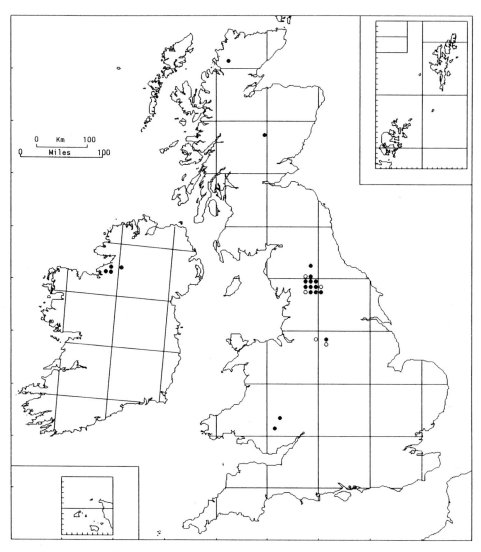

18/4. **Seligeria trifaria** (Brid.) Lindb.

Often the only bryophyte species present on lightly shaded, vertical or overhanging Carboniferous limestone rocks down which water seeps. An algal scum is usually present, and the moss is often heavily impregnated with tufa. It occasionally grows in similar situations on other hard limestones. Mostly at moderate or low altitudes, reaching 500 m on Nateby Common. GB 16+5*, IR 4.

Autoecious; capsules rather less frequent than in other members of the genus, but not uncommon.

North Spain, Alps, Tatra Mts, Hungary and Greece, extending eastwards to the Caucasus and Sayan Mts.

M. F. V. CORLEY

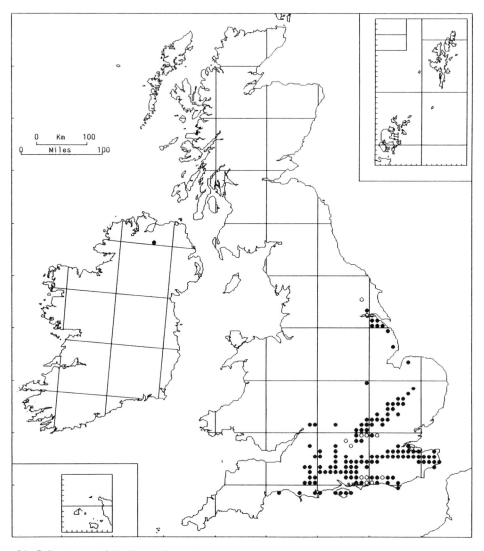

18/5. **Seligeria paucifolia** (Dicks.) Carruthers

Always found on fragments of chalk, of various sizes, normally on the ground in woodland, with such species as *Fissidens pusillus* var. *tenuifolius*, *Rhynchostegiella tenella*, *Taxiphyllum wissgrillii* and *Tortella inflexa*, but also in chalk pits and, rarely, in chalk grassland. Lowland, up to 250 m in Surrey. GB 132+13*, IR 1+1*.

Autoecious; capsules usually present, ripening in summer.

France, Belgium, Italy; a European endemic which is very rare outside Britain.

M. F. V. CORLEY

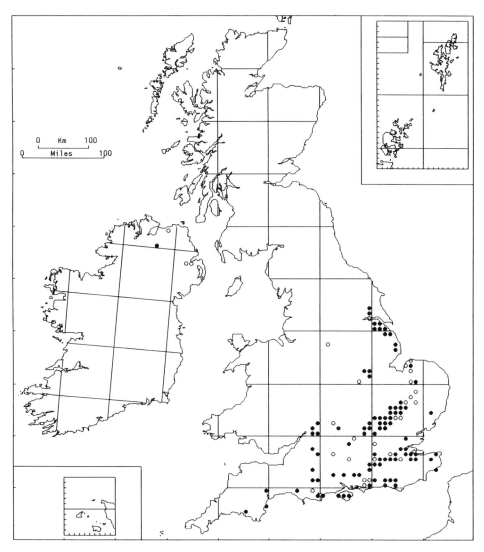

18/6. **Seligeria calcarea** (Hedw.) Br. Eur.

It grows on shaded, bare, calcareous vertical rock-faces, blocks and rubble. The substrate is chalk or oolite, rarely harder limestone. Disused chalk-pits are a frequent habitat, where it is reputedly a slow colonist, not becoming established until many years after the chalk is first exposed. Very rare on masonry, it has been found on a limestone balustrade. Associated species include *Leiocolea turbinata*, *Seligeria paucifolia* and *Tortula muralis*. Lowland, reaching 220 m in Surrey. GB 91+25*, IR 1+3*.

Autoecious; capsules are commonly present, ripening in summer.

Scattered through much of Europe north to Sweden and east to the Black Sea, but absent from the Arctic and most of the Mediterranean countries. Widespread but scattered in eastern N. America, from Tennessee north to Manitoba and Newfoundland; very rare in western N. America, but reaching 61°N in the Northwest Territories of Canada.

M. F. V. CORLEY

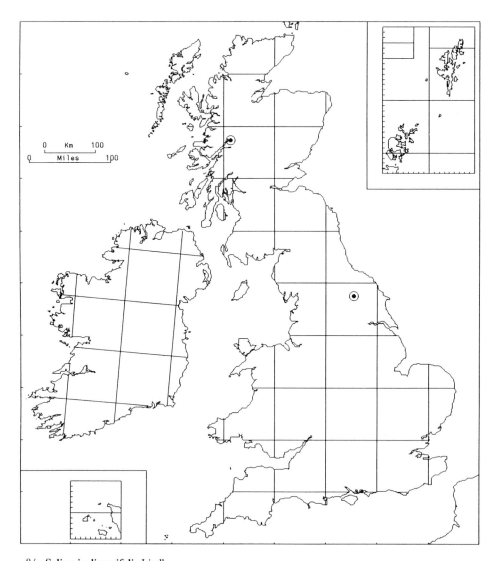

18/7. Seligeria diversifolia Lindb.

A plant of calcareous cliffs and crags, characteristically on the vertical surface of the rock. The Yorkshire locality is on a wooded calcareous cliff of the Upper Jurassic. The immediate associates at this site are not known but *S. recurvata* is plentiful and larger calcicoles such as *Tortella tortuosa*, *Neckera crispa*, *N. complanata* and *Rhynchostegiella tenella* also occur. In its Scottish locality, *S. diversifolia* has been found on a rock-face among crags. 150 m (Yorkshire) and 400 m (Beinn Riabhach). GB 2.

Autoecious; capsules common, ripening in spring (Yorkshire) and summer (Scotland).

N. and C. Europe, rather rare and with an eastern tendency. Sayan Mts, Arctic Alaska, Yukon, Quebec, New Brunswick.

First found in Britain in 1971 (Crundwell & Nyholm, 1973), it has been seen only once at each locality. It has possibly been overlooked because of confusion with other species of the genus, but is unlikely to be widespread in Britain.

T. L. BLOCKEEL

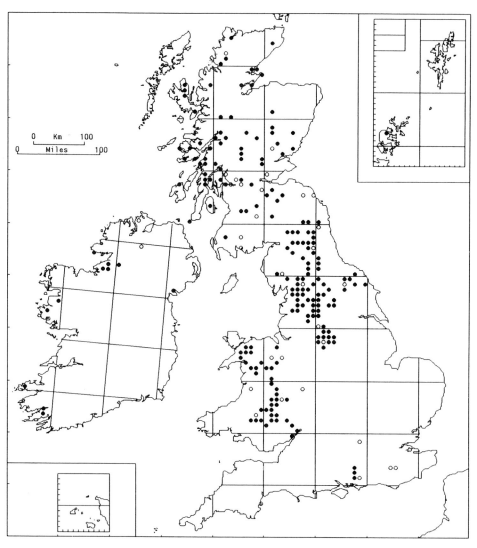

18/8. **Seligeria recurvata** (Hedw.) Br. Eur.

Less exclusively calcicolous than most of its congeners, this species occurs on a range of rock types including basalt, slate, limestone and sandstone. Its most frequent sites are at the foot of sheltered mountain-cliffs, in crevices and beneath overhangs, and on rocks by streams, often in dense shade. In S. England it is found on sandstone in hedgebanks. 0–700 m (Snowdon). GB 187+31*, IR 11+2*.

Autoecious; fruit is abundantly produced, ripe in early summer.

Widespread in Europe, especially C. Europe, becoming rare in N. Scandinavia and in the Mediterranean countries. Scattered through Asia and N. America.

M. F. V. CORLEY

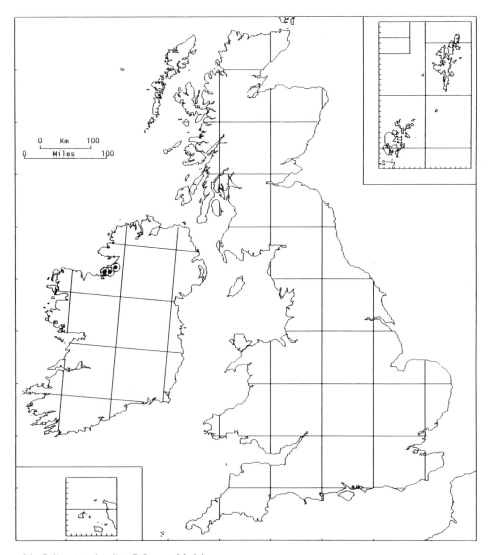

18/9. **Seligeria oelandica** C. Jens. & Medel.

This rare disjunct species is confined to N.E.-facing vertical Carboniferous limestone rock-faces kept constantly wet by water seepage; the plants tend to be encrusted with lime. The only associated bryophytes are *Jungermannia atrovirens* and *Seligeria trifaria*. 300 m. IR 3.

Autoecious; sporophytes are plentiful, maturing in summer.

In Europe known only from Ireland, Sweden (Öland, Gotland and Lapland), Norway and Svalbard, where it grows by hot springs. In northwest N. America it has been found in Arctic Alaska and in the Ogilvie Mts of the Yukon.

Its N. American localities are from the unglaciated portion of Alaska-Yukon, an area noted for its disjunct bryophytes, including eight out of the thirteen N. American *Seligeria* species (Vitt, 1976). Its presence in Ireland suggests that some Irish limestone was available as a habitat throughout the Pleistocene.

M. F. V. CORLEY

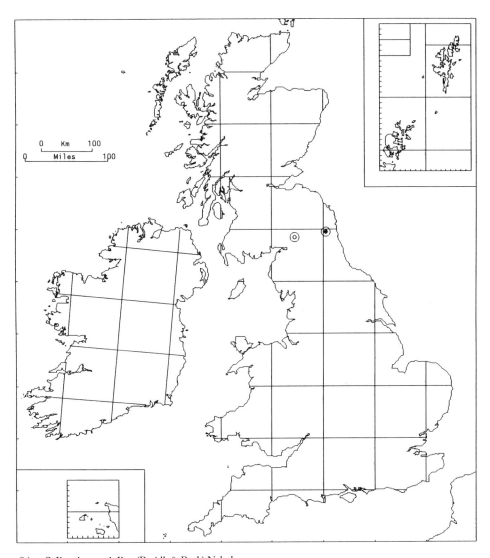

18/10. **Seligeria carniolica** (Breidl. & Beck) Nyholm

This very rare disjunct species is confined in Britain to rocks beside two streams in the Borders. Here it grows on damp calcareous sandstone or impure limestone, together with other small mosses including *Gyroweisia tenuis*, *Seligeria recurvata* and *Tetrodontium brownianum*. 250 m. GB 1+1*.

　　Autoecious; capsules frequent, ripening in summer.

　　Outside Britain confined to a very few localities in the Alps (Germany, Switzerland, Austria and Yugoslavia), France (Jura) and Scandinavia (Gotland and Norway). Its distribution is mapped by Dirske *et al.* (1990).

M. F. V. CORLEY

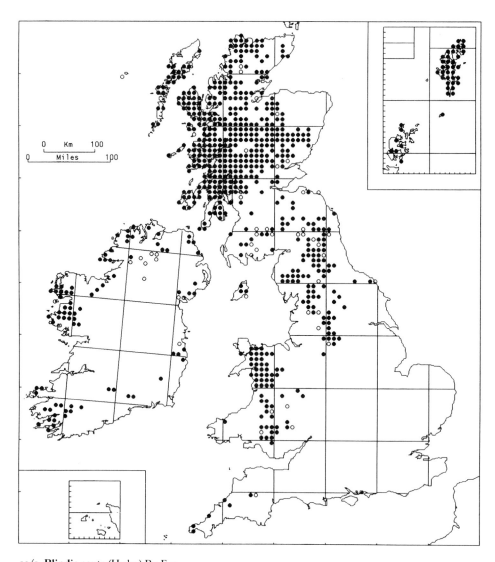

19/1. **Blindia acuta** (Hedw.) Br. Eur.

Often dominant on near-vertical, moderately basic rock-faces in the mountains where water seeps or drips from above, but also found by waterfalls, on stones at the edge of streams and flushes and occasionally submerged on stones in streams. In flushes it may grow on gravel, and it has been found on fixed sand-dunes. *Marsupella emarginata* and *Scapania undulata* are almost constant associates. Rare outside mountain districts; 0–1200 m (Ben Lawers). GB 598+39*, IR 81+11*.

Dioecious; sporophytes are frequent, maturing in summer.

Circumboreal, extending north to the High Arctic and south to Nepal and Guatemala. Almost throughout Europe north to Svalbard, but confined to the higher mountains in C. and S. Europe.

M. F. V. CORLEY

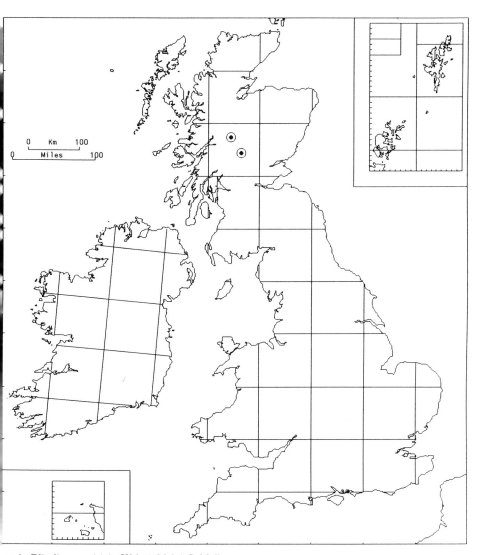

19/2. **Blindia caespiticia** (Web. & Mohr) C. Müll.

A calcicole of vertical rock-faces on N.- and E.-facing slopes, where it grows both on cliffs and on detached boulders of limestone or mica-schist. Associates include other rare montane calcicoles such as *Barbilophozia quadriloba*, *Campylium halleri*, *Encalypta alpina*, *Hypnum bambergeri*, *Lescuraea plicata* and *Myurella julacea*. 900 m (Ben Alder range) to 1175 m (Ben Lawers). GB 2.

   Autoecious; capsules are produced abundantly, maturing in summer.

   In all the main mountain ranges of C. and N. Europe. Possibly a European endemic; a record from C. Africa requires confirmation (Bartlett & Vitt, 1986).

M. F. V. CORLEY

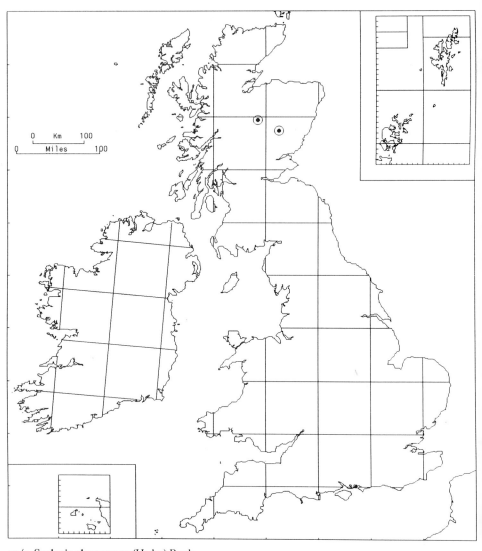

20/1. **Saelania glaucescens** (Hedw.) Broth.

A very rare but locally frequent plant of damp, basic montane rocks. It usually grows on damp soil in crevices of calcareous schists or on a thin covering of micaceous soil, always where shaded. In Glen Doll, at about 600 m, it is associated with *Oxytropis campestris*. GB 2.

Autoecious; capsules frequent, ripe in late summer.

Circumboreal with disjunct occurrences in the Southern Hemisphere. Widespread in N. Europe including the Arctic, north to Svalbard, south to the Pyrenees, Alps and Carpathians. Turkey, Caucasus, N. Asia, Himalaya, N. America, Greenland; southern Africa, Hawaii, New Zealand.

Said to be decreasing in its Clova locality (Coker, 1968). The characteristic glaucous coloration is largely due to an intimate covering of fungal hyphae, possibly symbiotic.

F. J. RUMSEY

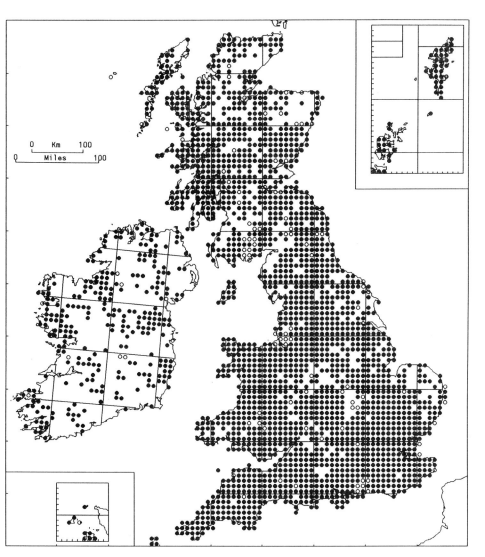

21/1a. **Ceratodon purpureus** (Hedw.) Brid. ssp. **purpureus**

A pioneer of open or disturbed, base-poor, usually dry, substrates. It is often abundant on peaty or sandy soil in moorlands, heathlands and acidic grasslands, on rocky and peaty banks, shingle beaches, leached sand-dunes, in sand- and gravel-pits, in arable fields and on railway ballast. It also grows in woodland rides and gardens, on walls and roofs, the flat tops of fence-posts, decaying thatch and wood, and as an epiphyte on bark, especially of elder and willow. Small tufts can turn up on almost any base-poor substrate, from old bracket-fungi to decaying boots. *C. purpureus* responds to nutrient-enrichment, and is often luxuriant on burnt ground and around rabbit warrens. It is also pollution-tolerant, extending into the centre of industrial conurbations. 0–1050 m (Glas Maol). GB 2120+92*, IR 272+9*.

Dioecious; sporophytes frequent and locally abundant, maturing in spring and early summer. Filamentous gemmae are rare; reports of rhizoid tubers are almost certainly erroneous.

Throughout Europe, from the melting glaciers of the Arctic southwards. *C. purpureus sensu lato* (including the tropical *C. stenocarpus* Br. Eur.) has a cosmopolitan distribution.

C. D. Preston

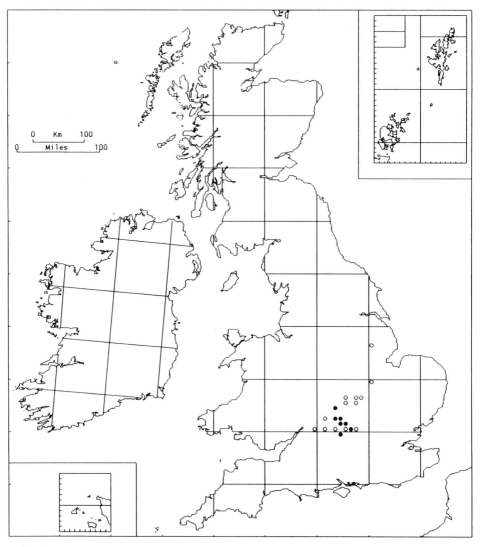

21/1b. **Ceratodon purpureus** (Hedw.) Brid. ssp. **conicus** (Hampe) Dix. (*C. conicus* (Hampe) Lindb.)

A plant of thin, dry earth over oolitic limestone. It was most often found on the top of earth-capped limestone walls, and in crevices in such walls. It is also known from paths (e.g. in cracks between paving-stones) and on hard, bare ground in stone quarries. Commonly found with *Barbula convoluta*, *B. fallax* and *B. hornschuchiana*; other associates on wall-tops were *Aloina* spp., *Encalypta vulgaris*, *Pottia lanceolata* and *Pterygoneurum ovatum*. Unlike ssp. *purpureus* it is a strict calcicole, and the two taxa only occasionally grow together. Lowland. GB 7+13*.

Dioecious; sporophytes infrequent, ripe May and June.

Norway, Sweden (Öland), Germany, the Alps and Crete. Canaries, N. Africa, S.W. Asia. Records from elsewhere require confirmation.

This account is based on a revision of British material in a recent generic monograph (Burley, 1986; Burley & Pritchard, 1990), which concluded that ssp. *conicus* merits specific rank. It has declined in Britain as the practice of earth-capping walls has ceased, and the walls themselves have been destroyed. It survives in small quantity on paths.

J. S. BURLEY & C. D. PRESTON

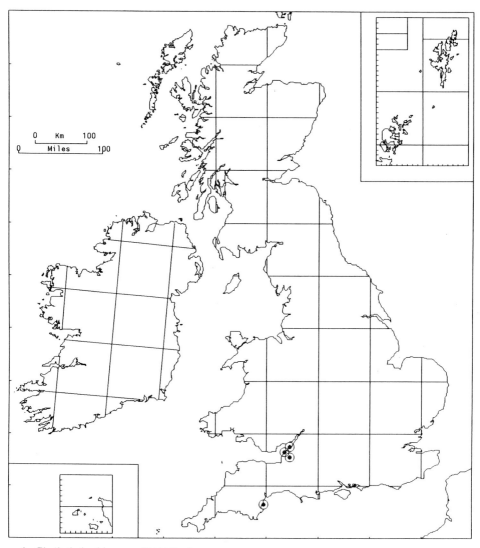

22/1. Cheilothela chloropus (Brid.) Broth.

On S.-facing limestone hills and sea-cliffs. It grows in loose turves or cushions or as scattered individuals in open communities on thin, compacted or friable soil around rock outcrops, where competition from coarse vascular plants is reduced by summer drought. It is also found in closed, short, rabbit-grazed *Festuca ovina–Koeleria* grassland. Associated species include *Barbula* spp., *Bryum caespiticium* var. *imbricatum*, *Pleurochaete squarrosa*, *Pottia recta*, *Scorpiurium circinatum*, *Trichostomum brachydontium*, *T. crispulum* and *Weissia microstoma*. Lowland. GB 4.

Dioecious; sporophytes have not been recorded in Britain. They are infrequent in the Mediterranean region, where they mature in spring.

A Mediterranean species which extends north along the Atlantic coast of Europe to the British Isles. Macaronesia, Algeria, S.W. Asia.

One of a small group of species whose European distribution is mainly southern and which in Britain is confined to limestone in S.W. England and S. Wales. Others include *Helianthemum apenninum*, *Koeleria vallesiana*, *Trinia glauca* and *Weissia levieri*.

R. D. Porley & C. D. Preston

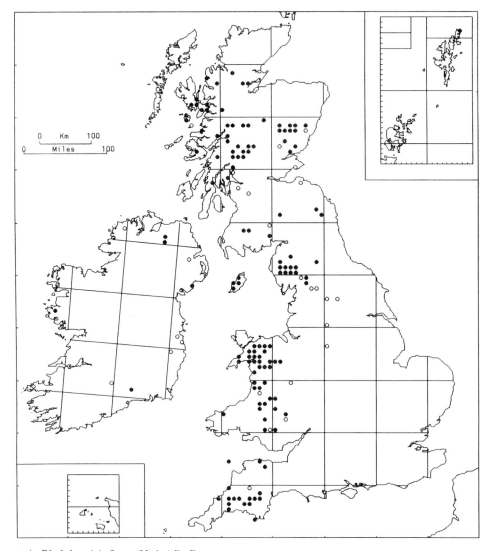

23/1. **Rhabdoweisia fugax** (Hedw.) Br. Eur.

In gritty rock-crevices, among stabilized scree and on cliff-ledges, favouring cool, shady, rather dry situations, sometimes spreading to shaded drystone walls. A strict calcifuge, it grows on siliceous rock without any enrichment with bases. Mostly at low and medium altitudes, often in woodland, ascending to 800 m on Stob Garbh. GB 125+22*, IR 5+9*.

Monoecious; capsules abundant, ripening irregularly through the year.

Widespread in Europe, north to the Arctic, becoming rare and scattered in the mountain ranges of the south. Macaronesia, Caucasus, S. Africa. Reported from C. and S. America.

M. O. Hill

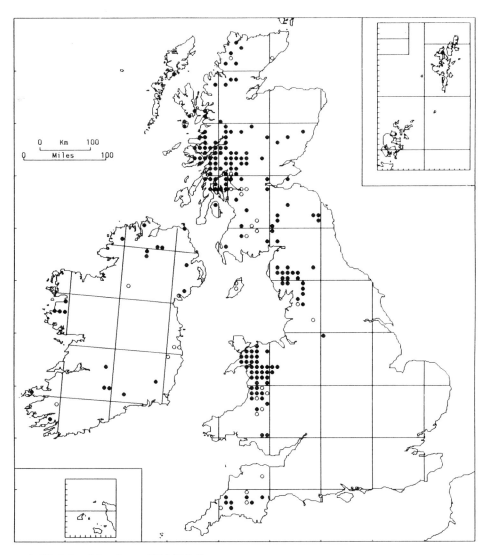

23/2. **Rhabdoweisia crispata** (With.) Lindb.

In crevices of siliceous rock, on cliff-ledges and under overhanging boulders, favouring sheltered and lightly shaded situations, typically in gullies, in wooded ravines, among scree and on crags. It sometimes colonizes quarry-waste. It is not restricted to acid rocks, occurring on basic schist in several parts of the Scottish Highlands. 0–950 m (Snowdon). GB 172+24*, IR 20+7*.

Monoecious; capsules abundant, ripening irregularly through the year.

Widespread in N. Europe and the mountains of C. and S. Europe. S. Africa, E. Asia south to Java, Hawaii, America from Arctic Alaska south to Bolivia, Greenland.

Intermediate between *R. fugax* and *R. crenulata* in morphology, its habitat overlaps that of both of them, especially *R. crenulata*.

M. O. Hill

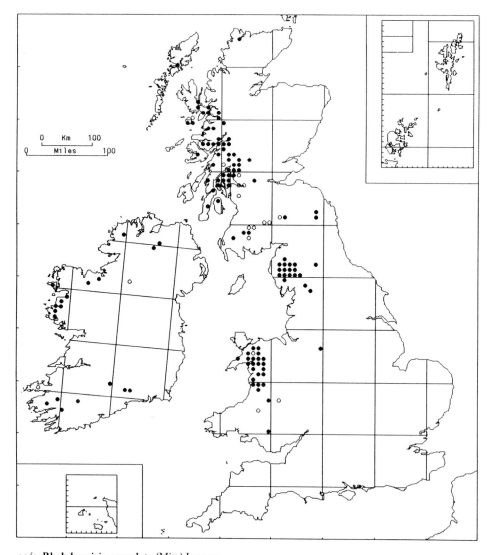

23/3. **Rhabdoweisia crenulata** (Mitt.) Jameson

On shaded ledges and crevices of siliceous rock in humid or damp situations, often under overhangs. It occurs typically on N.-facing crags, among boulders in scree, and in ravines, especially near waterfalls. Usually on acid substrata, it is also found occasionally on basic rock. 0–800 m (Clogwyn du'r Arddu). GB 98+15*, IR 19+2*.

Monoecious; capsules abundant, summer.

France, Germany, Belgium, Norway; very rare in Europe outside the British Isles. Himalaya, China, Hawaii, U.S.A. (N. Carolina), Colombia, Greenland.

The disjunct distribution of this species resembles that of several large liverworts such as *Pleurozia purpurea* and *Scapania ornithopodioides*. The liverwort disjunctions are sometimes attributed to fragmentation of range following failure of sexual reproduction, but such an explanation cannot possibly apply to the abundantly fertile *R. crenulata*.

M. O. HILL

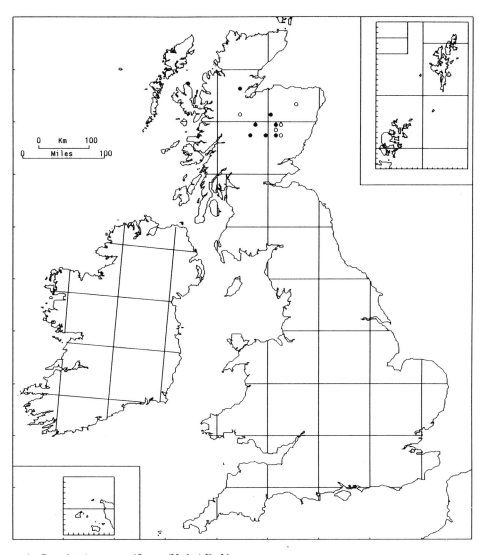

24/1. Cynodontium strumiferum (Hedw.) De Not.

A rare moss of acidic rocks and stones in sheltered screes, often in gullies. Sometimes associated with *Anastrophyllum saxicola* and *Cynodontium tenellum*. Mostly at 350–600 m (Sow of Atholl), but at only 30 m in Skye. GB 8+5*.

Autoecious; fruits in summer.

Circumpolar in boreal and arctic regions, extending north to the High Arctic and south to the mountains of Colorado (U.S.A.). Known from C. and E. Europe and from Scandinavia, where commonest in the east.

All records are based on fruiting plants: plants without sporophytes are hardly to be distinguished from *C. polycarpon*. Old British records were often erroneous, being based on other strumiferous mosses, notably *C. jenneri* and *Kiaeria blyttii*.

A. C. CRUNDWELL

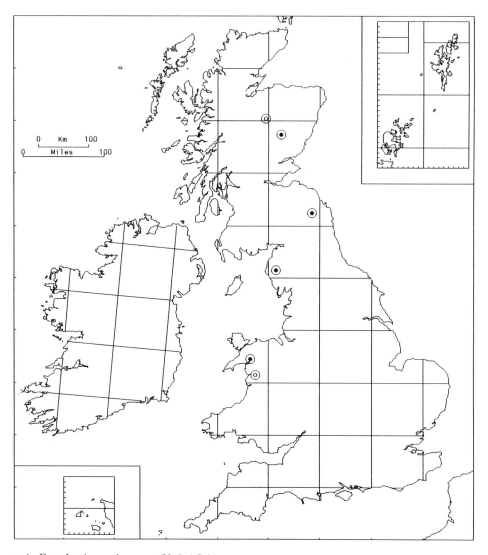

24/2. **Cynodontium polycarpon** (Hedw.) Schimp.

A rare species of sheltered or shaded upland rock-faces and screes, ascending to 880 m on Cader Idris. GB 4+2*.

Autoecious; fruits in summer.

Discontinuously circumboreal. E. and C. Europe, extending to N. Spain and Portugal; widespread in W. and C. Scandinavia, less frequent in the east. Absent from most of America; however the authors of a recent Canadian checklist (Ireland *et al.*, 1987) accept a record from the Northwest Territories.

Mapped records are based on fruiting specimens only. Old British records are unreliable, most of them being based on specimens of *C. jenneri*, *C. tenellum* and species of other genera.

A. C. CRUNDWELL

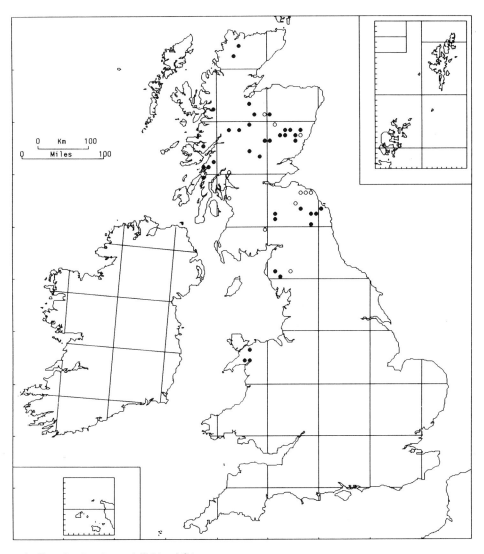

24/3. Cynodontium jenneri (Schimp.) Stirt.

A species of acid rocks, usually sheltered or in light shade, rarely on stone walls, occasionally on sandy soils in woodland. Frequent at low altitudes, but reaches about 730 m on Ben Loyal. GB 37+11*.

Autoecious; fruit common, summer.

In Europe known only from Westphalia (two localities) and from Scandinavia (from W. Finland westward). Very rare outside Europe; in Newfoundland and the Pacific Northwest of America.

Old British specimens were often named *C. polycarpon* or *C. strumiferum*.

A. C. CRUNDWELL

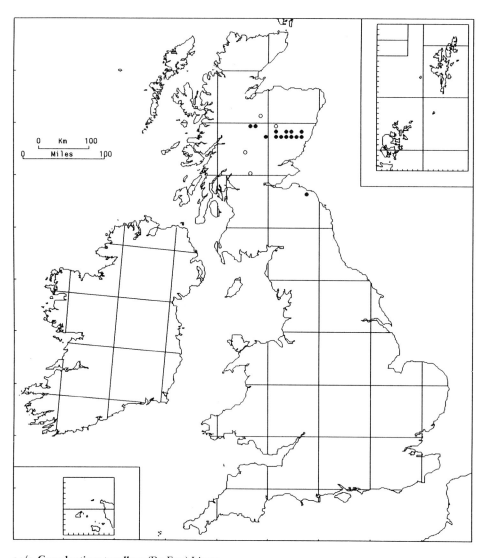

24/4. **Cynodontium tenellum** (Br. Eur.) Limpr.

A rare moss of acid rocks in screes and by streams; also found once on a wall. Not known from very high or very low altitudes, descending to c. 150 m in Berwickshire (Elba), reaching 600 m in Perthshire (Braigh Feith Ghiubhsachain). GB 14+4*.

Autoecious; fruits in summer.

Circumboreal, extending north to the High Arctic. Scandinavia (rare in the south), C. and E. Europe.

Mapped records are based on fruiting plants only. It was formerly confused in Britain with *C. polycarpon.*

A. C. CRUNDWELL

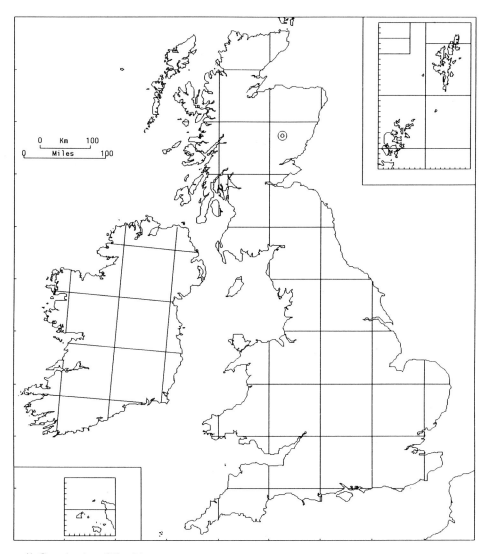

24/6. **Cynodontium fallax** Limpr.

Known in Britain only from a single specimen gathered in 1868 from Glen Fee, Clova, without ecological, altitudinal or phenological data. GB 1*.

Autoecious; sporophytes present.

Scattered in Sweden, Norway and in C. and E. Europe. N. Asia, east to the Altai Mts.

In continental Europe it fruits in summer and grows on shaded rocks in slightly damper situations than the other species of this genus. The British specimen was originally named *C. gracilescens* (Web. & Mohr) Schimp. by R. Braithwaite.

A. C. CRUNDWELL

127

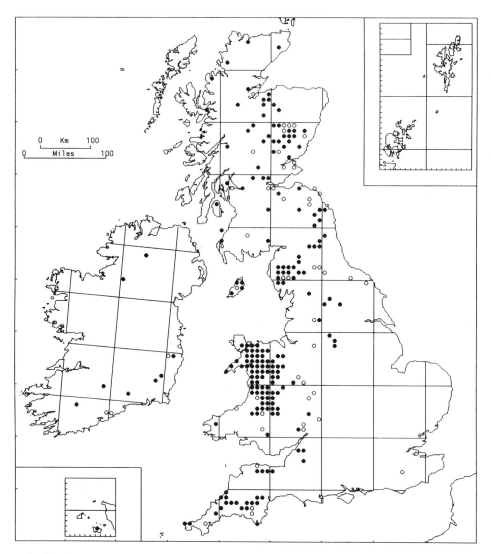

24/7. **Cynodontium bruntonii** (Sm.) Br. Eur.

On sheltered siliceous rocks and in rock crevices, frequently in light shade, sometimes in gullies or in woodland, very rarely on stone walls. The aspect is often N.-facing. In most districts it grows on acid rock, but in S.E. Scotland it is confined to slightly basic basaltic rocks. A plant predominantly of low altitudes but reaching 480 m (Creag Bhalg, Glen Quoich). GB 188+46*, IR 8+6*.

Autoecious; fruit common, summer. Rhizoidal tubers are produced frequently in plants from nearby parts of Europe (Arts, 1990) and doubtless occur in Britain and Ireland. Gemmae are produced on the protonema in culture.

In Europe most frequent in the west, extending eastwards to the Carpathians. Azores, Canaries, Turkey.

Its virtual absence from S.E. England is almost certainly related to the rarity of suitable rock rather than to climate. It is very similar in appearance in the field to *Dicranoweisia cirrata* and is no doubt occasionally overlooked for this reason.

A. C. Crundwell

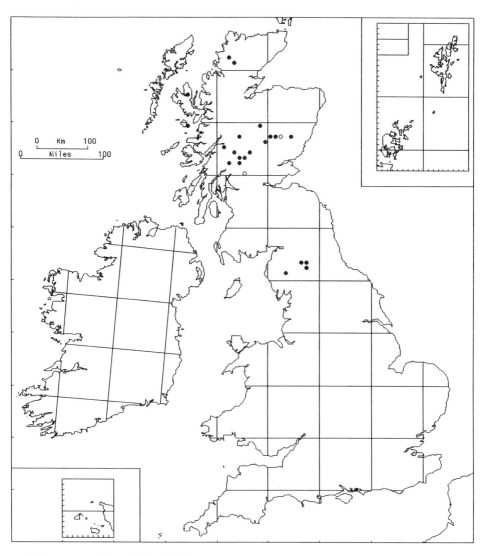

25/1. **Oncophorus virens** (Hedw.) Brid.

In damp calcareous turf, especially in flushes and close to streams, often mixed with species such as *Selaginella selaginoides, Thalictrum alpinum, Scapania undulata, Bryum pseudotriquetrum, Calliergon sarmentosum, Campylium stellatum, Drepanocladus revolvens* and *Philonotis fontana*. Rare associates in rich flushes include *Carex atrofusca, Tritomaria polita* and *Timmia norvegica*. Less often it is found on base-rich rock-faces, e.g. with *Orthothecium rufescens*. 640 m (Skye) to 970 m (Beinn Heasgarnich). GB 21+2*.

Monoecious; capsules frequent, summer.

Circumpolar arctic-alpine, extending from the Arctic south to N. Africa, Caucasus, Himalaya and California (U.S.A.).

D. G. LONG & M. O. HILL

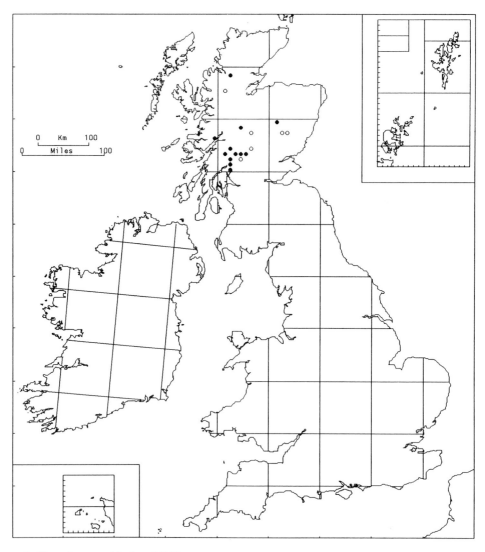

25/2. **Oncophorus wahlenbergii** Brid.

Apparently more restricted ecologically than *O. virens*, being found usually in relatively pure cushions on damp rocky hillsides and by springs and streamsides. Associates include *Calliergon sarmentosum* and *Fissidens adianthoides*. Mostly above 700 m, ascending to 1000 m (Creag Meagaidh). GB 12+6*.

Monoecious; capsules frequent, summer.

Circumpolar, from the High Arctic south to the Alps, Caucasus, Himalaya and Arizona (U.S.A.).

In boreal forest, this species is particularly characteristic of rotten logs, but this habitat is hardly available to it on British mountains.

D. G. Long & M. O. Hill

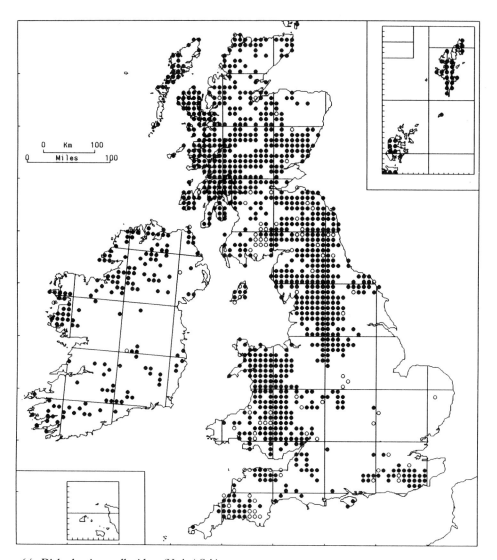

26/1. Dichodontium pellucidum (Hedw.) Schimp.

This typically forms pure turves on deposits of fine gravel or sand beside hill-streams and rivers, but also occurs on poorly drained soil elsewhere, in dune-slacks and, rarely, on periodically wet brickwork. It tends to be associated with basic but not specifically calcareous substrates. 0–1170 m (Ben Lawers). GB 1017+91*, IR 157+4*.

Dioecious and frequently fertile, although inflorescences are often of only one sex within a population; sporophytes occasional, maturing from late autumn to early spring. Gemmae are produced copiously on filaments originating from the stem (Newton, 1989), and on the protonema in culture (Whitehouse, 1987).

Mainly in Arctic and boreal regions, extending across N. and C. Europe, Siberia, China, Japan and N. America. In the north it reaches Greenland, Iceland and Svalbard; in the south it extends to the Pyrenees, Alps, Caucasus and Colorado (U.S.A.).

All non-fruiting *Dichodontium* records are mapped here. A few may be *D. flavescens*, which can be distinguished with certainty only by the form and inclination of the capsule. Gardiner (1981) suggests that the frequency of *D. pellucidum* in S.E. England has increased in recent years.

M. E. NEWTON

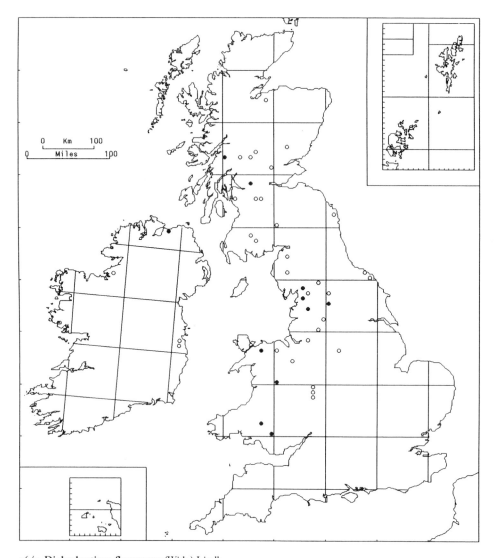

26/2. Dichodontium flavescens (With.) Lindb.

In similar habitats to *D. pellucidum* but tending to be restricted to more frequently inundated sites. Most localities are at low altitudes in hilly or mountainous areas. GB 10+30*, IR 1+3*.

Dioecious; identifiable only when bearing sporophytes, which mature from late autumn to early spring. Gemmae are not present on fertile stems.

The large proportion of old records mapped reflects the fact that the distributional data are mostly derived from herbarium specimens; it is not an indication that the species has declined. Distribution outside Britain is uncertain because other authorities have not emphasized the sporophyte characters regarded as essential by Smith (1978). Plants with erect capsules have been reported from C. Europe and N.W. America.

M. E. NEWTON

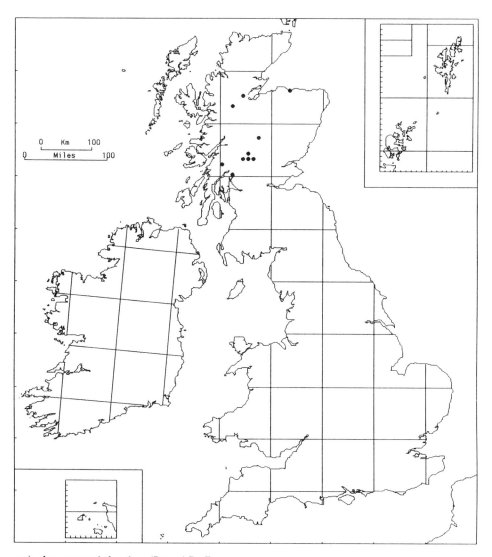

27/1. **Aongstroemia longipes** (Somm.) Br. Eur.

A plant of open communities on damp, slightly basic sandy soils, just below Highland dams, in quarries, gravel-pits and shingle-beds, sometimes associated with *Dicranella varia* and *Fossombronia incurva*. Ascends to 490 m (Lochan na Lairige). GB 10.

Dioecious; capsules known from only one British locality, ripe in summer. Vegetative propagation by means of fragile stems sometimes results in populations of one sex only.

Iceland, N. and C. Europe. Siberia, western N. America, Greenland.

First found in Britain in 1964 and undoubtedly much overlooked because of its small size and its resemblance in the field to depauperate plants of *Anomobryum* and *Pohlia*. All the published British records result from the finding of large patches, but Hesselbo (1918) reported that in Iceland it grew 'most frequently as single specimens scattered among other mosses' and that most specimens were not seen until after the material had been brought back from the field. British bryologists have yet to find such diminutive specimens.

A. C. CRUNDWELL

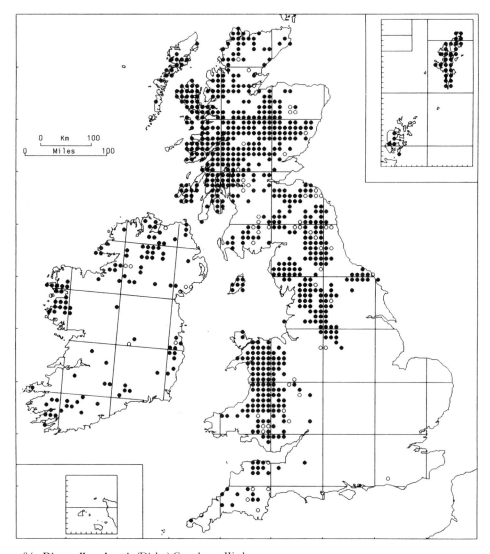

28/1. **Dicranella palustris** (Dicks.) Crundw. ex Warb.

A species of acidic flushes, streamsides, marshes, ditches and the sides of waterfalls, tolerant of light shade such as in ravines. It typically occurs where there is some water-movement such as splash-water from rocks or seepage through soil. It is highly characteristic of mountain streams or open flushes where it grows with *Scapania undulata*, *Bryum pseudotriquetrum* and *Philonotis fontana*. It occurs also on rocks and gravel, and on these substrates is often accompanied by *Dichodontium pellucidum*, which it superficially resembles. 0–1100 m (Ben Alder). GB 714+57*, IR 104+9*.

Dioecious; sporophytes rare. Rhizoidal tubers are almost always present but never abundant.

Circumboreal, reaching the Low Arctic; also in the subalpine and low-alpine zones of mountains farther south.

R. D. PORLEY

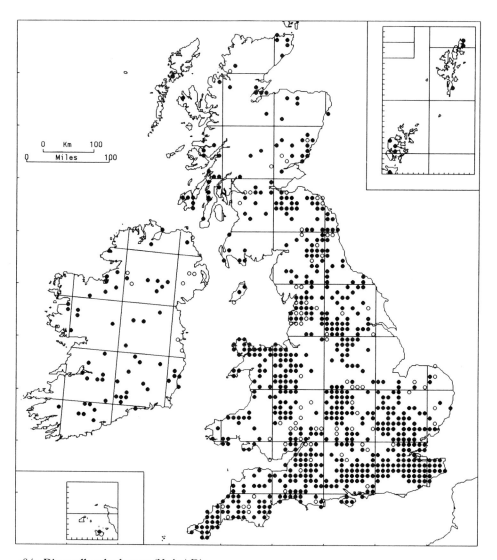

28/2. **Dicranella schreberana** (Hedw.) Dix.

On disturbed soil in a wide range of habitats including arable fields, ditchsides, river-banks, woods and moorland. It tolerates diverse neutral and basic soils, but is perhaps most frequent on calcareous clay. Lowland. GB 684+85*, IR 59+8*.

Dioecious; many plants are sterile, but sporophytes are occasional to frequent, maturing in autumn. Tubers are abundant on the rhizoids.

Circumboreal. Widespread in C. and N. Europe, reaching south to the Mediterranean mountains.

H. L. K. WHITEHOUSE

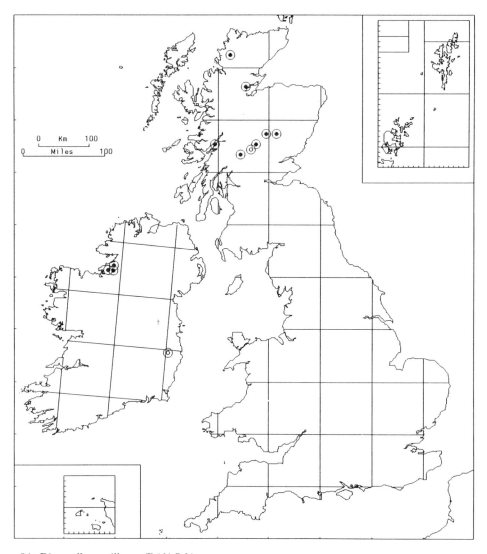

28/3. **Dicranella grevilleana** (Brid.) Schimp.

A montane species of basic, often disturbed, damp soils. It is found amongst turf on base-rich rock-ledges, in rock crevices, on moss-covered rocky banks and on damp soil in disused limestone quarries. Species that have been recorded as associates include *Eremonotus myriocarpus* and *Dicranella crispa*. Sometimes descending to near sea-level but mostly at higher altitudes, to at least 380 m on Ben Lawers. GB 7+1*, IR 3+1*.

Dioecious or autoecious; sporophytes frequent. Rhizoidal tubers normally present.

Circumboreal, mainly in the northern boreal zone and Arctic, extending north to the High Arctic and south in the mountains to the Pyrenees, Alps and Caucasus.

R. D. PORLEY

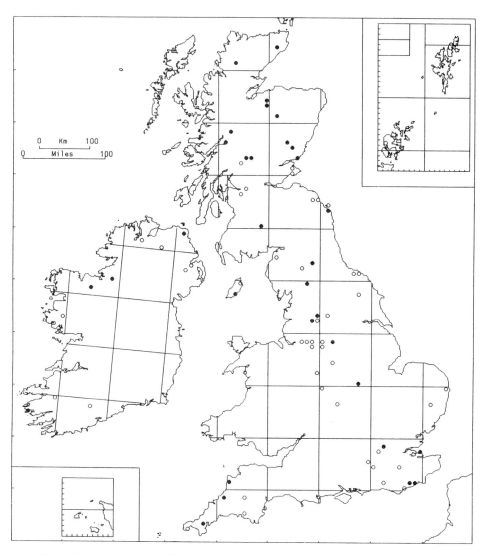

28/4. **Dicranella crispa** (Hedw.) Schimp.

An infrequently recorded species, found in a wide variety of habitats but apparently absent from both markedly acid and basic substrata. It is most typically found on disturbed open soil of clayey, sandy, loamy or gravelly banks by roadsides, rivers and streams. It is also found on rocks, especially vertical sandstone. It has been recorded in an arable field associated with *Anthoceros punctatus* and *Ephemerum serratum* var. *minutissimum*, and sometimes occurs in duneland. 0–550 m (Glen Coe). GB 28+40*, IR 4+7*.

Dioecious or autoecious; sporophytes common.

Circumboreal. North to the High Arctic, south to the Pyrenees and Alps.

The map suggests that this species is declining. However, it may have previously been recorded in error, or it may be currently overlooked.

R. D. PORLEY

137

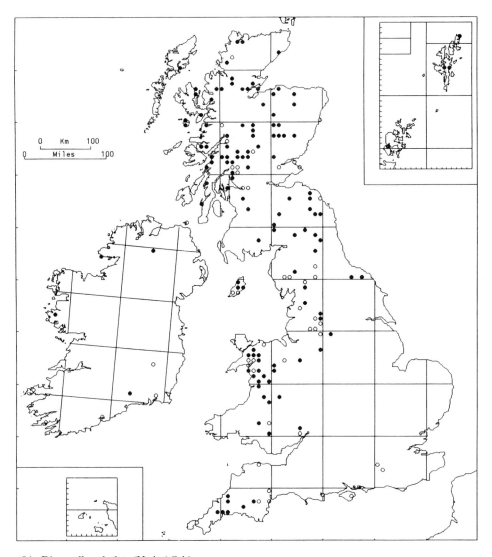

28/5. **Dicranella subulata** (Hedw.) Schimp.

A colonist typically found on damp, acid, sandy or stony ground on sloping or vertical banks, in crevices, in turf, along tracks and roads, and by streams. It also occurs on moist, peaty clay-banks, on wet sandstone rocks and in sandstone quarries. In these habitats it is often associated with *Diplophyllum albicans*, *Lophozia bicrenata*, *Nardia scalaris*, *Dicranella heteromalla* and *Pogonatum aloides*. At higher altitudes in Scotland it sometimes grows in localities with base-demanding bryophytes such as *Herbertus stramineus*, *Radula lindenbergiana*, *Bryoerythrophyllum ferruginascens*, *Leptodontium recurvifolium*, *Oxystegus hibernicus* and *Tetrodontium brownianum*. Whether it grows on basic substrata at these high altitudes, or whether it is on locally leached acid soil, is unclear. 0–950 m (Snowdon). GB 120+44*, IR 4+6*.

Dioecious; sporophytes common. Rhizoidal tubers often present.

Circumboreal. Widespread in the Arctic and the boreal zone, south in mountains to Pyrenees, Alps, Caucasus and California (U.S.A.).

R. D. PORLEY

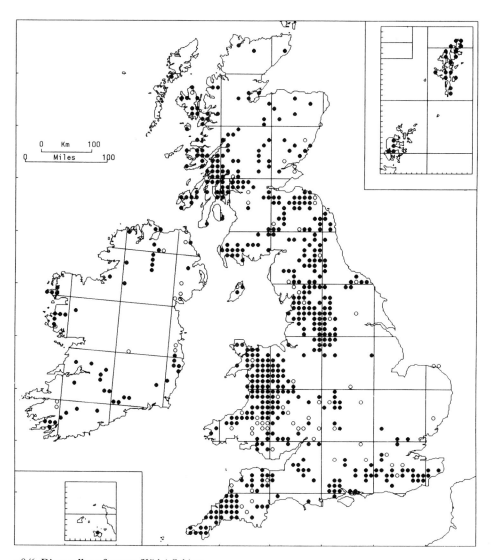

28/6. **Dicranella rufescens** (With.) Schimp.

A colonist that characteristically forms reddish tufts or scattered patches on damp acidic often alluvial or fine-textured, mineral soils by streams, ditches, pools and paths, in quarries, and on exposed margins of reservoirs and lakes. It is also found on woodland rides and paths, often in wheel-ruts. On Skye it occurs in oatfields. Amongst its many associates are *Fossombronia pusilla*, *F. wondraczekii*, *Bryum rubens*, *Dicranella staphylina*, *Discelium nudum*, *Pottia truncata* and *Pseudephemerum nitidum*. It occurs, rarely, on rocks by streams. 0–450 m (near Deiniolen). GB 532+66*, IR 51+12*.

Dioecious; sporophytes occasional to frequent. Rhizoidal tubers often present, especially on senescing plants.

Cool-temperate Europe, becoming rarer northwards, reaching just north of the Arctic Circle in W. Norway; rare and montane in S. Europe. Madeira, Caucasus, E. Asia, western and eastern N. America.

R. D. PORLEY

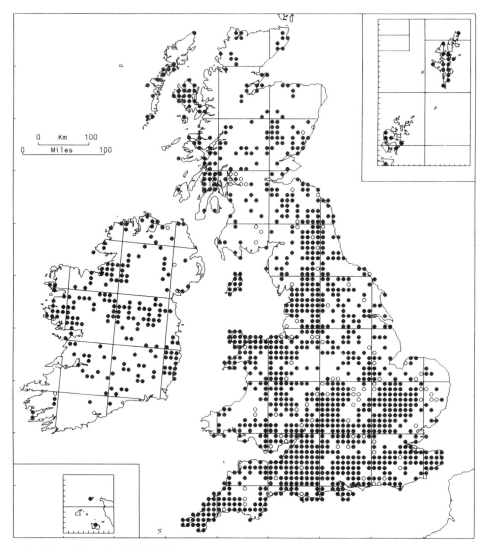

28/7. **Dicranella varia** (Hedw.) Schimp.

A common plant of open habitats on a wide range of neutral or basic soils, occurring particularly in disused chalk or sandstone quarries, on river-banks and on sandy or clay banks generally. It is uncommon in arable fields. Tolerant of high heavy-metal concentrations, it reaches its greatest abundance on metalliferous mine-waste. Mainly lowland but reaching at least 490 m as an associate of *Aongstroemia longipes* at Lochan na Lairige. GB 1128+99*, IR 193+5*.

Dioecious; sporophytes are common, maturing in winter. Tubers are frequent on the rhizoids on calcareous soils, but are otherwise often absent.

Europe from Spitsbergen south to the northern edge of the Mediterranean region. Siberia, Afghanistan, Himalaya, W. China (Yunnan), western and eastern N. America.

It is replaced in S. Europe by *D. howei* Ren. & Card. Plants that are intermediate between the two species and may be hybrids have been recorded in several places in Britain (Crundwell & Nyholm, 1977).

H. L. K. WHITEHOUSE

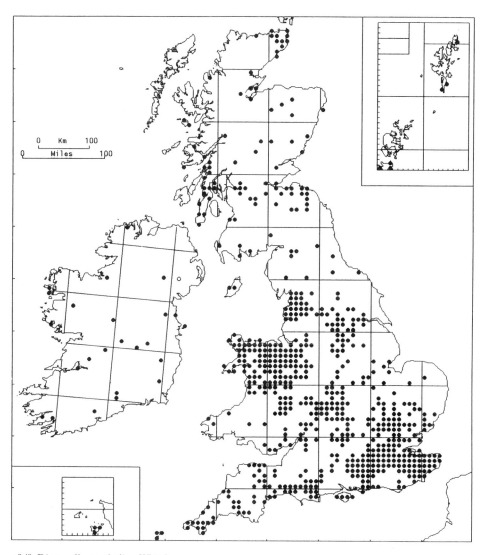

28/8. Dicranella staphylina Whitehouse

This species is abundant in arable fields on non-calcareous soils, but very rare on chalky soils (Preston & Whitehouse, 1986). It also occurs on disturbed acid soil in many other habitats, including woods, grassland and gravel-pits. 0–490 m (south of Newtown, Powys). GB 623, IR 24 + 1*.

Dioecious; most plants are female, but males have been seen at Midgley, S.W. Yorkshire. Sporophytes are unknown in Britain, but have been found in Luxemburg (Arts, 1985). Tubers are always abundant on the rhizoids, and filamentous protonemal gemmae have been observed in culture.

Widespread in W. and C. Europe, north to Finland.

The tubers appear to be capable of surviving for at least 48 years in damp soil (Whitehouse, 1984).

H. L. K. WHITEHOUSE

141

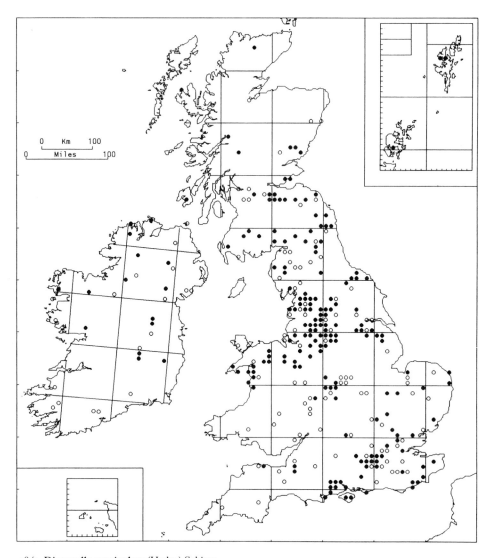

28/9. **Dicranella cerviculata** (Hedw.) Schimp.

A colonist characteristic of partially drained peat, especially peat-cuttings, in both ombrogenous and minerotrophic mires. It is typically associated with *Dicranella heteromalla*. Other habitats include ditch-banks and disturbed, damp gravelly or sandy soils, where it may be associated with *Ditrichum heteromallum*; in the Weald it commonly grows on loose sand derived from weathered sandstone rock. Rarely, it occurs on acid clay. Lowland. GB 178+102*, IR 15+20*.

Dioecious; sporophytes common.

Widespread in the boreal zone, becoming relatively uncommon in the Arctic, extending south to the mountains of C. Europe, southern Canada (British Columbia) and New England.

R. D. PORLEY

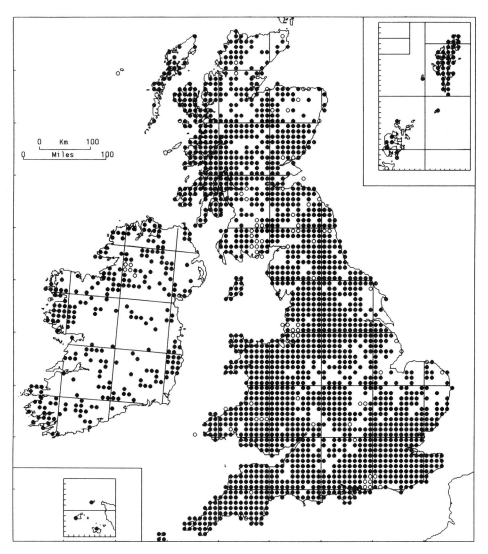

28/10. Dicranella heteromalla (Hedw.) Schimp.

This species is present in a wide variety of non-calcareous habitats and is one of the most frequent and widespread mosses of deciduous and coniferous woodland in the British Isles. It is a characteristic colonist of woodland banks, typically associated with *Calypogeia arguta, C. fissa, Diplophyllum albicans, Lepidozia reptans, Atrichum undulatum, Mnium hornum,* and, on peaty banks, *Tetraphis pellucida.* It also commonly grows on tree-bases, stumps, rotting wood, hedge-banks, sandstone rock, ditch sides, and, in the mountains, in rock crevices, on detritus and beside paths. On moors and heaths it occurs on dry peaty banks. Even in calcareous districts it occurs frequently on humus around the bases of trees. 0–1100 m (Ben Alder). GB 1940+79*, IR 255+7*.

Dioecious; sporophytes common, maturing in winter and spring.

Circumboreal. Mainly in the boreal and broadleaved deciduous-forest zones, scarcely reaching the Arctic, south to Macaronesia, N. Africa, Himalaya, Florida and Texas (U.S.A.). Recorded in the tropics from C. Africa, northern S. America.

R. D. PORLEY

143

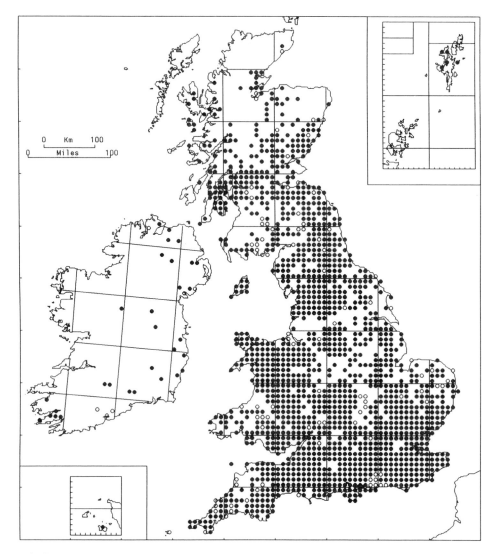

29/1. **Dicranoweisia cirrata** (Hedw.) Lindb. ex Milde

A common plant of tree-trunks, particularly in the more populated and industrial parts of Britain. It is infrequent in the north of Scotland and in Ireland, but even there it may be locally abundant close to built-up or industrial areas. It is common on fence-posts and old thatch, where thatch can still be found. It is relatively uncommon on rocks and walls, except in N. Wales and the Lake District. 0–625 m (Gragareth). GB 1473+96*, IR 30+6*.

Monoecious; usually with abundant sporophytes. Gemmae are often abundant on the leaves and occur on the protonema in culture (Whitehouse, 1987).

Europe north to S. Fennoscandia. Macaronesia, N. Africa, S.W. Asia, western N. America. Southern Hemisphere records are doubtful.

D. cirrata is one of the few mosses to have increased in the past 150 years. Like *Aulacomnium androgynum* and *Dicranum tauricum*, its spread can be attributed to acidification and the consequent disappearance of more sensitive corticolous species. Its habitat has changed. Of the thirty pre-1900 specimens in the Natural History Museum (BM), only two come from standing trees, most are from rocks, walls or thatch.

P. H. PITKIN

29/2. Dicranoweisia crispula (Hedw.) Milde

A local plant of dry acidic rocks in mountainous areas. It usually occurs in open sunny situations, often on boulders where it is sometimes the only bryophyte. At its few localities in N. Wales and the Lake District it grows with *Cryptogramma crispa*. It is occasional to frequent on the schists of the C. Highlands of Scotland and on the basalt of Skye. Braithwaite (1887–1905) reports a smaller blackish form 'growing exposed to the constant drip of snow water'. Mostly above 300 m, but down to 60 m on Skye, ascending to 1335 m on Ben Nevis. GB 45 + 11*.

Monoecious; usually recorded with sporophytes.

Circumboreal, reaching the High Arctic. Common in the Alps, where it occasionally grows on trees. South to the mountains of N. Africa, Himalaya and New Mexico (U.S.A.).

D. crispula is inconspicuous and difficult to recognize without capsules, so is possibly under-recorded. It may be declining in N. Wales and the Lake District.

P. H. PITKIN

145

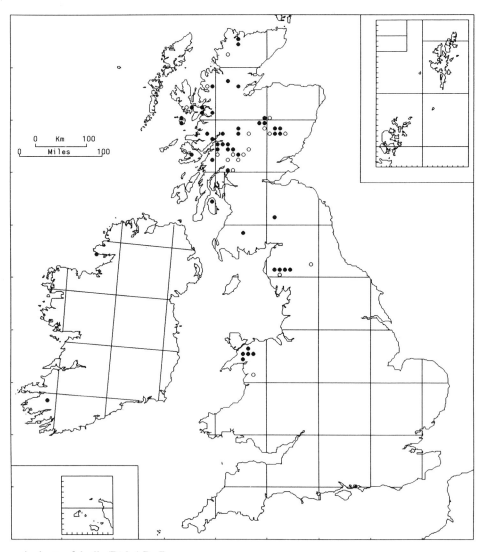

30/1. **Arctoa fulvella** (Dicks.) Br. Eur.

This species is found on summits and on alpine cliffs with a N. or E. aspect, on rocks, in rock crevices and occasionally in block-scree, rarely on soil. Most sites are in areas of late snow-lie. It is absent from limestone and from the most acid rocks, but is found on granite, andesite, basalt and mica-schist, where it occurs with a variety of other bryophytes, but most frequently with *Gymnomitrion concinnatum* and *G. obtusum*. 450 m (Argyll) to 1340 m (Ben Nevis). GB 45+20*, IR 2.

Autoecious; capsules are usually present, maturing in summer.

An arctic-alpine, found in nearly all European mountain ranges from the Pyrenees, Alps and Carpathians to the Arctic. Far East, western and eastern N. America.

M. F. V. CORLEY

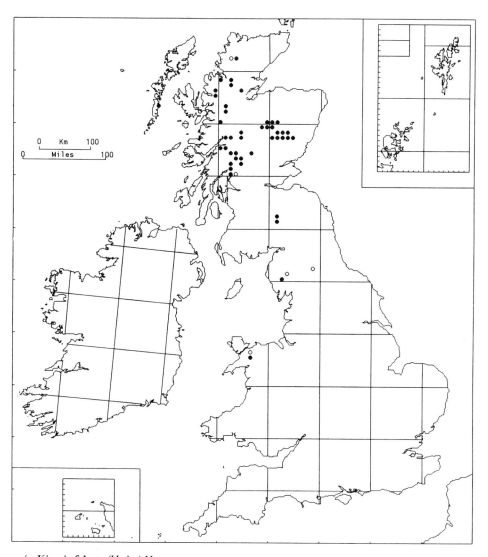

31/1. **Kiaeria falcata** (Hedw.) Hagen

Areas of late snow-lie are the principal habitat of this species. Most sites are high in corries among stone or on finer detritus on steep slopes with *Conostomum tetragonum*, *Kiaeria starkei* and *Pohlia ludwigii*, but it also occurs on and among boulders in similar places, on sheltered N.- or E.-facing banks on summit-ridges and, rarely, in tufts of other bryophytes such as *Racomitrium aquaticum* by high-level springs. It is almost indifferent to rock type. 950–1340 m (Ben Nevis). GB 44+6*.

Autoecious; fruit is frequent, ripening in summer.

Circumboreal arctic-alpine. In Europe from the mountains of N. Portugal, Spain, N. Italy and Romania northwards to the Arctic.

M. F. V. CORLEY

147

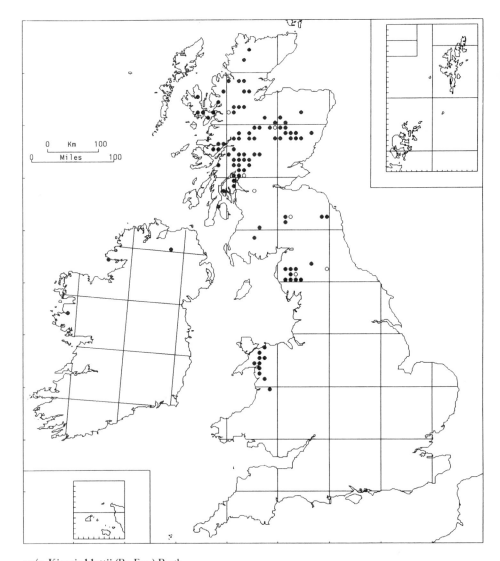

31/2. **Kiaeria blyttii** (Br. Eur.) Broth.

Less restricted to areas of late snow-lie than its congeners, this species is most frequently found on the sides and tops of boulders in lightly sheltered block-screes, and on large boulders in streams above normal flood-level. More rarely it is found on detrital gravel between boulders on summit-ridges. It avoids limestone and the hardest and most acid rocks. Commonly found with *Racomitrium sudeticum* and *R. fasciculare*, less often with *Dicranoweisia crispula*. Although it ranges from 100 m (Morvern) to 1344 m (Ben Nevis), it is most frequent above 500 m. GB 94+8*, IR 3.

Autoecious; sporophytes are commonly present, maturing in summer.

Circumboreal arctic-alpine. Mountain ranges of Europe north to the Arctic. Azores.

M. F. V. CORLEY

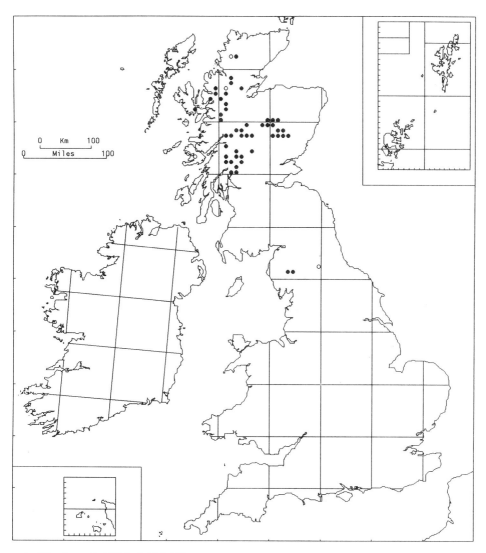

31/3. **Kiaeria starkei** (Web. & Mohr) Hagen

A characteristic plant of late-snow areas, where it may grow on near-horizontal slabs, but more often occurs on detrital soils, especially where a thin layer of grit lies over sloping rocks at the foot of crags; also found between larger stones and boulders in gullies and in block-screes. It occurs on most rock types except limestone and the most acid rocks. Among the most frequently associated species are *Barbilophozia floerkei*, *Diplophyllum albicans*, *Nardia scalaris*, *Conostomum tetragonum*, *Kiaeria falcata* and *Pohlia ludwigii*. 580–1340 m (Ben Nevis). GB 48+4*.

Autoecious; fruit is frequent, maturing in late summer.

Circumboreal arctic-alpine. Mountains of C. and N. Europe north to the Arctic. Southern Hemisphere records are doubtful.

M. F. V. CORLEY

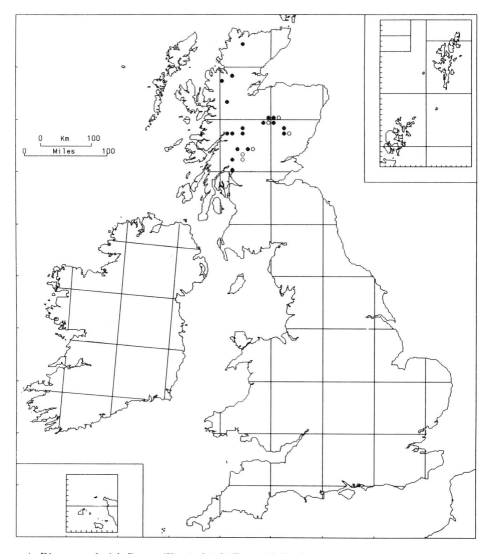

32/1. **Dicranum glaciale** Berggr. (*Kiaeria glacialis* (Berggr.) I. Hag.)

It is found at high altitude in N.- and E.-facing corries on stony or rocky substrates among boulders in late-snow areas. It also grows in sheltered spots among boulders on summit-ridges and cols. Associates are mostly common species such as *Diplophyllum albicans* and *Rhytidiadelphus loreus*. 900 m (Beinn Dearg) to 1330 m (Ben Nevis). GB 18+6*.

Autoecious; fruit moderately frequent.

A circumpolar species of the Arctic and of mountains in the boreal zone. In Europe, widespread in Fennoscandia and Iceland but, farther south, only in Britain and Czechoslovakia.

Likely to be under-recorded as it is easily overlooked if not fruiting, and few British bryologists are familiar with it. Although superficially *Dicranum*-like, this species has the habitat and sexuality of a *Kiaeria* (cf. Corley, 1979).

M. F. V. Corley

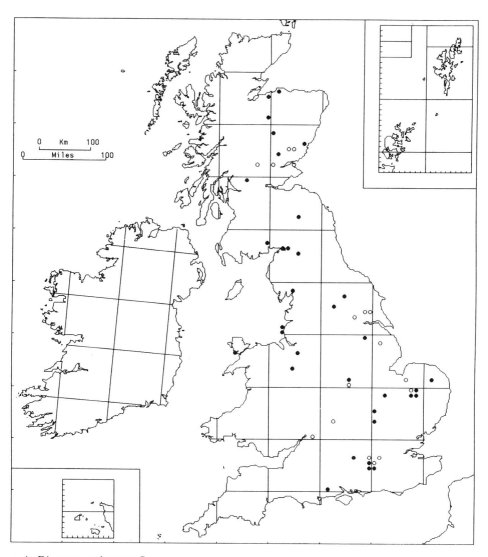

32/2. Dicranum polysetum Sw.

A lowland species found principally on the ground in coniferous (especially *Pinus*) woods and plantations, with *Hypnum jutlandicum*, but also in birch woodland, on heaths and in raised bogs. First collected near Kinnordy (Angus) in 1848, and at Wolford Heath (Warwickshire) in 1887, it appears to be spreading. Lowland. GB 34+17*.

Dioecious, with minute male plants; sporophytes are unknown in Britain, but plentiful in Scandinavia, which could be the source from which British plants derive.

A boreal species, rare or absent in S. and W. Europe. Asia, N. America.

M. F. V. CORLEY

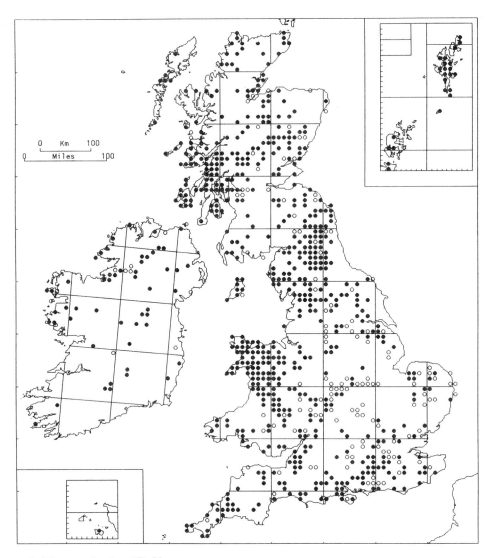

32/3. **Dicranum bonjeanii** De Not.

Chiefly a plant of bryophyte-rich turf in damp places, including fens, both rich and poor, particularly around lakes where it grows with *Aulacomnium palustre*, *Sphagnum auriculatum* and *S. teres*, in flushes and marshy ground with *Plagiomnium elatum*, in damp heathland with *Pleurozium schreberi* and in valley bogs. It also occurs in drier grassland on chalk, limestone, and, more rarely, dunes and sea-cliffs. Most habitats are open, but it sometimes grows in damp open woodland or in woodland rides. 0–430 m (Skye). GB 619+137*, IR 52+11*.

Dioecious, with minute male plants; sporophytes very rare. Flagelliferous shoots are occasionally present and presumably serve for vegetative propagation.

Circumboreal. Throughout C. and N. Europe, rare or absent in the Mediterranean region.

M. F. V. CORLEY

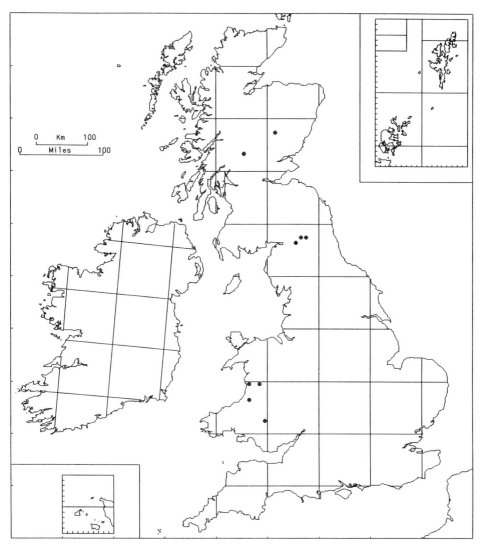

32/4. **Dicranum leioneuron** Kindb.

Hitherto considered to be confined to sphagnum hummocks on blanket and raised bogs, sometimes mixed with hepatics such as *Odontoschisma sphagni*, but recently it has been found under heather on a bank at the top of a ravine, and in thin turf on top of an erratic boulder. Mainly lowland, ascending to at least 800 m (Meall nan Tarmachan). GB 9.

Dioecious; fruit unknown in Britain. Fragile deciduous apical shoots sometimes present.

Fennoscandia. N.E. Asia, N. America.

Closely resembling *D. bonjeanii* and *D. scoparium*, and doubtless much overlooked.

M. F. V. CORLEY

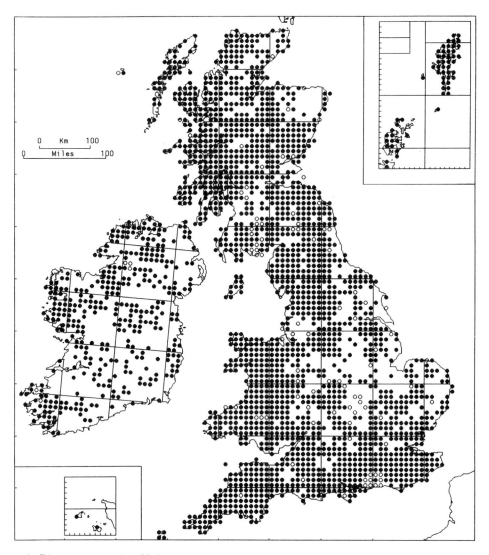

32/5. Dicranum scoparium Hedw.

A common calcifuge with an extremely wide ecological range. It grows on soil, rocks, trees and stumps in open or shaded habitats, being particularly abundant on the ground in acid woodland and on heaths and moors, often under tall *Calluna*. It also occurs in marshes, in leached calcareous grassland, on fixed dunes, in turf on the top of boulders, on rocks in woods, on tree-trunks and branches, on rotten stumps, among sphagnum in bogs, among rocks on mountains, in late-snow areas and in mountain-top turf. With such a wide range of habitats it has a great range of associates, but among the most frequent are *Polytrichum formosum* in acid woodland, *P. juniperinum* on dry boulder-tops, *Hypnum jutlandicum* on heaths and moors and under conifers, and *H. cupressiforme* on trees such as oak and hawthorn. 0–1220 m (Cairngorms). GB 1948+91*, IR 364+7*.

Dioecious, with dwarf male plants; sporophytes rare in the drier south-east, more frequent in humid sites and high-rainfall areas, maturing late summer (Argyll) or winter (Cornwall).

Circumboreal. Almost throughout Europe, but less frequent in the south and absent from some Mediterranean islands.

M. F. V. CORLEY

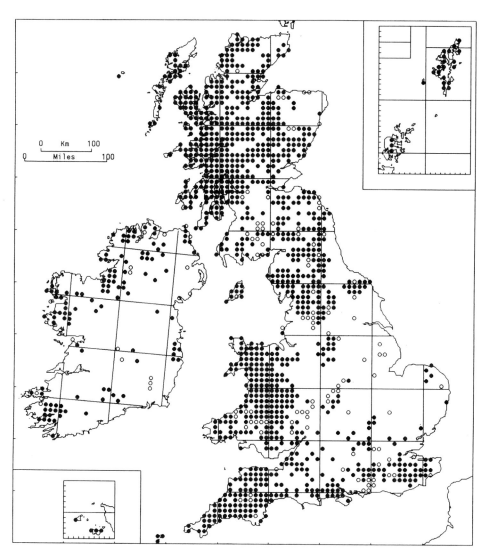

32/6. Dicranum majus Sm.

A characteristic species of sheltered acidic to mildly basic turf on the ground and on turf-covered boulders in oak, birch or coniferous woodland, in hedge-banks, and on N.- to E.-facing banks on mountains. Less frequently it grows on the ground among boulders on moorland, and on heathy banks in quarries. Frequent woodland associates are *Mnium hornum*, *Polytrichum formosum*, *Rhytidiadelphus loreus* and *Thuidium tamariscinum*. On mountains it grows with *Polytrichum alpinum* and *R. loreus*. 0–1080 m (Aonach Beag). GB 1055+103*, IR 115+19*.

Dioecious, with dwarf male plants; sporophytes are frequent in high-rainfall areas, ripening in winter.

Circumboreal. C. and N. Europe, rare in the south and absent from some Balkan countries and all the Mediterranean islands.

M. F. V. CORLEY

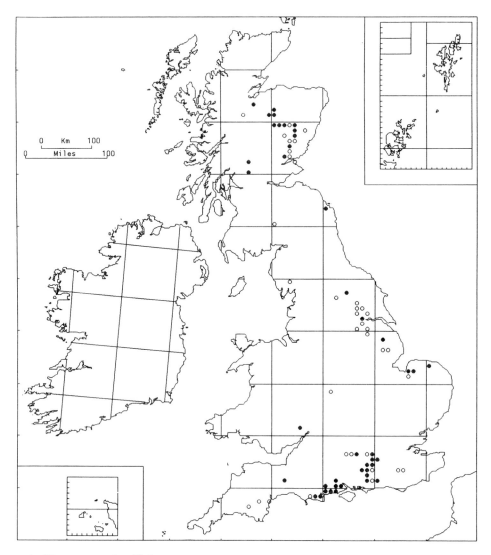

32/7. **Dicranum spurium** Hedw.

Typically a plant of heathland, where it occurs under both wet and dry conditions, often with *Pleurozium schreberi*, but in Scotland also found on warm S.-facing heathy banks beneath scattered Scots pines. It has decreased in England largely through the ploughing up or afforestation of heathland. It is rather slow to come back after burning. Lowland. GB 43+36*.

Dioecious, with minute male plants; sporophytes very rare.

A boreal moss, rare or absent in S. and W. Europe. Siberia, Himalaya, eastern N. America.

M. F. V. CORLEY

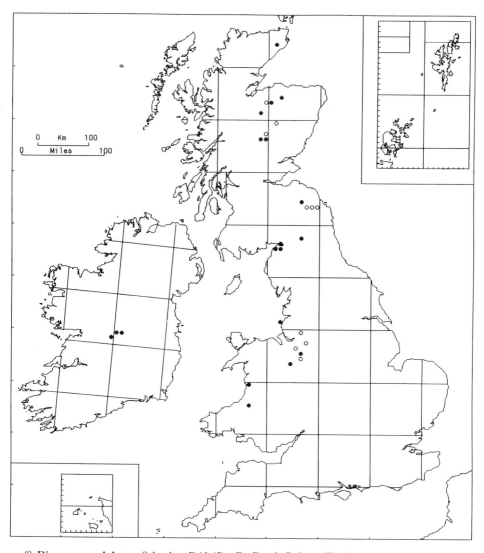

32/8. **Dicranum undulatum** Schrad. ex Brid. (*D. affine* Funck, *D. bergeri* Bland.)

It is confined to bogs, typically raised bogs. Associates include *Odontoschisma sphagni, Pleurozium schreberi, Polytrichum alpestre* and *Sphagnum* species, especially *S. tenellum*. 0–350 m (Blair Atholl). GB 16+11*, IR 3.

Dioecious, with minute male plants; sporophytes rare, ripening in summer.

Circumboreal. C. and N. Europe, rare in the south and absent from Portugal, Greece and the Mediterranean islands.

Subfossil records of this species indicate that it was formerly more widespread than at present.

M. F. V. CORLEY

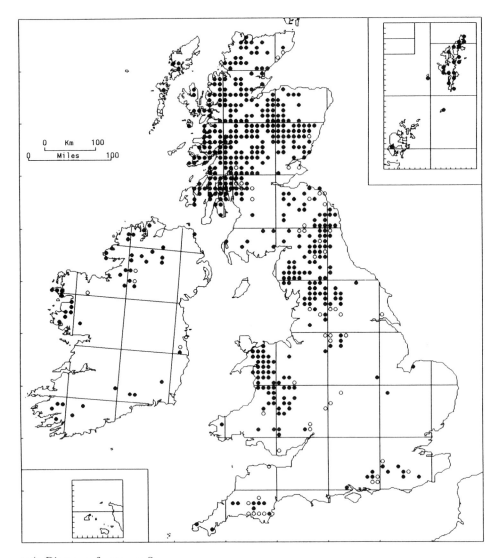

32/9. Dicranum fuscescens Sm.

Occurs in a wide range of well-drained acid habitats. In woodland, it is found on tree-bases, trunks and branches, especially of oak, ash and birch, together with *Hypnum cupressiforme* or *H. mammillatum*, on stumps, particularly of conifers, with *Campylopus paradoxus* and *Pohlia nutans*, on boulders, and on vertical rock-faces. In more or less open ravines it occupies similar habitats. In the hills, it occurs on exposed rocky bluffs and boulders, on sheltered boulders and cliffs in corries, on stony ground or in turf on summits and exposed mountain sides with *Racomitrium lanuginosum* or *R. sudeticum*, and, rarely, on dry peaty ground on moorland. 0–1205 m (Ben Lawers). GB 489+55*, IR 45+6*.

Dioecious; fruit is occasional, ripe summer.

Circumboreal, occurring widely in the Arctic as well as further south. Most of Europe except Hungary, becoming montane in the south and absent from the Mediterranean islands.

Many, but not all, plants from exposed alpine sites are referable to var. *congestum* (Brid.) Husn.

M. F. V. CORLEY

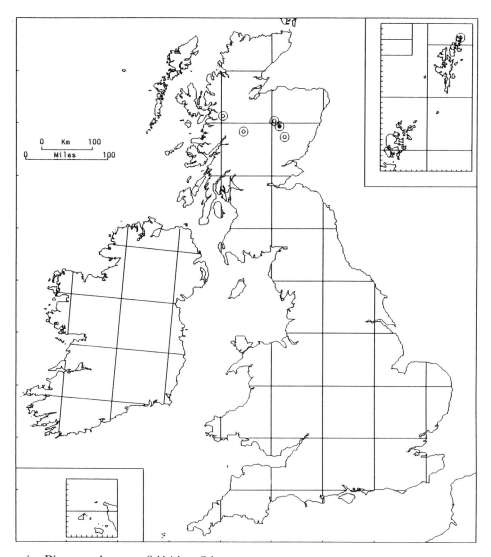

32/10. **Dicranum elongatum** Schleich. ex Schwaegr.

On peaty soil and on peat overlying mountain rocks. In 1987 it was found on peaty soil in a cleared pine-plantation near Braemar, where it appeared to be a recent colonist. From near sea-level (Unst) to 1150 m (Cairn Gorm). GB 1+5*.

Dioecious; capsules not found in Britain.

Circumpolar arctic-alpine, very common in the tundra, extending north to the High Arctic and south to the Alps, Caucasus, C. Asia and Colorado (U.S.A.).

According to Steere (1978), in the Alaskan tundra it often forms large tufts and high mounds on which snowy owls (*Nyctea scandiaca*) perch.

D. G. LONG & M. O. HILL

159

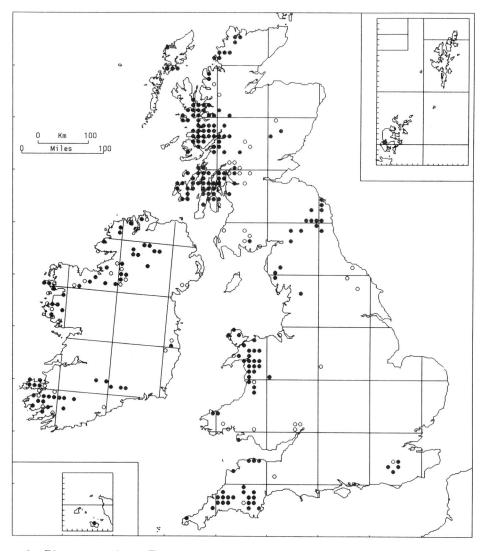

32/11. **Dicranum scottianum** Turn.

Largely confined to dry, acidic, near-vertical rock-faces of boulders, cliffs and ravines, including sheltered sites by the sea such as the boulders and cliffs of raised beaches. Sites may be open or shaded by trees, but it only rarely grows on trees. In the Weald it grows on sandstone rocks, but elsewhere it is found on a wide variety of harder rock types including andesite and Lewisian gneiss. Among the more frequently associated species are *Bazzania trilobata*, *Frullania tamarisci*, *Lepidozia cupressina* and *Isothecium myosuroides*. 0–430 m (Galtee Mts). GB 176+38*, IR 55+24*.

Dioecious; fruit moderately frequent, ripening in summer.

W. Europe from Portugal to S. Norway and Sweden. Macaronesia.

M. F. V. CORLEY

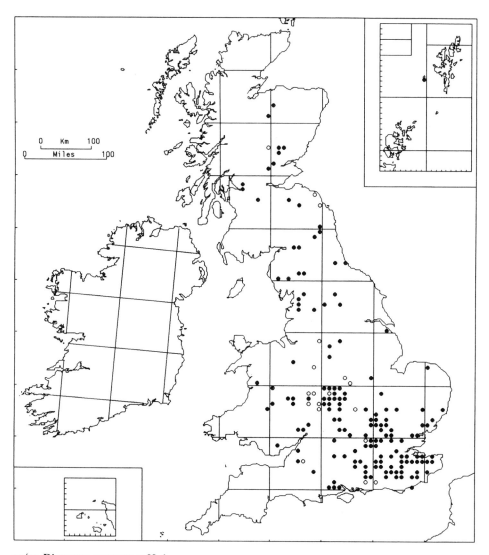

32/12. **Dicranum montanum** Hedw.

In woodland and parks this species is found on exposed roots, trunks and branches of trees such as oak, birch and ash, often with *Hypnum cupressiforme*. It also occurs on logs, stumps and, rarely, sandstone rocks. In a treeless locality on Foula in Shetland it probably occurred on rock. Lowland. GB 155+18*.

Dioecious; fruit unknown in Britain. Deciduous leaves are normally present in abundance.

Circumboreal. C. and N. Europe, reaching the Arctic in Fennoscandia, rare in the south; absent from Portugal, Greece and the Mediterranean islands, and from Iceland. Azores, Madeira.

This species has undoubtedly increased in recent years, possibly as a response to 'acid rain'.

M. F. V. CORLEY

161

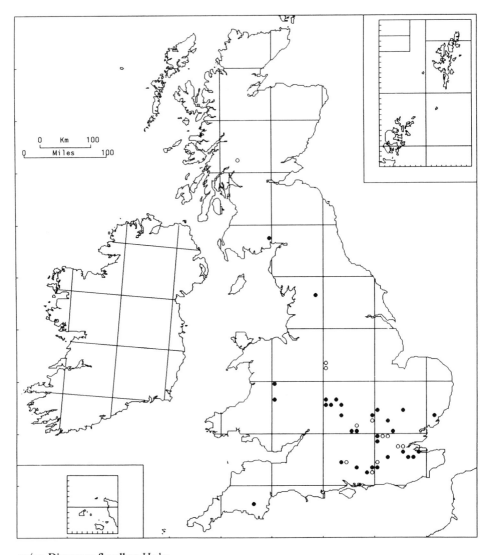

32/13. Dicranum flagellare Hedw.

In Britain this species occurs only on decaying wood, being found on logs and stumps in woods. It is associated with common dead-wood species such as *Lophocolea heterophylla*. Lowland, but reaching 370 m (Malham Tarn). GB 31+12*.

Dioecious; fruit very rare, ripe in summer. Flagelliferous branches are often present.

Circumboreal, extending south to Macaronesia and C. America. In Europe, most frequent in C. Europe and S. Fennoscandia, becoming rare or absent in the N., W., S. and S.E. of the Continent.

In N. America this species, although commonest on decaying wood, also occurs on humus, peaty soil, bark at the base of trees and on rocks (Crum & Anderson, 1981).

M. F. V. CORLEY

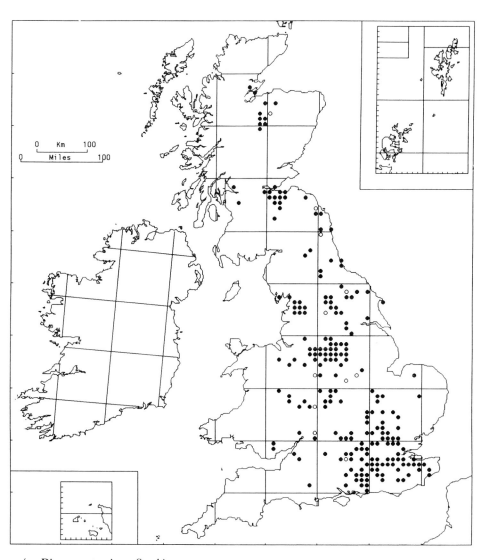

32/14. Dicranum tauricum Sapehin

Mainly occurring on decaying wood of stumps, logs and fallen branches in woodland, but also found on trunks and branches of trees such as ash and elder. It has been recorded from rocks. The most frequent associates are *Lophocolea heterophylla* and *Hypnum cupressiforme* including var. *resupinatum*. Lowland. GB 213 + 11*.

Dioecious; fruit recorded only once in Britain, ripening in summer. The tips of many of the older leaves break off, serving as a means of vegetative propagation.

Widespread in Europe, but rare and montane in the south. N. Africa, Turkey, Kamchatka, western N. America.

This species has spread and greatly increased in abundance in recent years, particularly in areas of high air pollution. It was found, new to Britain, in Staffordshire in 1864, but remained very rare up to about 1930.

M. F. V. CORLEY

163

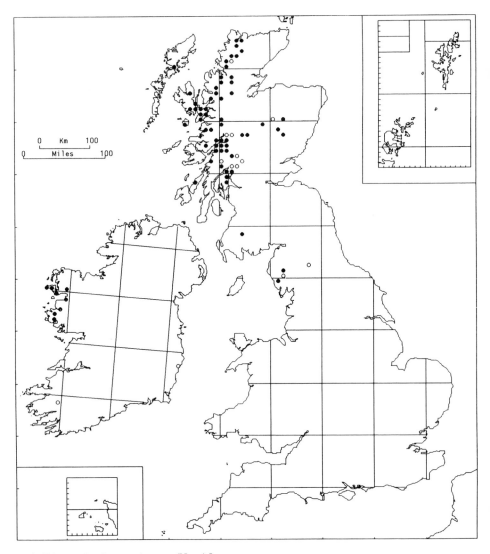

33/1. Dicranodontium uncinatum (Harv.) Jaeg.

A species of acid humus in sheltered moist N.- to E.-facing sites, such as under heather on rocky slopes with *Mylia taylorii*, *Scapania ornithopodioides* and other large oceanic hepatic species, and on mountain cliff-ledges with similar species including *Bazzania tricrenata*. At lower altitudes it is locally abundant on N.-facing periodically irrigated near-vertical rock-faces and in similar sites in wooded ravines and in crevices on raised-beach cliffs. 0–750 m (Ben Alder range). GB 59+11*, IR 8+2*.

Dioecious; fruit unknown in Britain.

N. Spain, Alps, Tatra Mts, Norway. Himalaya, Java, Philippines, Taiwan, Japan, W. Canada.

M. F. V. CORLEY

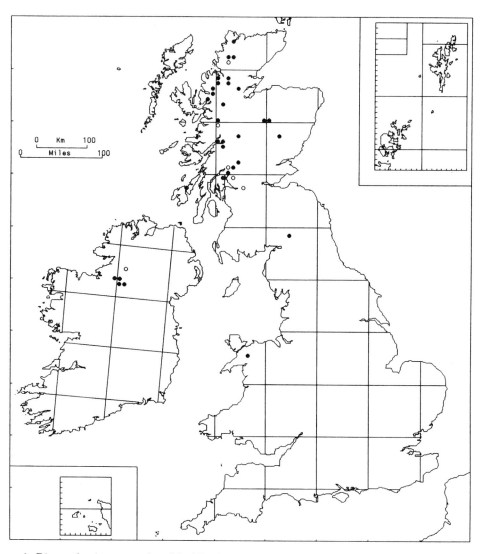

33/2. Dicranodontium asperulum (Mitt.) Broth.

It resembles *D. uncinatum* in its preference for sheltered moist N.- to E.-facing sites, but is perhaps a little more tolerant of exposure as it occasionally occurs on W.-facing slopes, and is less strongly associated with large oceanic hepatic species. It is most frequently found on acid humus among rocks and on low vertical rock-faces on steep mountain slopes and on rock ledges. In such habitats *Diplophyllum albicans* is an almost constant associate, often with *Bazzania tricrenata*. More rarely it grows on gravelly soil among boulders near mountain streams. 330–850 m (Ben Vorlich, Dunbartonshire). GB 27+5*, IR 4+1*.

Dioecious; fruit unknown in Britain. Deciduous leaves are shed by most plants, occasionally in large numbers.

Alps, Tatra Mts, Norway. Himalaya, W. China (Yunnan), Taiwan, Japan, N. America (W. Canada, Appalachian Mts).

M. F. V. CORLEY

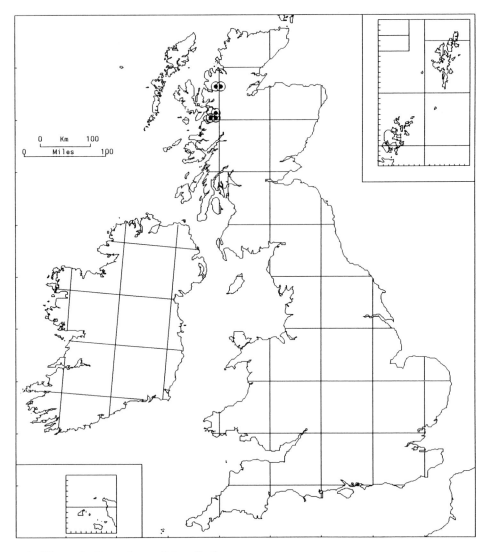

33/3. Dicranodontium subporodictyon Broth.

Rare but locally abundant on acid rock in areas of very high rainfall. Suitable sites are on steeply sloping granite or Torridonian sandstone with permanent seepage or with a small stream glissading down over a wide slab of smooth rock or down a narrow chute. Here it grows in patches, sometimes of considerable area, as down the whole length of a chute, often along crevices, usually associated with *Narthecium ossifragum*, which serves as an anchor. The most frequent bryophyte associates are *Marsupella emarginata* and *Campylopus atrovirens*. It is also found on sloping rock-slabs in open Caledonian pine forest and in lightly wooded ravines. 10–180 m (Kinloch Hourn). GB 5.

Dioecious; fruit unknown in Britain.

Himalaya, W. China (Yunnan), W. Canada.

D. subporodictyon, like several other bryophytes of cool-temperate high-rainfall regions, has a remarkably disjunct distribution. It has only recently been added to the British flora (Corley & Wallace, 1974), and is likely to be found in additional localities.

M. F. V. CORLEY

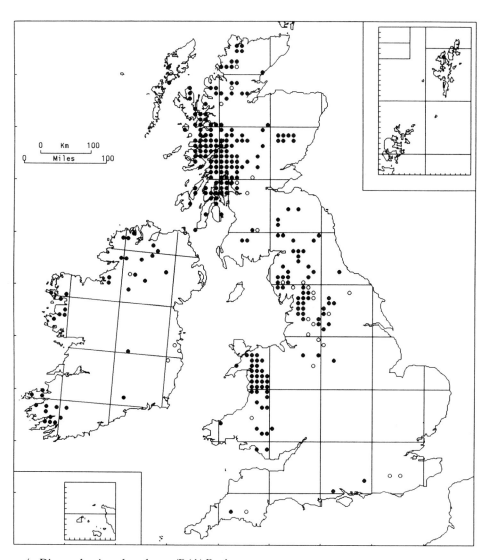

33/4. **Dicranodontium denudatum** (Brid.) Broth.

Rarely found outside mountain districts, this species occurs on decaying wood of stumps and fallen branches and on mossy boulders and peaty banks in woods. Among its many associates in woodland are *Calypogeia* spp., *Lepidozia reptans*, *Hypnum jutlandicum* and *Mnium hornum*. It is also found on sheltered peaty banks under heather or among rocks, mainly on N.- or E.-facing slopes and on rock ledges. Occasionally it grows in blanket bog, usually in the middle of a tussock of *Leucobryum* or *Sphagnum*. 0–950 m (Foel Grach). GB 239+27*, IR 37+6*.

Dioecious; fruit very rare. Some plants have the majority of their leaves deciduous, leaving long sections of their stems bare.

Throughout the central latitudes of Europe, becoming rare in the south and north. Caucasus, Turkey, Himalaya, China, Taiwan, Korea, Japan, Hawaii, N. America.

The upper altitudinal limit refers to var. *alpinum* (Schimp.) Hagen, an unsatisfactory taxon which is not mapped separately.

M. F. V. CORLEY

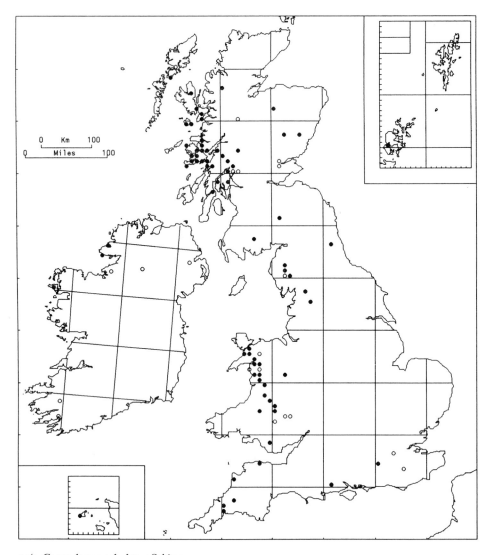

34/1. **Campylopus subulatus** Schimp.

A locally common species of unshaded, acid, gravelly and sandy places. Habitats include tracks, paths, lay-bys, old tarmac on roadsides, and stony or gravelly river-margins. More rarely it is found on sandstone rocks, rock crevices by sheltered coastal inlets, stony reservoir banks and fine mountain-scree. Common associates in gravelly places are *Pohlia drummondii* and *Racomitrium ericoides*. On little-used tracks it also grows with *Scapania irrigua, Archidium alternifolium, Calliergon cuspidatum, Hypnum lindbergii* and *Philonotis fontana*. Mainly lowland, but reaching 850 m in Angus. GB 67 + 16*, IR 3 + 6*.

Dioecious; sporophytes very rare, recorded only once in Britain, and equally rare elsewhere. Vegetative propagation by means of deciduous shoot-tips.

Suboceanic in Europe, extending eastwards as far as Yugoslavia, Austria, Germany and Sweden. Turkey, Bhutan, W. China (Yunnan), U.S.A. (California).

Reports from outside Europe are often unreliable; those listed here have been confirmed by Frahm (1980) or by D. G. Long.

M. F. V. CORLEY

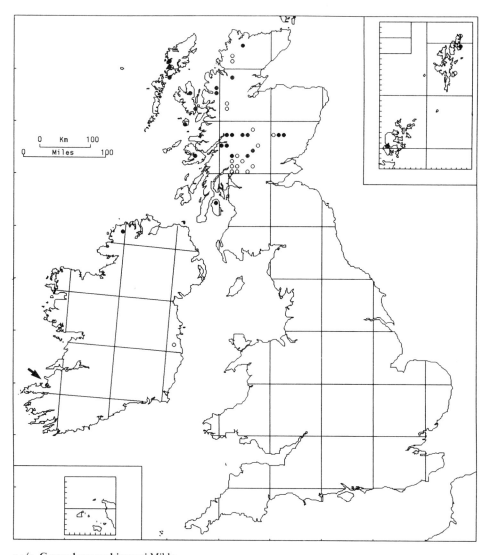

34/2. **Campylopus schimperi** Milde

Ecologically distinct from *C. subulatus*, being found mainly in areas of relatively late snow-lie in sheltered turf on acid humus-enriched detrital soils on steep N.W.- to E.-facing slopes in corries, on the floor of wide gullies with *Oligotrichum hercynicum*, or on banks below cliffs, where it may form extensive patches. It also grows in slightly sheltered pockets on summit ridges with a northerly inclination. 60 m (Fetlar) to 850 m (Ben Vorlich, Dunbartonshire), but rarely below 400 m. GB 24+19*, IR 1+2*.

Dioecious; fruit unknown in Britain. Vegetative propagation by means of fragile leaf-tips, and sometimes deciduous leaves or shoot-tips.

Circumboreal arctic-alpine. Alps, Carpathians including the Tatra Mts, Norway, Sweden, Iceland, Svalbard. Caucasus, C. Asia, Himalaya, Japan, N. America, Greenland.

Probably under-recorded.

M. F. V. CORLEY

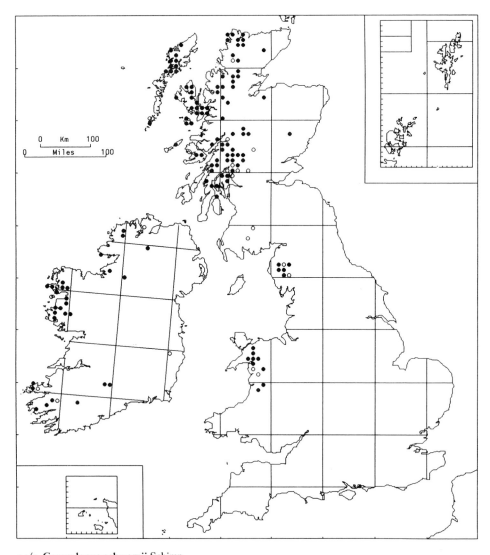

34/3. **Campylopus schwarzii** Schimp.

A locally common species of flushes, of steep, acid, humus-rich banks among rocks, especially where there is some water-seepage, and of flushed rock-slabs, often in exposed moorland sites, where it is usually mixed with *C. atrovirens*. It can also be found on dripping vertical rock-faces in open ravines and on mountain cliffs. More rarely it is found in bogs, on the edge of peaty pools and in turf in late-snow areas. 0–900 m (Aonach Beag). GB 99+16*, IR 28+5*.

Dioecious; fruit unknown. Vegetative propagation by fragile leaf-tips and shoot-tips.

Alps, Czechoslovakia, Romania, Faeroes, Norway. Himalaya, Taiwan, Korea, Japan, Kamchatka, Canada (British Columbia).

M. F. V. CORLEY

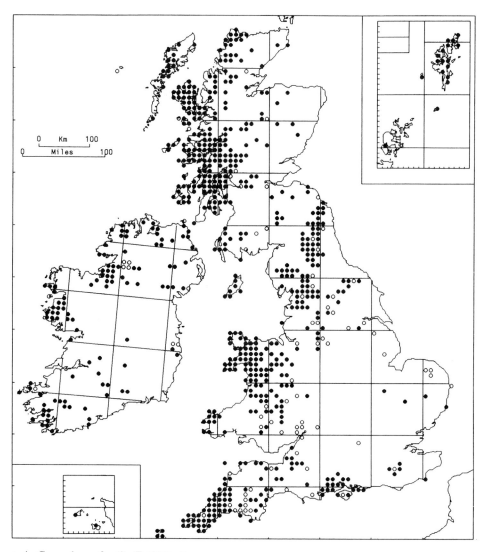

34/4. Campylopus fragilis (Brid.) Br. Eur.

More tolerant of basic conditions, including salt-spray, than other *Campylopus* species, nevertheless its most frequent habitats are peat on moorland banks, peaty soil on rocks or rocky banks, and heathy cliff-tops. Other habitats include slightly sheltered rock crevices in woods and near the sea, and walls and hedgebanks. Occasional habitats include rotten wood and tree-boles, limestone turf and dunes. In recent years it appears to have become increasingly frequent on the ground in fen carr in E. England. 0–760 m (Beinn Odhar Bheag). GB 564+71*, IR 106+10*.

Dioecious; fruit occasional, ripe summer. Vegetative propagation by deciduous leaves, which may be produced in abundance at the shoot tips. Tubers have been found in an Irish specimen and are probably frequent (Arts, 1989).

W. and C. Europe, becoming rarer in the S., E. and N. of the Continent; absent from Iceland, Finland, Hungary and most of the Mediterranean area. Macaronesia, E. Africa, Turkey, Himalaya, Japan, Canada (British Columbia), U.S.A. (Arkansas), C. and S. America.

M. F. V. Corley

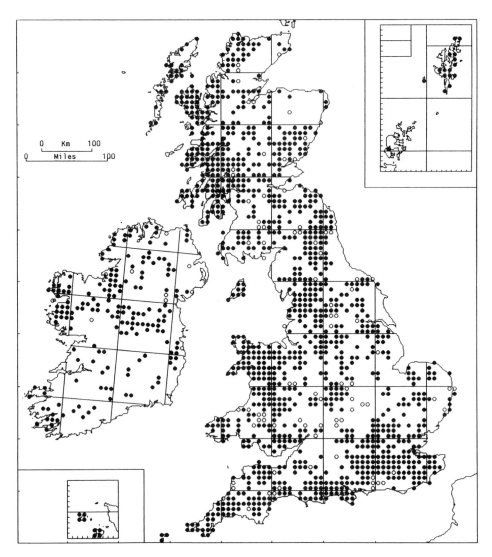

34/5. **Campylopus pyriformis** (Schultz) Brid.

A plant of bare, acid, sandy or peaty soil in oak, birch or coniferous woodland, and on heaths and moors, where it occurs on banks, sides of drainage ditches, and burnt or disturbed ground. It is less common on decaying stumps and logs and on decaying tussocks of *Molinia caerulea*. Mainly lowland but reaching 580 m (Glen Duror). GB 1121+91*, IR 175+10*.

Dioecious, male plants normal-sized or occasionally minute and gemmiform, fruit occasional to frequent, spring and early summer; vegetative propagation by deciduous leaves of normal or reduced size. Tubers have been found in specimens from Belgium and numerous places outside Europe (Arts & Frahm, 1990).

W. and C. Europe, becoming rare in the S., E. and N. of the Continent; absent from Finland, Bulgaria, Greece, Portugal and some Mediterranean islands. Azores, S. Africa, south-eastern U.S.A., S. America, Australia, New Zealand.

On the basis of its world distribution, which resembles that of *C. introflexus*, Corley & Frahm (1982) suggest that *C. pyriformis* may also be an introduction to Europe, having arrived before the beginning of the 19th century. There is no direct evidence to support this theory.

M. F. V. CORLEY

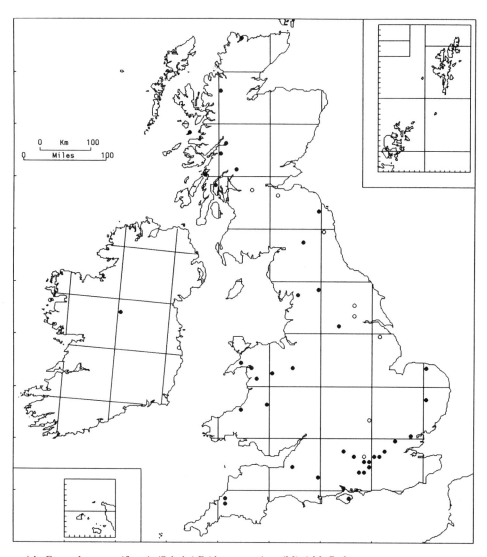

34/5b. **Campylopus pyriformis** (Schultz) Brid. var. **azoricus** (Mitt.) M. Corley

This variety favours moister habitats than var. *pyriformis*, being particularly frequent on decaying tussocks of *Molinia caerulea* and Cyperaceae, but it also occurs on boggy ground under trees at the edge of valley bogs, on peaty soil and on rotten logs lying on wet ground. 0–580 m (Glen Nevis). GB 40+8*, IR 1+2*.

Dioecious; fruit frequent, maturing in summer. Vegetative propagation by means of broken-off leaf-tips.

The distribution outside the British Isles is unclear, because var. *azoricus* is not always considered distinct from var. *pyriformis*. It certainly occurs in France, Germany and the Azores. Var. *fallaciosus* (Thér.) M. Corley is of uncertain taxonomic value and is not mapped separately.

M. F. V. Corley

173

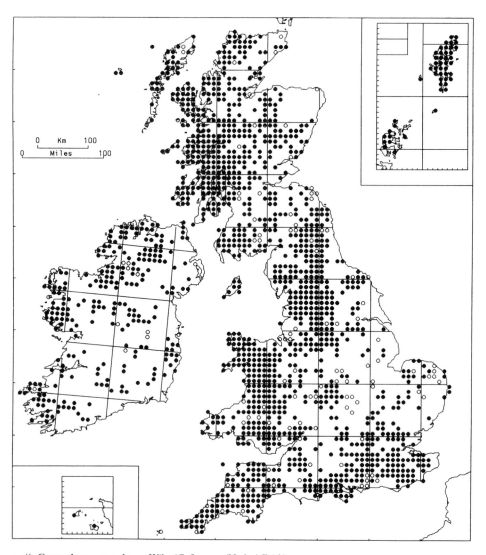

34/6. **Campylopus paradoxus** Wils. (*C. flexuosus* (Hedw.) Brid.)

A common calcifuge of shaded or unshaded habitats, including decaying stumps and logs, peat on moorland banks and burnt heather-moors, peat-cuttings, edges of bog-pools, sandy humus in woods and plantations, peaty soil overlying boulders, and wall-tops, rocks and cliffs in woods, in open ravines and on mountains. More rarely, it grows low on tree-trunks. 0–800 m (Clova). GB 1279+105*, IR 240+12*.

Dioecious; fruit occasional to frequent, ripening in summer. Vegetative propagation by deciduous shoot-tips. Tubers have been found in an Irish specimen but appear to be rare (Arts, 1989).

A suboceanic species found in W. Europe as far east as Yugoslavia, Poland and Sweden. Macaronesia, Africa, Mauritius, India, N., C. and S. America.

M. F. V. CORLEY

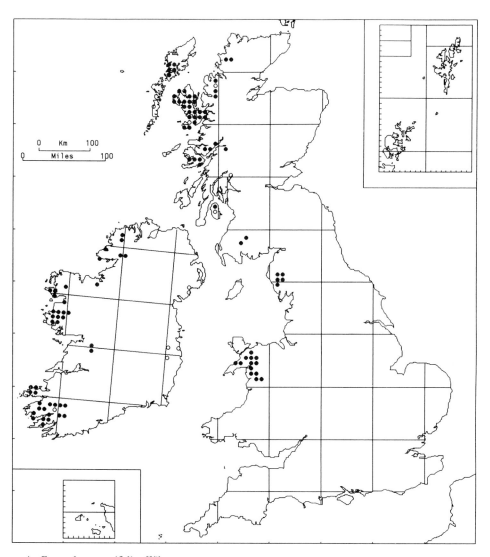

34/7. **Campylopus setifolius** Wils.

An oceanic species with a curiously patchy distribution, found in a variety of acid habitats. It often grows amongst tall *Calluna vulgaris* on N. to E. aspects in steeply sloping turf with *Mastigophora woodsii*, *Scapania ornithopodioides* and *Breutelia chrysocoma*; also on rock ledges with water dripping from above, in the spray zone of waterfalls, in block-screes and at the edge of streams and flushes. It is sometimes found in bogs. All localities have a humid microclimate, and it appears to be more frequent in the warmer parts of its range. Sea-level (Skye) to 800 m (Borrowdale). GB 62+6*, IR 42+6*.

Dioecious, male plants very scarce and known only from Skye and the adjacent mainland; fruit has been recorded in Britain, but all fruiting records are dubious.

Outside the British Isles known only from N. Spain.

M. F. V. CORLEY

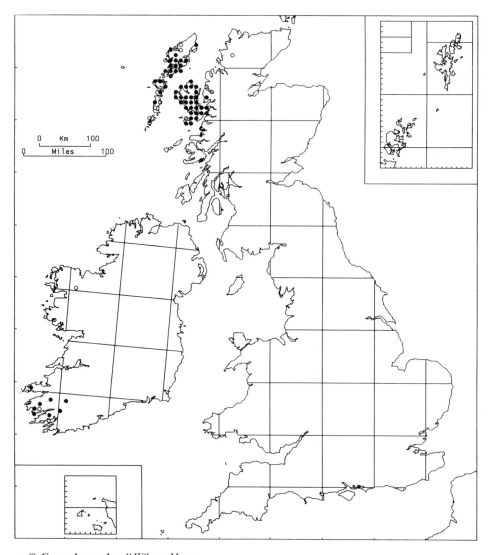

34/8. Campylopus shawii Wils. ex Hunt

It favours moist or wet acid ground where there is some water movement, such as flush bogs, turf in *Carex demissa*–*C. echinata* flushes, and under *Calluna vulgaris* on sloping boggy hillsides. It also occurs at the edge of hummocks in shallow bog-pools and in peat-cuttings. It has been found on rock ledges with water dripping from above. 0–440 m (N. Harris). GB 51+12*, IR 8+3*.

Dioecious; fruit unknown in the British Isles, where only female plants have been recorded. Vegetative propagation by fragile leaf-tips, or occasionally whole leaves.

Azores, Caribbean Is. (Jamaica, Cuba, Hispaniola, Puerto Rico).

Although this species is locally abundant at low to moderate altitudes in cool oceanic districts, it has not been recorded from some areas where it would be expected to occur. *C. shawii* was until recently thought to be endemic to the British Isles, but is now recognized as conspecific with plants from the Azores and Caribbean (Frahm, 1985).

M. F. V. CORLEY

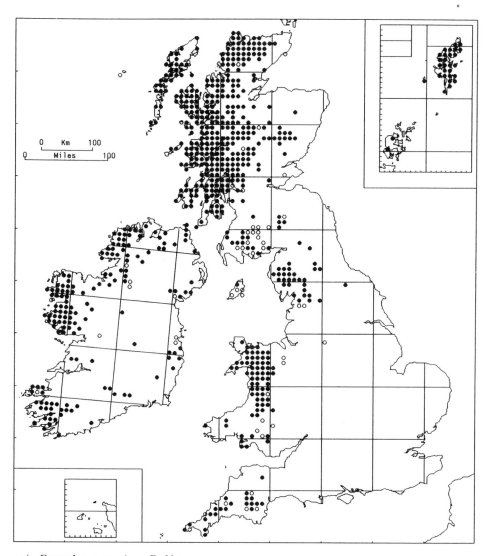

34/9. **Campylopus atrovirens** De Not.

This is a conspicuous species of upland moors and acid, mountain rocks. It is found in a wide variety of peaty sites including shallow bog-pools, flushes, and sloping boggy hillsides. It also occurs on moist rock-ledges, flushed sloping rock slabs and wet soil among rocks. 0–940 m (Aonach Beag). GB 512+43*, IR 152+9*.

Dioecious; fruit unknown in the British Isles. Vegetative propagation by deciduous shoot-tips and fragile leaf-tips.

W. Europe, from Portugal north to 65°N in Norway, and Alps. Azores, Turkey, Caucasus, Nepal, China, Japan, N. America, New Zealand.

Except for var. *falcatus*, the varieties are of doubtful taxonomic value and are not mapped separately.

M. F. V. CORLEY

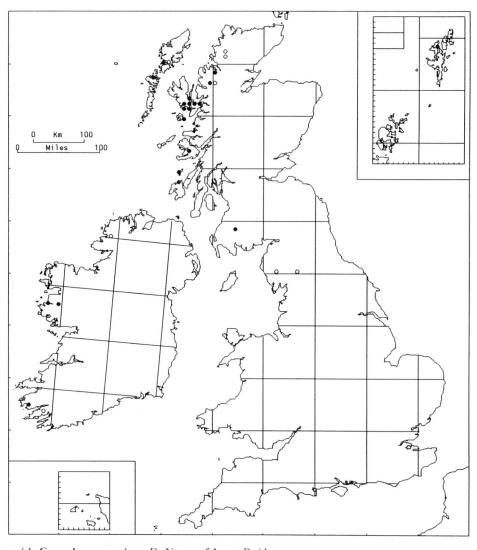

34/9b. **Campylopus atrovirens** De Not. var. **falcatus** Braithw.

In similar habitats to the type-variety, but confined to the districts of highest rainfall. GB 17+7*, IR 3+2*.
Dioecious; fruit unknown.
Norway.

M. F. V. CORLEY

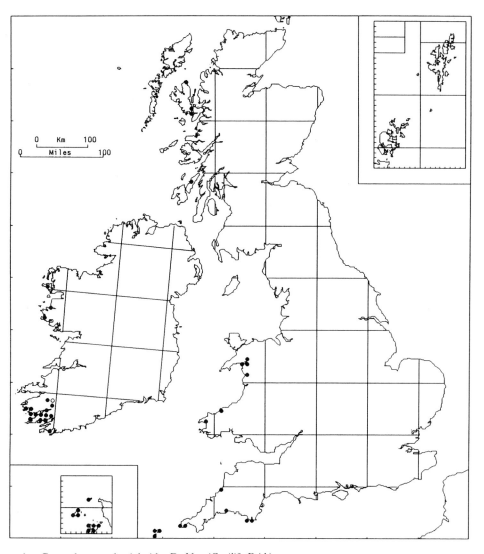

34/10. **Campylopus polytrichoides** De Not. (*C. pilifer* Brid.)

A thermophilous species found on dry, insolated acid rocks or soil thinly overlying them, or on stony ground, mainly on the coast. Associates include *Sedum anglicum*, *Grimmia laevigata* and *Polytrichum juniperinum*. Lowland. GB 19+1*, IR 17+1*.

Dioecious; fruit very rare. Vegetative propagation by deciduous leaves and shoot-tips.

W. Europe east to Italy and Belgium. Macaronesia, Africa, India to Java, southern U.S.A. and Mexico south to Venezuela, Galapagos Islands.

M. F. V. CORLEY

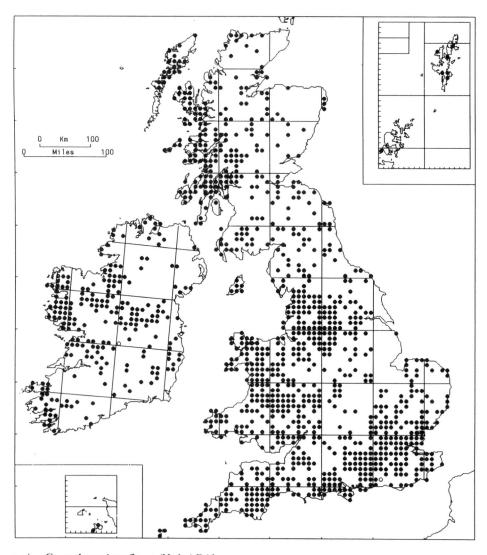

34/11. **Campylopus introflexus** (Hedw.) Brid.

A colonist of bare peat following burning, peat cutting, ploughing or digging of drainage ditches. It is also common on decaying logs and stumps, but rare on trees. Another major habitat is acid sandy or gritty ground at the edge of forestry tracks, on heathland, on railway ballast and on mine-waste. Occasionally it grows on acid rocks and roof tiles. 0–400 m (near Llangollen). GB 979+1*, IR 223+2*.

Dioecious; fruit frequent in wetter districts, often in great quantity, ripening early summer. Vegetative propagation by deciduous leaves or shoot-tips.

Introduced into N.W. Europe and now known from N. Spain and Switzerland to Sweden. S. Africa, U.S.A. (California, probably introduced), S. America, New Guinea, Australia, New Zealand.

Although a comparatively recent introduction into Europe, this species is now found all over the British Isles. The first records were from Washington, Sussex in 1941 and Howth, Co. Dublin in 1942. Richards & Smith (1975) give details of its spread in Britain and Ireland up to 1973.

M. F. V. CORLEY

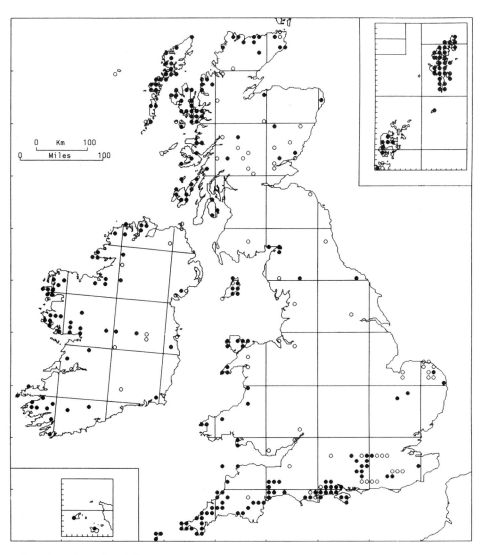

34/12. **Campylopus brevipilus** Br. Eur.

Locally common on heath and moorland, but surprisingly rare in the areas of highest rainfall. It is found in wet blanket and valley bogs, often with *Gymnocolea inflata*, and on peaty lake-margins. In heathland, it occurs on the edge of seasonal pools and, occasionally, on dry gritty soils. More rarely it grows on acid fixed dunes. Lowland. GB 219+64*, IR 55+13*.

Dioecious; fruit very rare. Vegetative propagation by deciduous leaves and fragile leaf-tips.

W. Europe east to Italy, Switzerland, Germany and Norway. Azores, Algeria.

M. F. V. CORLEY

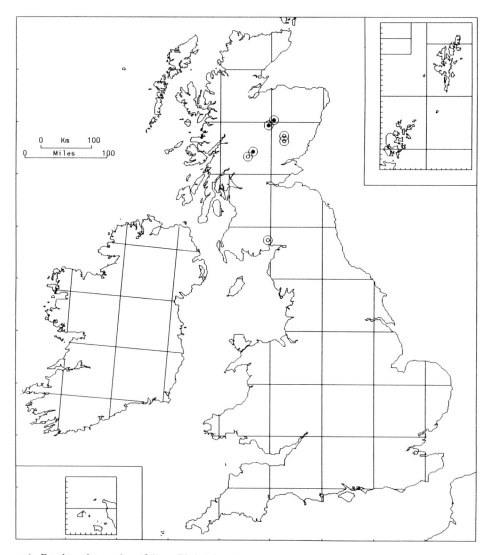

35/1. **Paraleucobryum longifolium** (Hedw.) Loeske

A species until recently considered extinct in the British Isles, but now found in several new Cairngorm sites and rediscovered on Ben Lawers in 1990. On the latter mountain it grows on the sides of massive base-rich blocks in a humid, boulder-filled gully; associates include *Barbilophozia lycopodioides*, *Gymnomitrion obtusum*, *Plagiochila porelloides*, *Dicranum scoparium*, *Mnium spinosum*, *Plagiothecium denticulatum* var. *obtusifolium* and *Tortella tortuosa*. In the Cairngorms it is restricted to the sides of sheltered granite boulders close to corrie lochans, in pure small cushions, sometimes associated with *Dicranum fuscescens*. 730 m (Loch Avon) to 1000 m (Coire an Lochain). GB 3+4*.

Dioecious; sporophytes not found in Britain. Crum & Anderson (1981) report fragile leaves on some N. American populations, suggesting vegetative propagation.

Circumboreal. In Europe, it is common in the boreal forest, extending north to Arctic Fennoscandia and south to the Alps and other high mountain ranges. Extends south to Caucasus, Himalaya, Japan and southern U.S.A. (New Mexico and N. Carolina).

D. G. LONG

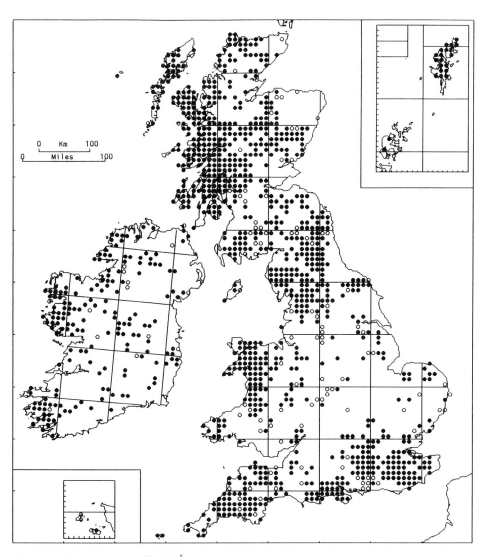

36/1. Leucobryum glaucum (Hedw.) Ångstr.

The wide distribution of *L. glaucum* reflects its broad ecological range. The main habitats include acidic woodland, damp and wet heathland, moorland, and various types of mire from lowland valley bogs and fens to upland blanket bog. Although patchy in its occurrence, it may be locally plentiful and an important structural component of the ground vegetation, forming massive hummocks which become colonized by other bryophytes and vascular plants. It does not grow directly on base-rich outcrops, but is very occasionally found in grass-heath on acid soil overlying limestone. 0–1030 m (Beinn a'Chaorrainn). GB 924+102*, IR 174+14*.

Dioecious, with males epiphytic on females; sporophytes rare (probably in part due to scarcity of males (Blackstock, 1987)), but reported from scattered localities, mostly in the south, ripe in winter. Vegetative regeneration via leaf- and shoot-fragments provides a locally effective means of dispersal, supplemented occasionally by the formation of unattached moss-balls.

Widely distributed in temperate Europe, reaching 66°N in W. Norway, rare and montane in the south. Macaronesia, Turkey, Caucasus, Japan, eastern N. America.

Large cushions, up to 2 m in diameter in a Berkshire woodland, are thought to be about 70 years old (Bates, 1989).

T. H. BLACKSTOCK

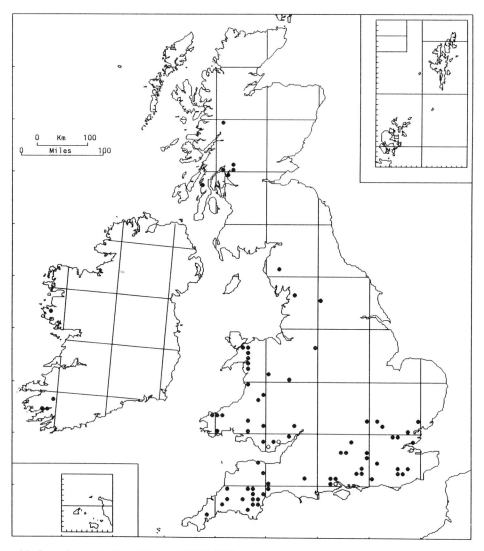

36/2. **Leucobryum juniperoideum** (Brid.) C. Müll.

Most records are from acidic oak- and beech-woodland. It is sometimes locally abundant, forming dense cushions at soil level and also extending on to rock outcrops, tree-bases and, very occasionally, stumps, trunks and branches. *L. glaucum* may also be present, and the two species can co-exist in close association. *L. juniperoideum* has also been found in wooded ravines, chestnut coppice, conifer woodland and, rarely, in more open habitats such as heathland. Poorly drained soils are generally avoided. Mainly lowland, up to 400 m (Artle Beck). GB 76+3*, IR 4.

Dioecious, with males epiphytic on females; sporophytes rare, but locally frequent in some woodlands, ripe in winter. Vegetative propagation by shoot-fragments and deciduous leaves.

L. juniperoideum has now been recorded widely in W. and C. Europe, but is absent from Fennoscandia (except Denmark) and from most of S. Europe. Macaronesia, Turkey, Caucasus, E. Asia, including Japan, China and Taiwan; it is also reported in the Southern Hemisphere from Madagascar and nearby islands.

The status of *L. juniperoideum* in the British Isles has been unclear until quite recently (Crundwell, 1972); it is still under-recorded.

T. H. Blackstock

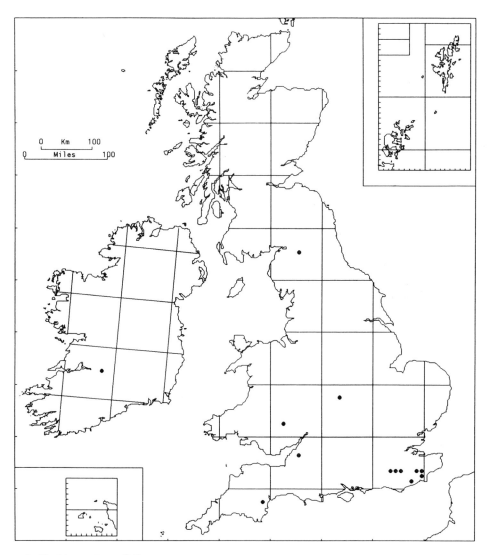

37/1. Fissidens exiguus Sull.

The smallest British aquatic *Fissidens*, *F. exiguus* occurs in small quantity, sometimes mixed with other larger *Fissidens* spp., and is possibly overlooked on this account. It is found on shaded, acidic, wet or submerged sandstone rocks in the beds of streams or small rivers in ravines and woods. Ascends to 305 m (south of Brampton). GB 12, IR 1.

Dioecious; capsules common, spring.

Very rare in continental Europe, where it is known with certainty only from Denmark and Germany (extinct). N. America.

Although reported from England by Mitten (1885), the existence of this plant in Europe was not definitely established until it was found by Mrs J. A. Paton (at that time Miss J. A. Comyn) in Kent in 1952 (Potier de la Varde, 1953). It may occur more widely in Europe, but has been much confused with *F. viridulus* var. *lylei* Wils., which is merely a small form of *F. viridulus*.

A. J. E. SMITH

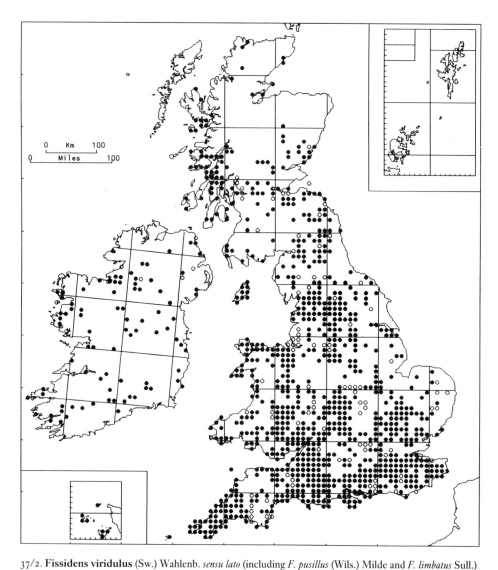

37/2. **Fissidens viridulus** (Sw.) Wahlenb. *sensu lato* (including *F. pusillus* (Wils.) Milde and *F. limbatus* Sull.)

F. viridulus is a frequent or common species in lowland areas forming small patches on basic to slightly acidic soil that is not permanently wet. It is found on banks, on streamsides, in woodland, amongst rocks, and in rock crevices. Although it may occur on thin soil overlying rock it very rarely occurs directly on rock. *F. pusillus* occurs mainly on rock, frequently in fast-flowing streams and rivers as well as in terrestrial habitats. Mainly at low altitudes but ascending to 700 m on Snowdon (*F. viridulus*). GB 873+95*, IR 86+7*.

Autoecious, synoecious or dioecious; capsules common, mainly in winter but ripening sporadically until June.

Throughout Europe, extending to about 68°N in Scandinavia. Macaronesia, Africa, Asia, N. and C. America, Australia.

Because of disagreement and confusion over the taxonomy of *F. viridulus* and *F. pusillus* it has not been possible to map them separately. *F. limbatus* was not distinguished by most recorders during the Mapping Scheme and is also included in the aggregate map. There is controversy over the taxonomic status of the first six *Fissidens* taxa mapped here, some authorities (e.g. Crum & Anderson, 1981) regarding them merely as forms of *F. bryoides*, whilst others consider them to be good species. For the distinction between *F. pusillus* and *F. viridulus*, refer to Corley (1980).

A. J. E. SMITH

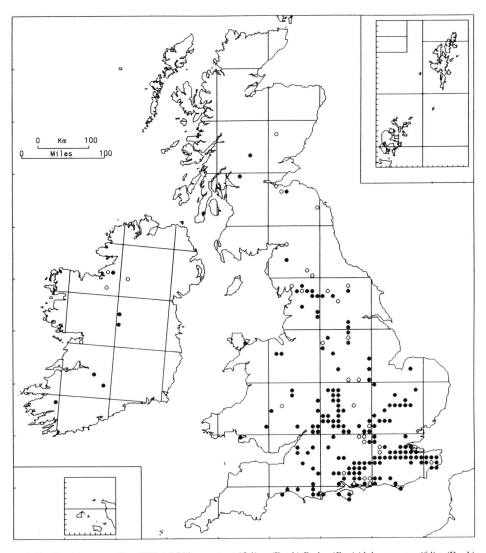

37/2Ab. **Fissidens pusillus** (Wils.) Milde var. **tenuifolius** (Boul.) Podp. (*F. viridulus* var. *tenuifolius* (Boul.) A. J. E. Smith, *F. gracilifolius* Brugg.-Nann & Nyh.)

A minute plant, gregarious or forming small patches on shaded, dry chalk, limestone or calcareous sandstone in woodland, road-cuttings, quarries and occasionally on limestone walls. It is found especially on chalk stones amongst leaf-litter in S. England. Lowland. GB 166+30*, IR 6+4*.

Dioecious or autoecious; capsules common to abundant, late summer to early spring.

Widespread but rare in Europe, extending from the Mediterranean to S. Scandinavia. Madeira, Turkey, Japan (Hokkaido).

In Britain, var. *tenuifolius* is a poorly delimited variety but the taxon is treated as a distinct species by some European authors (e.g. Nyholm, [1987]).

A. J. E. SMITH

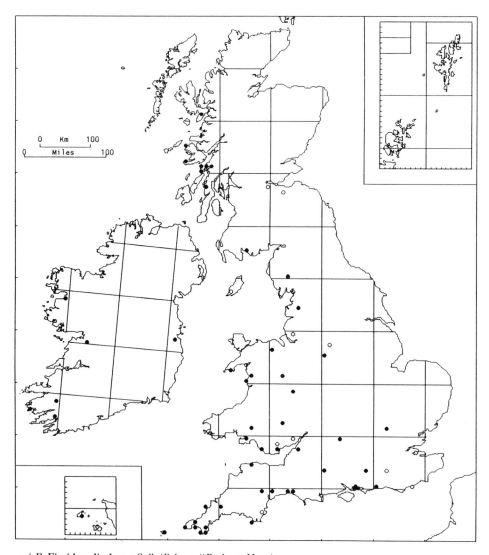

37/2B. Fissidens limbatus Sull. (*F. herzogii* Ruthe ex Herz.)

On calcareous to slightly acid substrata in sheltered but not wet habitats, growing on compacted soil or on thin soil over rock. It is most frequent on banks in woods and hedgerows or by streams and ditches, especially near the coast. It also occurs on shaded boulders, on mortar in stone-walls, on soil between limestone rocks, and in rabbit-holes. Lowland. GB 45 + 13*, IR 6.

Dioecious, autoecious or synoecious; capsules abundant, autumn to spring.

W. and S. Europe, Switzerland. Azores, Canaries, Turkey, Israel, Iran, N. America.

Probably much under-recorded in the British Isles, being very similar to *F. pusillus* and *F. viridulus* and not distinguished from them until recently (Corley, 1980).

A. J. E. SMITH

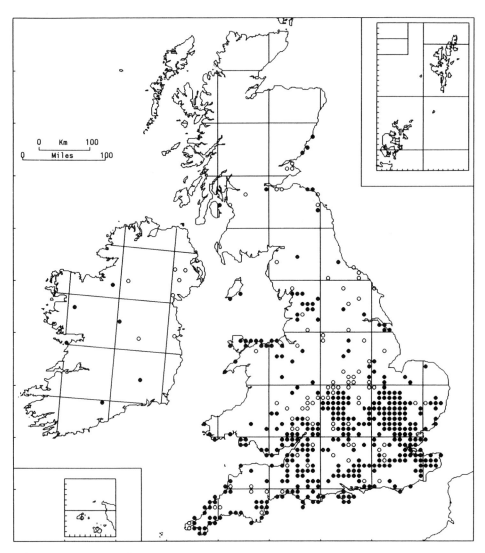

37/3. **Fissidens incurvus** Starke ex Röhl.

Plants gregarious, on basic to slightly acid soil in shady or open habitats in fallow fields, in bare patches in turf, on banks, on streamsides, in woodland and on coastal cliffs. Lowland. GB 382+90*, IR 6+8*.

Dioecious or autoecious; capsules abundant, autumn to early spring.

Europe from the Mediterranean region north to S. Sweden and S. Russia. Macaronesia, Africa, Asia, N. America, Australia.

A. J. E. SMITH

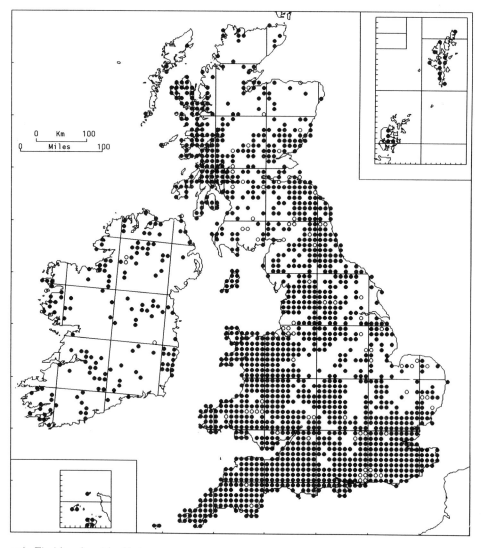

37/4. Fissidens bryoides Hedw.

Usually forming small patches on neutral to acidic, sandy, loamy or clayey soils in open or shaded sites in fallow fields, in pasture-leys, in gardens, on roadsides, in woods, on cliffs, in moorland, on banks, by streams and by ditches. It rarely occurs on wood or stones. Mainly lowland, ascending to 580 m (Beinn Sgulaird). GB 1513+83*, IR 150+6*.

Autoecious, rarely synoecious; capsules abundant, winter.

Throughout Europe, rare in the Arctic but north to Spitsbergen. Macaronesia, N. Africa, Cameroon, Madagascar, N. America, southern S. America, New Zealand, Prince Edward Island.

A. J. E. SMITH

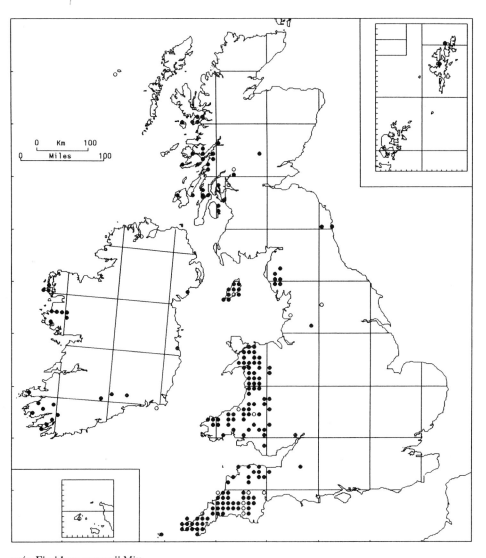

37/5. Fissidens curnovii Mitt.

Occurring as scattered plants to extensive patches on moist acidic soil, flushed rock-faces and moist crevices in slatey rock. It is found beside oligotrophic streams and rivers at or near normal water-level, in ditches, on moist road-cuttings and on coastal cliff-faces. Lowland. GB 166+18*, IR 23+2*.

Autoecious; capsules common, late autumn to summer.

Spain, Portugal, France, Belgium, Italy, Sicily, Switzerland, Germany, Czechoslovakia. Macaronesia, N. Africa.

A. J. E. SMITH

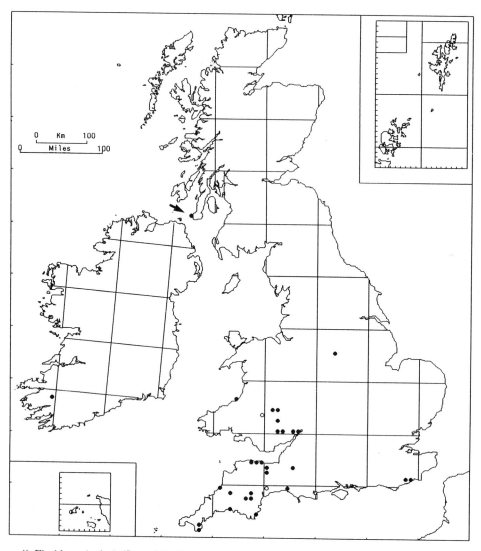

37/6. **Fissidens rivularis** (Spruce) Br. Eur.

It occurs as scattered plants or patches on shaded, moist or submerged, neutral to acidic rocks in streams, in rivers and on lakesides. Very rare on limestone. The plants may be encrusted with alluvium and in some localities appear tolerant of a considerable degree of aquatic pollution. Lowland. GB 28+2*, IR 1.

Autoecious; capsules occasional, autumn.

Widely dispersed but rare in Europe, from the Iberian Peninsula north to Belgium and Germany, east to Crete and the Balkans. Macaronesia, Africa, Turkey, Caucasus.

In S. Europe it is often found on limestone.

A. J. E. SMITH

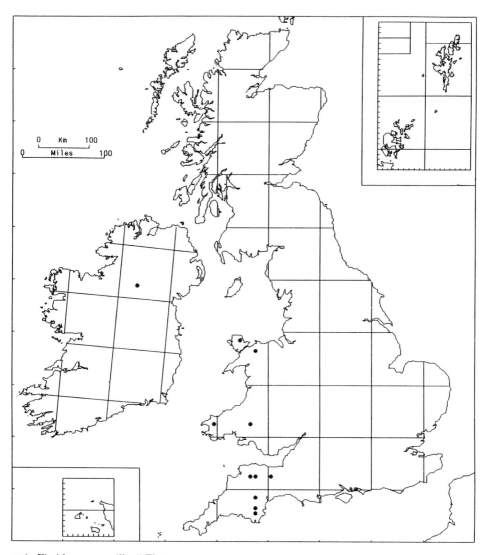

37/7. Fissidens monguillonii Thér.

The rarest of the large aquatic *Fissidens* in Britain and Ireland, occurring as scattered plants or small patches on silt-covered rocks and rock ledges, on mud-banks at or below the flood-level of slow-flowing rivers, and on the bases of *Phragmites australis* stems in fens. Lowland. GB 10, IR 1.

   Autoecious; capsules occasional, winter.

   Spain, France, Belgium. Macaronesia, Africa.

A. J. E. Sᴍɪᴛʜ

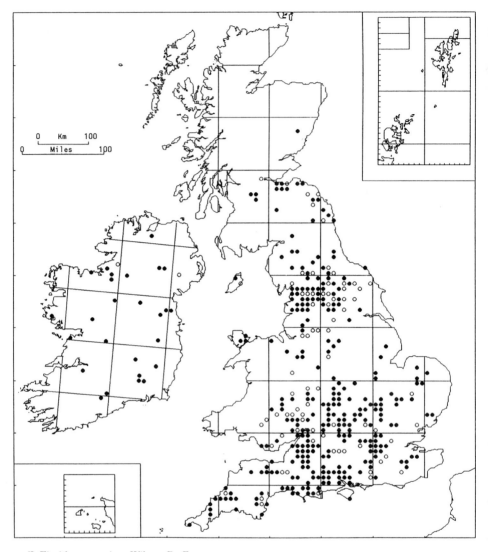

37/8. **Fissidens crassipes** Wils. ex Br. Eur.

Occurring as gregarious plants or patches at or below normal water-level on shaded or exposed, usually but not invariably basic rocks and stones in eutrophic streams and rivers, on bridge abutments, on sides of canals and on concrete culverts and sluices. This species is able to tolerate some degree of aquatic pollution. Lowland. GB 283+76*, IR 26+2*.

Dioecious, autoecious or rarely synoecious; capsules frequent, late summer to winter.

Throughout Europe north to S. Sweden. Morocco, Algeria, Turkey, Kashmir, Australia.

A. J. E. Smith

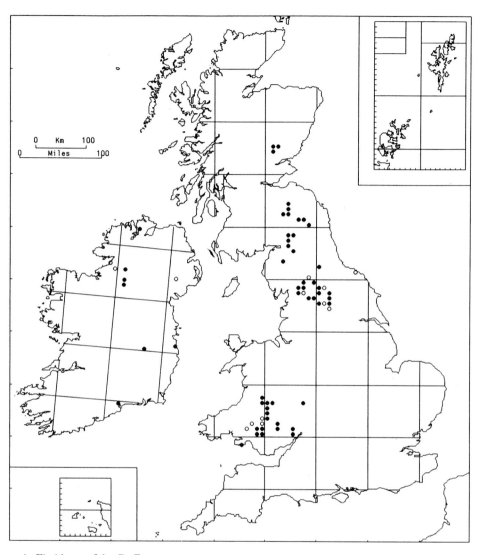

37/9. **Fissidens rufulus** Br. Eur.

F. rufulus forms patches on limestone, neutral or acidic rocks at or below normal water-level in fast-flowing unpolluted streams and rivers. In the Craven Pennines, where it is locally abundant, it grows on stony and rather sandy stream-beds, usually where submerged for the greater part of the year (Proctor, 1960). 0–300 m (Water-break-its-neck). GB 47+10*, IR 7+3*.

Dioecious; capsules occasional, winter.

Rare in W., C. and S. Europe. Turkey.

A. J. E. SMITH

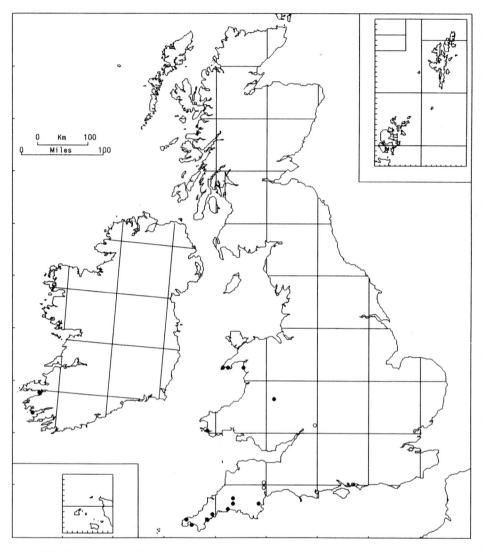

37/10. **Fissidens algarvicus** Solms

A rare but probably overlooked species of acidic clayey soils on shaded banks, sometimes associated with *Epipterygium tozeri*. Lowland. GB 13+3*, IR 2.

Dioecious; capsules occasional, ripe winter to spring.

Widely distributed but rare in S. and W. Europe from the Iberian Peninsula north to Luxemburg and France and east to Yugoslavia and Greece. Macaronesia, N. Africa, Turkey.

A. J. E. SMITH

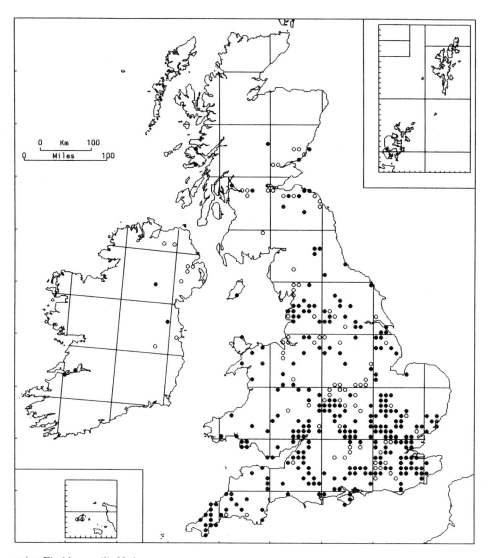

37/11. **Fissidens exilis** Hedw.

A winter ephemeral plant, occurring as scattered shoots or in thin patches on neutral or acidic loam or clayey soil in woods, on woodland rides, on sheltered banks, on streamsides and in damper parts of fallow fields or grass-leys. In established woodland it is confined to sites that are free of leaf-litter, notably molehills. It may occur mixed with *F. bryoides* and *F. taxifolius*. Lowland. GB 284+81*, IR 4+8*.

Dioecious or autoecious; capsules abundant, mainly in winter, but the season of mature capsules extends from autumn to spring.

Widespread but rather rare in Europe, from Italy and Yugoslavia north to about 62°N in Fennoscandia and from the Iberian Peninsula east to Russia. Canary Islands (Gomera), Algeria, Turkey, N. Asia, Japan (Hokkaido), eastern N. America.

Except when capsules are present, this species may be overlooked because of its minute size.

A. J. E. SMITH

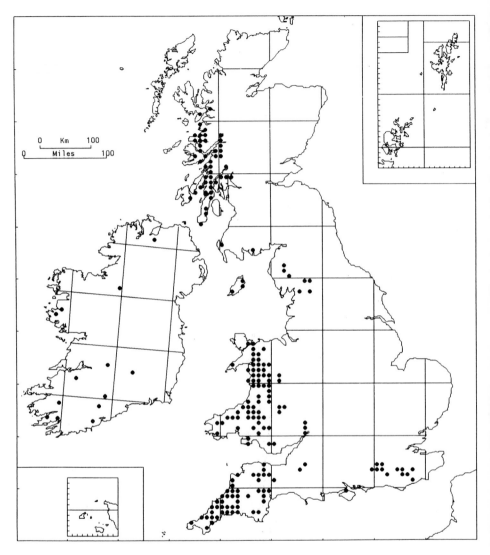

37/12. **Fissidens celticus** Paton

Occurring as scattered or gregarious plants or small patches on unstable substrata in shade. It grows on disturbed, fine-textured, acidic mineral soil and on soil-covered rocks, mainly on steep banks by streams and rivers or in woods and ravines. Frequent associates are *Calypogeia arguta* and *Diplophyllum albicans*. Lowland. GB 189, IR 13.

Dioecious, only female plants are known.

Apparently endemic to the British Isles.

F. celticus was first discovered in Pembrokeshire in 1958 by A. H. Norkett and was described as a new species by Paton (1965). It may be under-recorded, occurring in habitats not usually regarded as profitable bryologically. Endemism to the British Isles must be regarded as apparent as the species is still likely to be found in oceanic parts of continental Europe.

A. J. E. SMITH

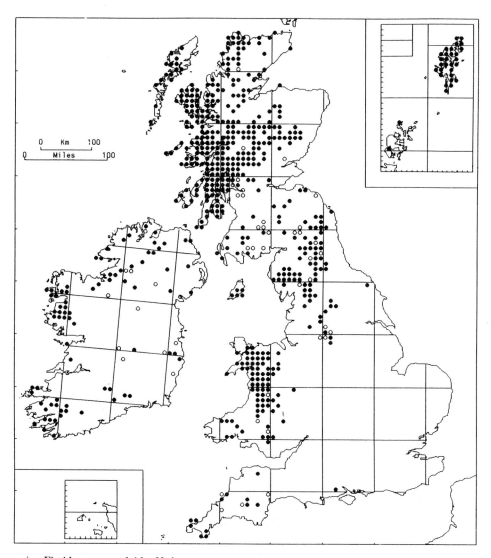

37/13. **Fissidens osmundoides** Hedw.

In dense tufts on moist ledges and crevices of cliffs and ravines, especially where slightly basic, often associated with *Amphidium mougeotii*; also beside moorland streams, and in flushes and damp montane grassland associated with species such as *Campylium stellatum* and *Drepanocladus revolvens*. Usually in montane and submontane habitats but descending to sea-level along W. coasts, also on the bank of a streamlet in the New Forest. 0–920 m (Ben Lawers). GB 503+35*, IR 77+13*.

Dioecious; capsules occasional or rare. Tubers have been found in an Irish specimen (Arts, 1988b) and may be frequent.

Circumboreal, reaching the High Arctic. Most of Europe north to Iceland and Spitsbergen, becoming rare and montane in the south, and absent from several Mediterranean countries.

A. J. E. Smith

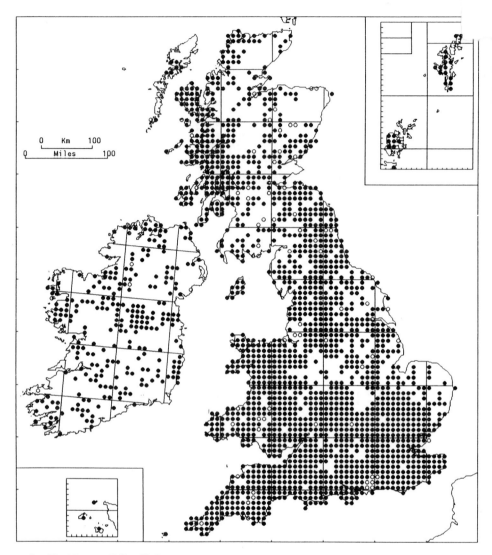

37/14. Fissidens taxifolius Hedw.

It grows on damp basic to slightly acidic soil, especially where clayey, on rocks, in rock crevices and rarely on tree-roots, in woods, in arable fields, on banks, by ditches and streams, on disturbed ground, in garden-beds and on sea-cliffs. *F. taxifolius* is able to withstand a considerable degree of atmospheric pollution. 0–460 m (Skye). GB 1744+69*, IR 257+5*.

Autoecious; capsules occasional to frequent, winter to early spring. Rhizoidal tubers sometimes present, plants possessing them apparently not producing capsules.

Europe north to 64°N. Macaronesia, Tunisia, Asia, N. and C. America, southern S. America, New Zealand.

The map shows the distribution of ssp. *taxifolius*. Ssp. *pallidicaulis* (Mitt.) Mönk. is rare and rather poorly marked in the British Isles. It is not mapped separately.

A. J. E. Smith

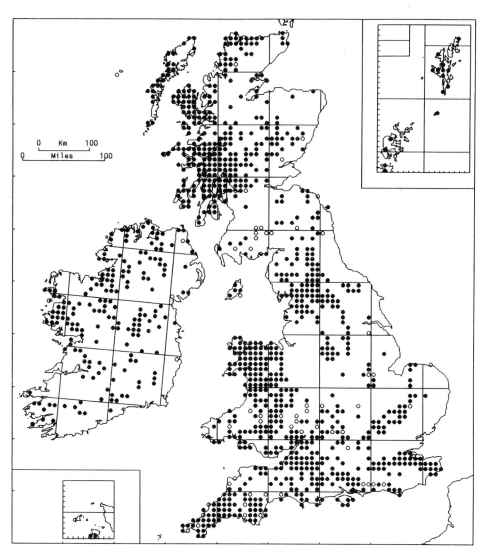

37/15. **Fissidens cristatus** Wils. ex Mitt. (*F. dubius* P. Beauv.)

A plant of open to sheltered, well-drained or intermittently moist habitats, forming dense tufts. It is found in chalk and limestone grassland where the grass is short, on sand-dunes, on rocky banks, amongst rocks and mine-waste, in limestone grikes, on ledges and in crevices of cliffs and ravines, on mortared walls and bridges, and on streamside rocks. *F. cristatus* is usually a calcicole but also grows on non-calcareous substrata, including tree-trunks, in the wetter parts of the country. 0–760 m (Skye). GB 871+59*, IR 183+5*.

Autoecious or dioecious; capsules occasional to frequent, winter to spring. Rhizoidal tubers, which remain viable for up to 20 months in dried herbarium specimens, have been found in non-fruiting Belgian, French and Swiss specimens (Arts, 1986); they are probably frequent in Britain and Ireland. As with *F. taxifolius*, gemmae and capsules appear to be mutually exclusive.

Europe north to Iceland and W. Norway, where it reaches 68°N, extending east to Ukraine and Crimea. Macaronesia, N. Africa, Turkey, Caucasus, Himalaya, N.E. Asia, S.E. Asia, N. America, Mexico, Haiti.

A. J. E. SMITH

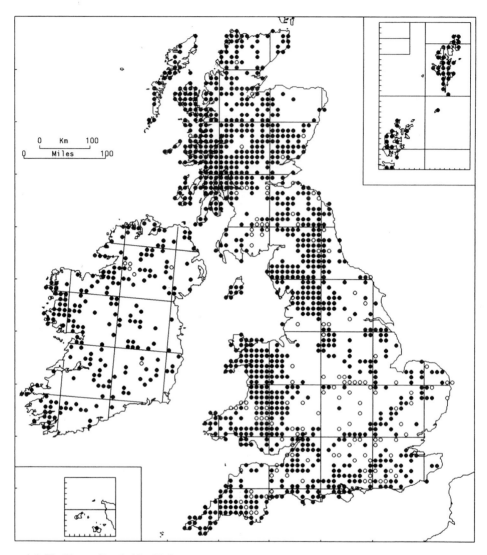

37/16. Fissidens adianthoides Hedw.

A robust species forming dense to lax, sometimes extensive, tufts up to 10 cm high. It occurs mainly in damp or wet habitats, especially where the water is basic, being found on wet or dripping rock-faces, on ledges and crevices of cliffs, in flushes, drainage runnels, marshes, fens and dune-slacks, and rarely on rotting wood or tree-trunks. It is also found in chalk and limestone grassland but in moister places than *F. cristatus*, where the grass is longer. In its less wet habitats *F. adianthoides* is a calcicole. 0–950 m (Ben Lawers). GB 1106+130*, IR 215+8*.

Autoecious or dioecious; capsules occasional to frequent. Rhizoidal tubers unknown but should be looked for in specimens from drier habitats, where it tends to be stunted and sterile.

Circumboreal, reaching north to the High Arctic and south to N. and E. Africa and southern N. America (Louisiana and Florida). Disjunct in the Southern Hemisphere; known from Tierra del Fuego, Tasmania and New Zealand.

A. J. E. Sᴍɪᴛʜ

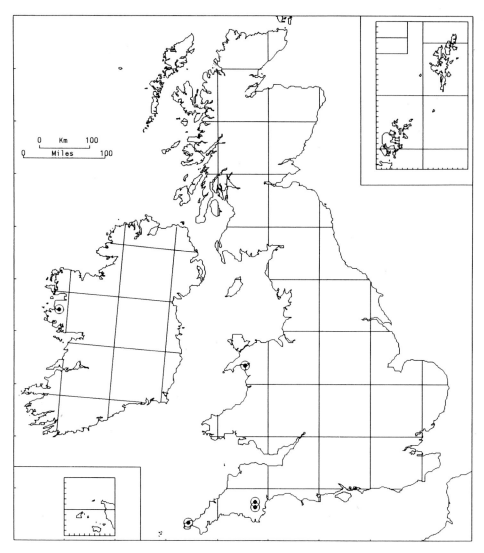

37/17. **Fissidens serrulatus** Brid.

Lax patches on alluvial sand and gravel, and on rocks below flood-level of deeply shaded streams; also on wet shaded rocks in a cave. Lowland. GB 4, IR 1.

Dioecious, only male plants known in Britain and Ireland. Rhizoidal tubers have been found in a specimen from Devon (Arts, 1988b) and are probably frequent.

Spain, France, Luxemburg, Corsica, Italy, Yugoslavia, Greece (Rhodes). Macaronesia, Tunisia, Algeria.

A. J. E. SMITH

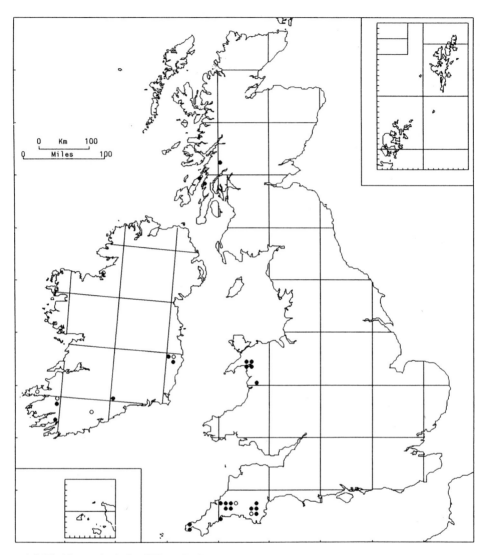

37/18. Fissidens polyphyllus Wils. ex Br. Eur.

Dark green patches, sometimes extensive, on usually deeply shaded rocks and hard-packed soil by streams and waterfalls, rarely in exposed habitats; also found inside a cave at sea-level and on dripping rocks by mine-workings. Associated species may include *Marsupella emarginata*, *Heterocladium heteropterum* and *Hyocomium armoricum*. Lowland. GB 20+2*, IR 5+5*.

Dioecious, only male plants known in Britain and Ireland. Rhizoidal tubers have been found in a Welsh specimen and may be frequent (Arts, 1988b).

Spain, Portugal, W. France, Italy, extreme S.W. Norway. Azores, Madeira, Canaries.

A. J. E. Smith

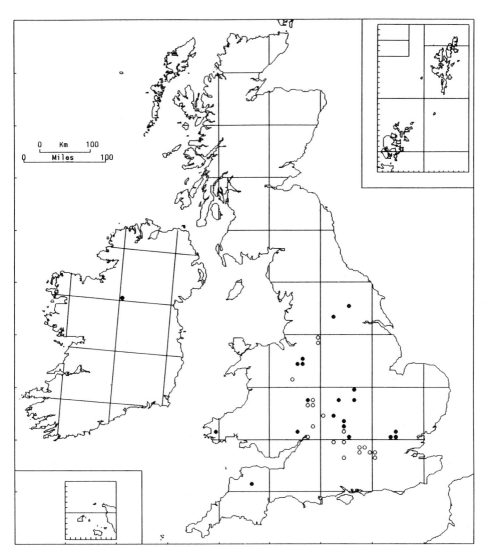

38/1. **Octodiceras fontanum** (La Pyl.) Lindb. (*Fissidens fontanus* (La Pyl.) Steud.)

An aquatic moss which grows submersed in clean or somewhat polluted water of rivers and canals, from near the surface to a depth of 80 cm. It is undiscriminating in its choice of substrate, colonizing wood, stones, rocks, concrete embankments, lock-gates, iron girders, floating pontoons and even the surface of freshwater sponges (Sowter, 1972). It has also been recorded from stones in reed-swamp. *Fissidens crassipes*, *Fontinalis antipyretica* and *Rhynchostegium riparioides* are sometimes mixed with it. Lowland. GB 19+19*, IR 1.

Autoecious; capsules have been observed in C. Europe but not in the British Isles. Vegetative propagation by detached leafy branches has also been reported in C. Europe.

Europe, north to S. Scandinavia, with isolated localities in N. Sweden and Finland. Madeira, N., E. and S. Africa, N. America, Mexico, Chile, Australia, New Zealand.

O. fontanum has become extinct at some localities, possibly because of increased pollution. It may be particularly palatable to freshwater snails; Lohammar (1954) found that in the aquarium it was eaten in preference to *Drepanocladus aduncus* and *Fontinalis antipyretica* and in the wild it was absent from water-bodies where snails were numerous.

C. D. PRESTON & A. J. E. SMITH

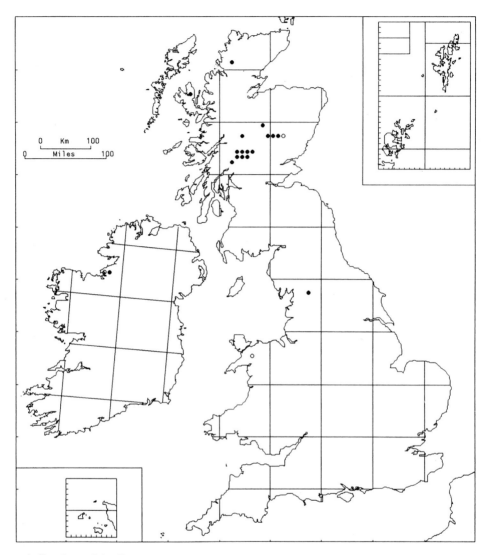

39/1. **Encalypta alpina** Sm.

Occurs as small pure tufts in dry or seasonally moist rock crevices and cliff-faces of limestone, calcareous schist, or crumbling calcareous basalt, commonly associated with montane calcicoles such as *Polystichum lonchitis, Cirriphyllum cirrosum, Hypnum bambergeri, Mnium spinosum, M. thomsonii* and *Pohlia cruda*. Favours, but is not restricted to N.- or E.-facing cliffs. Always in small quantity. 350–1205 m (Ben Lawers). GB 16+2*, IR 1+1*.

Autoecious; sporophytes common.

Circumpolar arctic-alpine, extending from the High Arctic south to the Alps, Caucasus, Altai Mts, China and Colorado (U.S.A.).

H. J. B. BIRKS

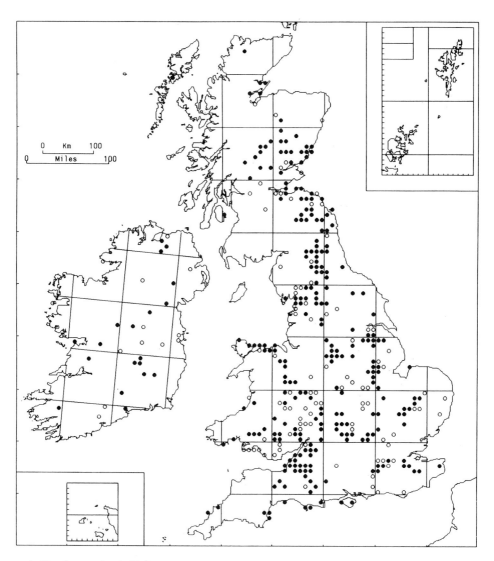

39/2. **Encalypta vulgaris** Hedw.

A strict calcicole, almost exclusively confined to chalk, limestone or calcareous schists; very rare on basic igneous rocks such as dolerite. It occurs as pure tufts on dry, well-drained rocks and soil in a range of habitats. These include open soil in short calcareous grassland, especially on steep rocky slopes; bare chalk soil or limestone rocks in woods and along wooded lanes; shaded chalk banks; bare soil in quarries; bare soil in crevices of sea-cliffs; and soil-capped ledges and crevices in limestone walls and inland cliffs, including upland cliffs. Common associates include *Leiocolea badensis*, *L. turbinata*, *Aloina* spp., *Ctenidium molluscum*, *Encalypta streptocarpa*, *Neckera complanata* and *N. crispa*. 0–730 m (Mickle Fell). GB 252+96*, IR 19+13*.

Autoecious; sporophytes abundant.

Europe except the far north. Canaries, N. and S. Africa, W. and C. Asia, Tibet, Kashmir, Himalaya, N. America, New Guinea, Tasmania, New Zealand.

In North America, *E. vulgaris* and *E. rhaptocarpa* intergrade so extensively that they could be regarded as subspecies of a single species (Horton, 1983), but in Britain they are normally easy to distinguish.

H. J. B. BIRKS

207

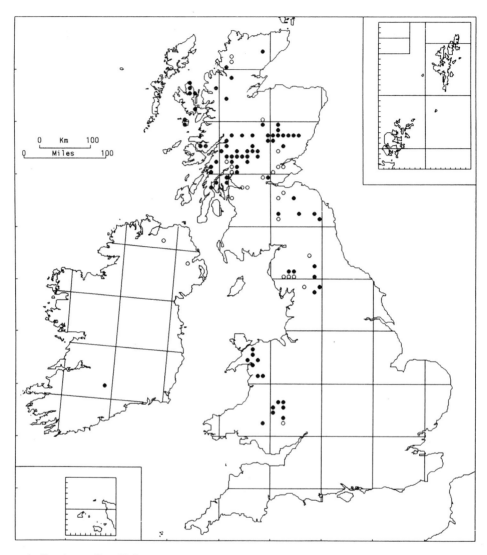

39/3. Encalypta ciliata Hedw.

A local species of dry or periodically damp, shaded crevices or cliff-faces of calcareous montane cliffs, usually of limestone, calcareous schist, basalt, volcanic tuff, or other basic igneous rocks, growing with other montane calcicoles such as *Asplenium viride, Mnium marginatum, M. stellare* and *Pohlia cruda*. Also occurs, more rarely, in shaded crevices on rock-walls of Carboniferous limestone ravines and gills with *Orthothecium intricatum* and *Plagiopus oederi*. Always in small quantity and seemingly indifferent to aspect. 300–1100 m (Ben Lawers). GB 81+23*, IR 1+2*.

Autoecious; sporophytes abundant.

Northern and montane parts of Europe. Algeria, Ethiopia, C. and S. Africa, Caucasus, N., C. and E. Asia, Kashmir, Himalaya, Hawaii, N., C. and S. America, New Guinea, Australia, New Zealand.

H. J. B. Birks

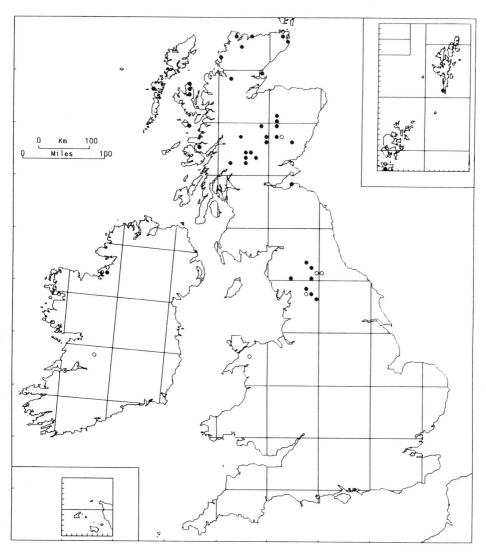

39/4. Encalypta rhaptocarpa Schwaegr.

Always in small quantity on dry or seasonally moist crevices and crumbling earthy ledges on limestones, highly calcareous schists or basalts, or other soft, basic igneous rocks, associated with calcicoles such as *Pseudoleskeella catenulata*, *Schistidium apocarpum*, *S. strictum* and *Tortula subulata*. Favours but is not confined to E.-, S.- or W.-facing cliffs. Also occurs on calcareous sand-dunes in E. and N. Scotland. 0–1190 m (Ben Lawers). GB 38+9*, IR 1+4*.

Autoecious; sporophytes abundant. Protonemal gemmae are produced in culture (Whitehouse, 1987).

Widespread in Arctic and montane parts of Europe. N. Africa, N., W. and C. Asia, Tibet, Himalaya, Kashmir, Hawaii, N. America, Greenland.

H. J. B. Birks

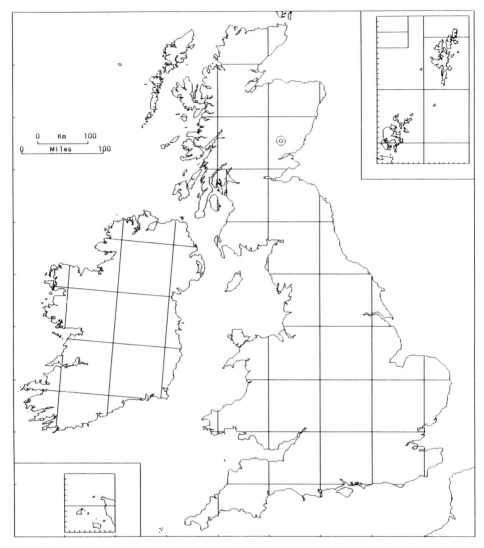

39/4A. Encalypta brevicollis (Br. Eur.) Ångstr.

In a steep river valley, mixed with *E. ciliata*. Lowland. GB 1*.

    Autoecious; sporophytes present.

    Fennoscandia north to Svalbard. Urals, N. Siberia, N. America, Greenland.

    The occurrence of this taxon in the British Isles is inferred from a specimen in W. Mitten's herbarium, labelled 'Reeky Lynn Aug. 1871 W. M.' This was interpreted by Horton (1980) as Reekie Linn, a low waterfall of the River Isla in the Den of Airlie, Angus. It has not been seen recently in the British Isles. Horton argues that it should be looked for 'where the rock type is acidic'. Its close association with *E. ciliata* suggests, however, that *E. brevicollis* is a calcicole, at least in Scotland.

H. J. B. BIRKS

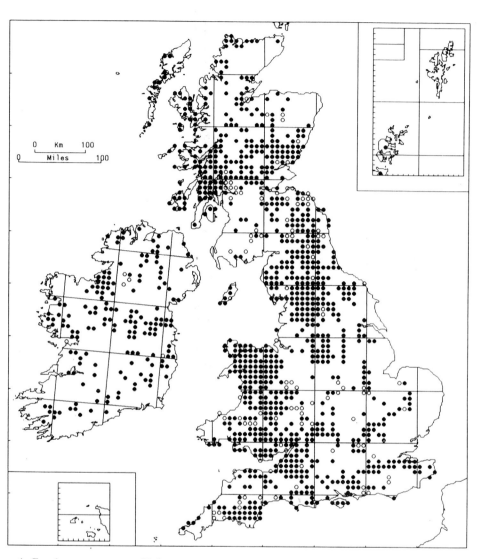

39/5. Encalypta streptocarpa Hedw.

A calcicole occurring on a wide range of substrata. Natural or semi-natural habitats include dry, often shaded calcareous rock-walls of low-lying wooded ravines; shaded limestone rocks in woods; bare chalk soil and rubble in beech-woods, particularly on steep slopes; crevices in limestone outcrops, especially in areas of bare limestone pavement; limestone scree; crevices and small ledges on damp, base-rich cliffs; bare patches in calcareous grassland; and calcareous sand-dunes. Artificial habitats include limestone walls and old crumbling mortar of other walls, bridges and old buildings, often where there is some shade; loose rubble in old limestone quarries; and bare chalk banks and hedgerows. Unlike other *Encalypta* species, it is not confined to calcareous areas because of its ability to grow on mortar of walls and other artificial habitats. 0–760 m (Glen Callater). GB 884+108*, IR 164+6*.

Dioecious; sporophytes very rare. Filamentous gemmae abundant in leaf axils; gemmae are also produced on the protonema in culture (Whitehouse, 1987).

Throughout Europe. Canaries, N. Africa, Turkey, Iran. Reports from N. America are due to confusion with *E. procera* Bruch (Horton, 1983).

H. J. B. Birks

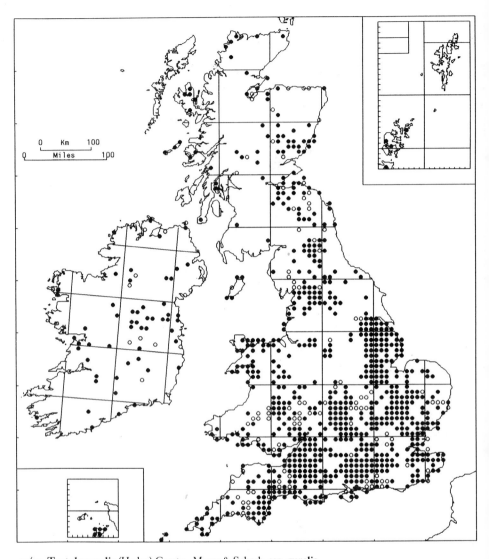

40/1a. **Tortula ruralis** (Hedw.) Gaertn., Meyer & Scherb. ssp. **ruralis**

This taxon occurs in a wide variety of habitats, usually in base-rich situations: in some lowland parts of the country it is very common on tiles and thatch on rooftops, but it also occurs widely on natural rock-outcrops and walls, especially of limestone, on stony and sandy ground, and on sand-dunes. It may also grow on artificial substrata such as asbestos, concrete and asphalt. The sites are usually exposed and sunny, and the species is tolerant of prolonged periods of drought. 0–350 m (Malham Tarn). GB 738+106*, IR 54+13*.

Dioecious; capsules infrequent, maturing in spring and early summer. Rhizoidal tubers occur in southern Africa but have not been reported in Europe.

Throughout most of Europe. Widespread in the Northern Hemisphere but its range is not accurately known because of taxonomic difficulties; southern Africa, southern S. America.

Many British populations are of the form recognized on the Continent as *T. calcicolens* Kramer. In Britain, intermediates between this and *T. ruralis sensu stricto* are too common for the taxa to be distinguished, let alone mapped separately.

T. L. BLOCKEEL

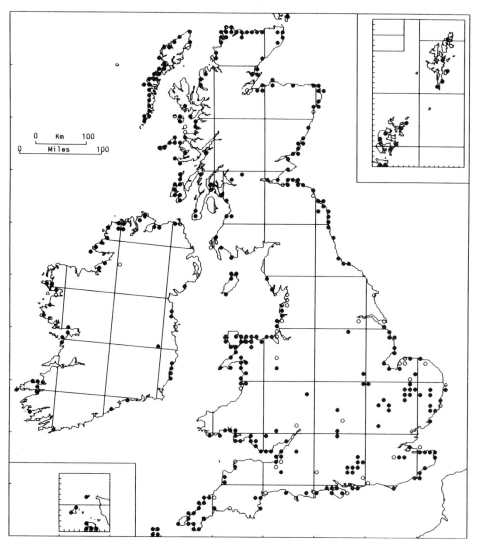

40/1b. **Tortula ruralis** (Hedw.) Gaertn., Meyer & Scherb. ssp. **ruraliformis** (Besch.) Dix. (*T. ruraliformis* (Besch.) Grout)

This plant grows in similar habitats to ssp. *ruralis* but is much commoner on sand-dunes, where it is a characteristic colonist of loose sand on unstable dunes and persists on the stabilized sand. It also occurs on sandy coastal banks and cliffs. It is less common inland, occurring on open sandy ground and old sand-pits, and rather rarely on stones, stony ground, rock outcrops and rooftops. Lowland. GB 305+44*, IR 42+5*.

Dioecious; capsules are rather rare, maturing in spring and summer.

Throughout much of Europe, north to S. Scandinavia. Macaronesia, N. Africa, W. Asia, western N. America.

T. L. BLOCKEEL

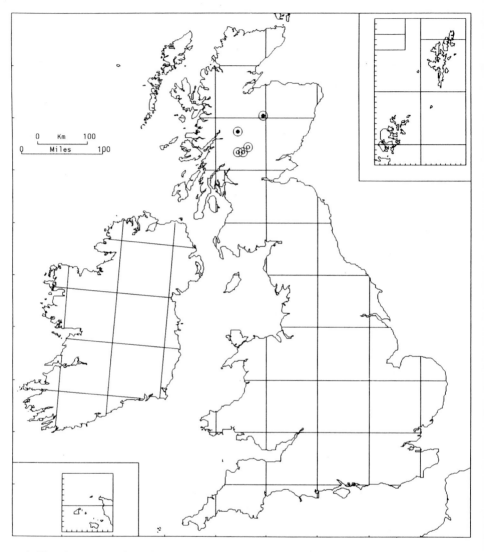

40/2. **Tortula norvegica** (Web. f.) Wahlenb. ex Lindb.

A rare montane species of rock crevices and soil among rocks on strongly basic substrata, especially limestones and mica-schist. Confined to high altitudes, reaching 1050 m (Ben Alder range). GB 2+3*.

Dioecious; capsules have not been recorded in Britain.

Circumboreal. N. Europe and widespread on mountains further south, but absent from most of the Mediterranean region.

T. L. BLOCKEEL

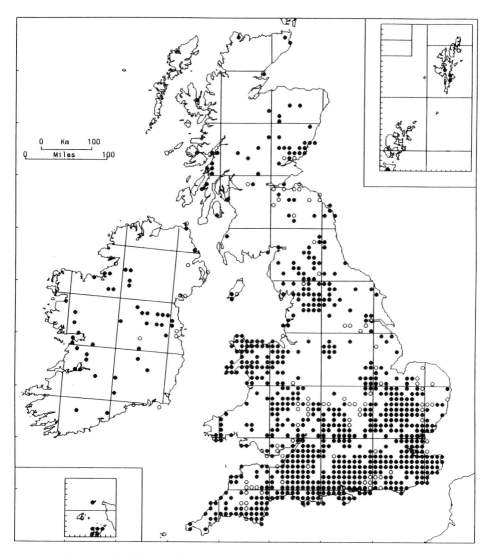

40/3. **Tortula intermedia** (Brid.) De Not.

Often abundant on limestone rocks and walls, growing in tufts or cushions. More rarely it grows on non-calcareous rocks, but always where there is some base content. It is frequent on artificial substrata such as brick walls, wall-mortar, roof-tiles, asbestos, concrete and asphalt, and on old thatch. It occurs rarely on trees and on stony ground and soil. Sites are often exposed and the species is tolerant of prolonged drought. Mainly lowland, ascending to 625 m on limestone of Gragareth. GB 736+89*, IR 44+12*.

Dioecious; capsules occasional, maturing in spring and summer.

Most of Europe, north to S. Scandinavia. Macaronesia, N. Africa, W. Asia, N. America.

T. L. Blockeel

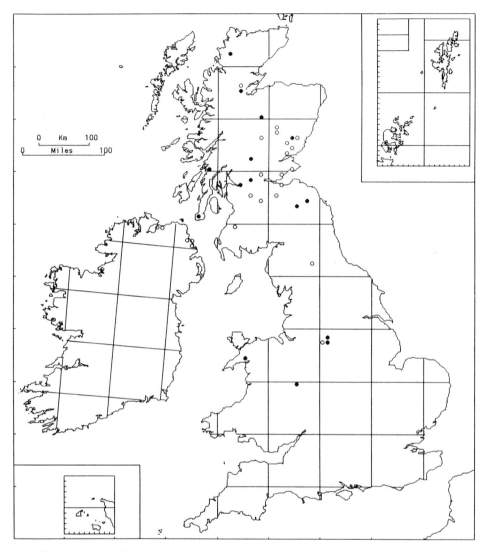

40/4. **Tortula princeps** De Not.

A species of base-rich outcrops and cliffs, especially on limestones and basic sandstone, in well-illuminated situations and in open woodland. It often occurs on soil among rocks, on thin soil on ledges and in rock crevices, more rarely on walls and trees. Though mainly confined to the upland districts of Britain, it is normally found at moderate rather than high elevations. 100 m (Kirkcaldy) to 500 m (Moel Hebog). GB 15 + 18*, IR 6*.

Synoecious; capsules frequent, maturing in summer.

W. and S. Europe. Macaronesia, N. Africa, Asia, western N. America; very widespread in temperate and Antarctic regions of the Southern Hemisphere.

T. L. BLOCKEEL

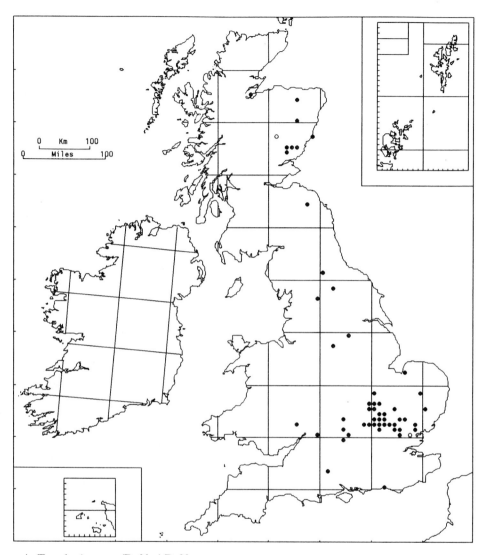

40/5. Tortula virescens (De Not.) De Not.

This species commonly occurs on the boles and bases of mature trees in open situations, less often on smaller trees such as elder. It is also widespread in man-made habitats, growing on stones, walls, asphalt paths and concrete, and more rarely on thatched roofs. Lowland. GB 52+2*.

Dioecious; capsules very rare, found only once in Britain. Gemmae have been reported in Continental material but not yet in Britain.

W., C. and E. Europe, north to C. Scandinavia; absent in much of the Mediterranean region. Turkey.

It was not recognized in Britain until 1958 and is probably under-recorded. It had been collected in Scotland in 1886, so there is no evidence that it is an introduction.

T. L. BLOCKEEL

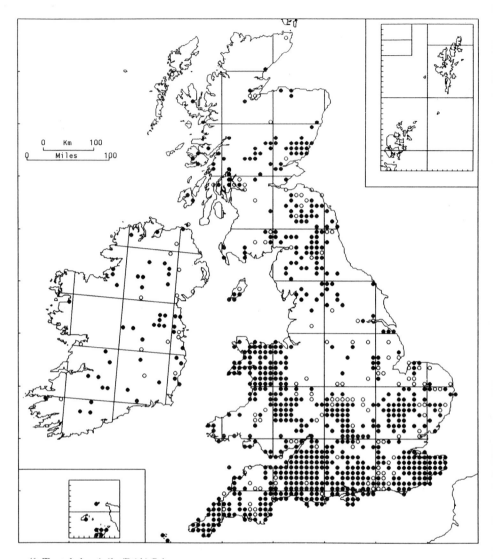

40/6. **Tortula laevipila** (Brid.) Schwaegr.

This species occurs as an epiphyte on the trunks and branches of a wide variety of trees, especially elm, ash, willows and old elders, but also oak, alder, beech and other species. It prefers well-illuminated situations and often occurs on isolated trees, as on roadsides. It also grows occasionally on the roots of trees by streams and rivers, and rarely on non-acidic rocks and walls. Lowland. GB 704+125*, IR 47+15*.

Synoecious or dioecious; capsules frequent, maturing in summer. Axillary leaf-like gemmae occur occasionally among the upper leaves.

Widespread in S. and C. Europe, common in the Mediterranean region and extending north to the British Isles and S. Sweden. Macaronesia, N. Africa, W. Asia, N. America.

Forms with leaf-like gemmae are sometimes distinguished as var. *laevipiliformis* (De Not.) Limpr., but recent observations suggest that this is not a distinct taxon (M. M. Yeo, pers. comm.).

T. L. Blockeel

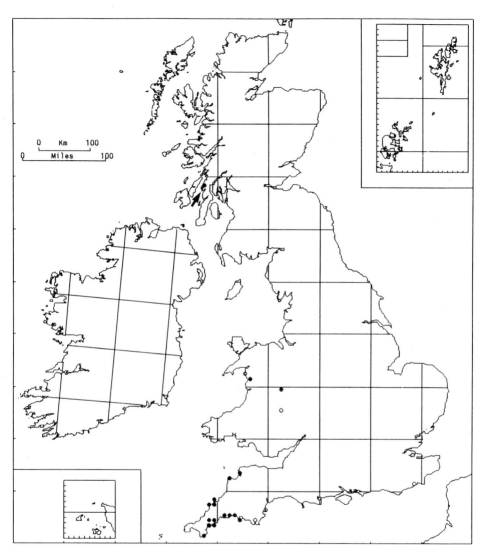

40/7. **Tortula canescens** Mont.

A species of soil in turf, on rocky and shaly banks and in crevices of walls, chiefly on the coast. Inland, near the Welsh border, it occurs on dolerite. It has a marked preference for sunny S.-facing slopes. Lowland. GB 17+5*.

Autoecious; capsules are frequent, maturing in spring.

S., W. and C. Europe. Macaronesia, N. Africa, W. Asia. The European distribution is mapped by Häusler (1984).

T. L. BLOCKEEL

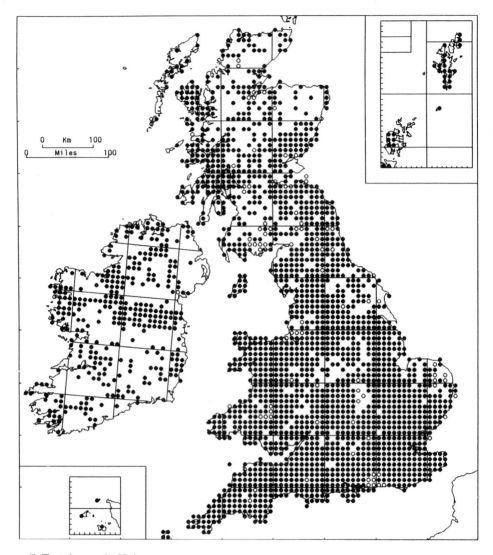

40/8. **Tortula muralis** Hedw.

An abundant species in all but the most sparsely populated parts of the country, occurring very commonly on brick and stone walls, roof-tiles, concrete and other man-made substrata. It is tolerant of air pollution and is common even in city centres and industrial areas. It occurs also on natural rock-outcrops, especially limestones, and on trees and wood, and is absent only from the most acid substrata. Sites are usually well illuminated or only lightly shaded. 0–950 m (Ben More, Mull). GB 1925+93*, IR 333+3*.

Autoecious; capsules abundant, maturing in spring, summer and autumn.

Cosmopolitan. Throughout Europe, except the extreme north.

Forms with a short hair-point are sometimes distinguished as var. *aestiva* Hedw., but this may not be a good taxon; it has certainly not been recorded consistently enough to permit mapping.

T. L. BLOCKEEL

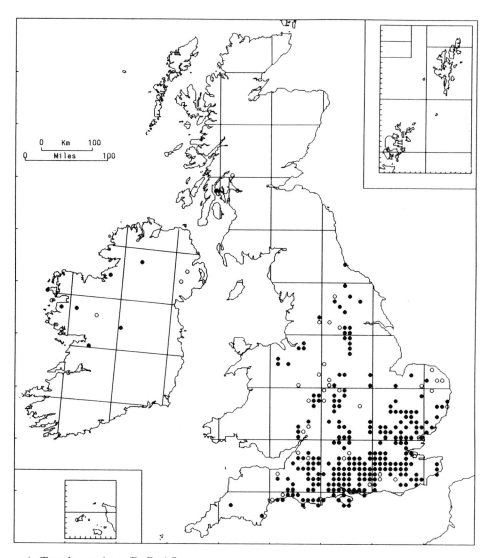

40/9. **Tortula marginata** (Br. Eur.) Spruce

On moist, base-rich rocks and walls. It occurs naturally on shaded limestones and sandstones, especially in woodland, but is also widespread in man-made sites on walls and buildings, commonly near ground-level, both on stonework and on old mortar and plaster. The habitat is often heavily shaded or at least of sheltered aspect. Lowland. GB 245+35*, IR 7+4*.

Dioecious; capsules common, maturing in late spring and summer.

A Mediterranean-Atlantic species ranging north in Europe to the British Isles and Germany. Macaronesia, N. Africa, W. Asia.

T. L. BLOCKEEL

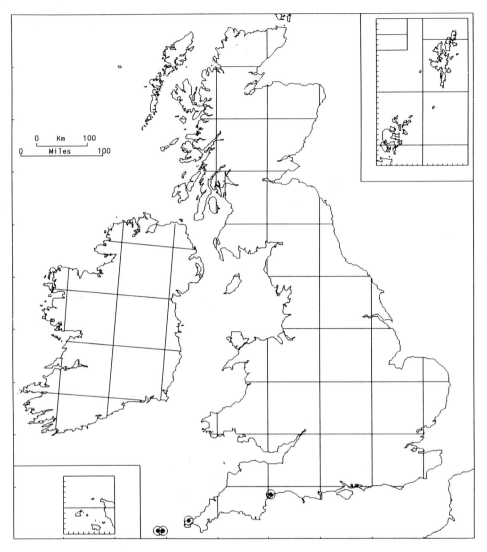

40/9A. **Tortula solmsii** (Schimp.) Limpr.

On slightly basic or neutral substrata on or near the coast. In Devon, it grows on friable rock on New Red Sandstone cliffs; in Cornwall, it is known from crevices of granite walls and from moist soil-banks. The species is tolerant of shade; most British plants belong to an atypical shade form. Lowland. GB 4.

Dioecious; capsules occasional in Britain, maturing in spring.

A Mediterranean-Atlantic species reaching its northern limit in Britain. Macaronesia, N. Africa, Turkey.

It was first collected in Britain in 1956 but, because of its atypical appearance, was not identified until 1980 (Long & Hill, 1982).

T. L. Blockeel

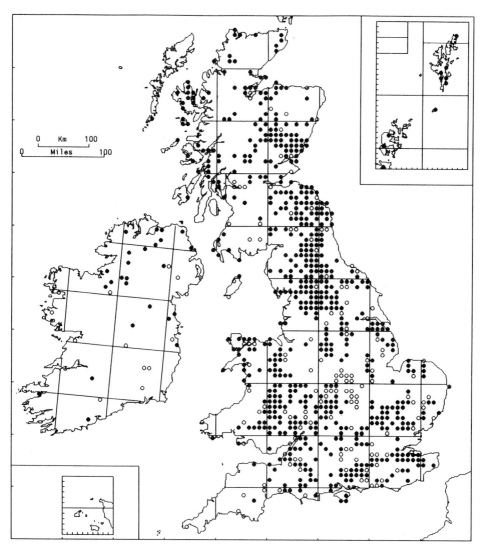

40/10. **Tortula subulata** Hedw.

A species of neutral and calcareous habitats on soil and about tree-bases, growing on roadside and woodland banks, on sea-cliffs, on ledges and crevices of rocks, and on old walls. It is also common on tree-bases and roots in the flood-zone of rivers and streams, usually in association with *T. latifolia*. It often occurs on sandy base-poor soil and on leached ground, as on chalk ridges, but is also widespread on limestone and other base-rich rocks. 0–500 m (North Wales, Skye). GB 627+141*, IR 25+13*.

Autoecious; capsules frequent, maturing in spring and summer.

Widespread in Europe, but rare in the north. Macaronesia, N. Africa, western N. America.

Of the varieties recognized by Smith (1978), var. *angustata* (Schimp.) Limpr. and *subinermis* (Brid.) Wils. are often hard to distinguish from var. *subulata*. Var. *graeffii* Warnst. is normally well marked, but is not mapped separately.

T. L. Blockeel

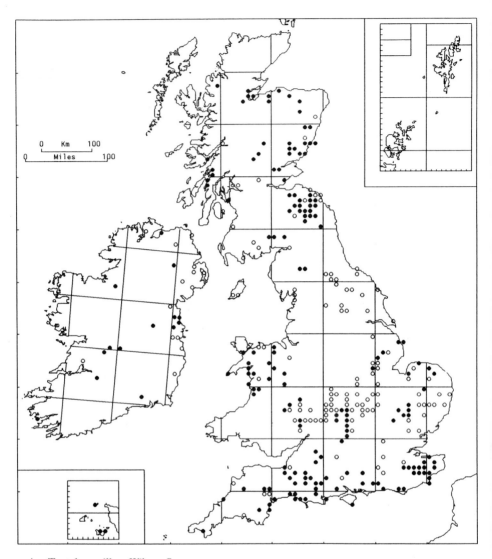

40/11. **Tortula papillosa** Wils. ex Spruce

Usually on the boles of mature trees in hedges and by roads, but also on smaller trees and bushes, especially elder. More rarely it occurs on walls, stones and tarred surfaces. Though sometimes found in pure patches, it is usually found in small quantity mixed with other bryophytes. It has a tendency to occur in roadside situations and near ground level; the species is probably nitrophilous. It is however sensitive to air pollution and has decreased over much of C. and E. Britain. Lowland. GB 164+120*, IR 14+14*.

Dioecious; capsules and inflorescences are unknown in Britain. Gemmae are constantly present.

Most of Europe except N. Scandinavia and much of the Mediterranean region. C. Asia, N., C. and S. America; widespread in the temperate Southern Hemisphere.

T. L. Blockeel

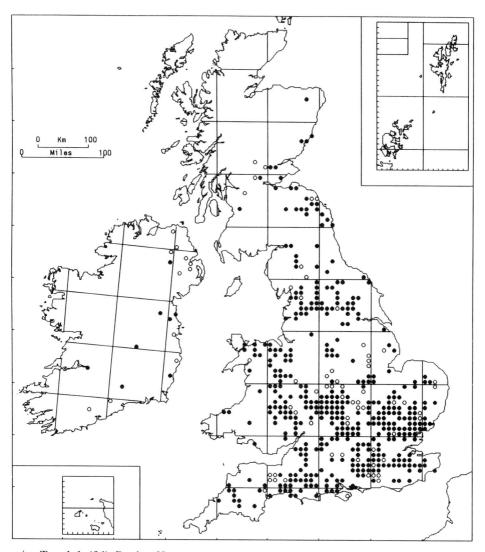

40/12. **Tortula latifolia** Bruch ex Hartm.

This species is characteristic of the banks of streams and rivers which are subject to periodic flooding. It is usually confined to the flood-zone and occurs both on the roots and boles of trees and on rock and stone, including walls, bridges and embankments; it is often embedded in sand or silty detritus. The most frequently colonized trees are alder, willows and sycamore. It is very commonly associated with *Leskea polycarpa* and in some districts with *Orthotrichum sprucei*. More rarely it occurs away from water, on tree-bases, moist stonework and concrete, and on old asphalt. It is probably nitrophilous. Lowland. GB 406+60*, IR 8+11*.

Dioecious; capsules very rare, maturing in spring. Gemmae are common.

W. and C. Europe, north to S. Scandinavia. W. Asia, western N. America.

T. L. Blockeel

225

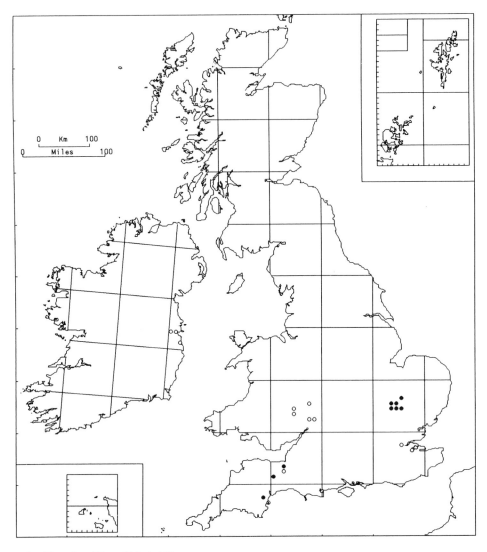

40/13. **Tortula vahliana** (Schultz) Mont.

Most occurrences are on chalk, where it grows chiefly as pure patches on deeply shaded, moist chalky-clay banks along roadsides, in road-cuttings and in disused chalk-pits. Very rare on other limestones. Associated species are few, but can include *Leiocolea turbinata*, *Seligeria calcarea*, *Tortula marginata* and *T. muralis*. Lowland. GB 9+9*, IR 3*.

Autoecious; capsules are frequent, but very few appear to mature, being adversely affected by late spring frosts.

Mediterranean region and Atlantic seaboard of Europe north to Britain and Ireland. Macaronesia, N. Africa.
The map is based on herbarium records revised by C. D. Preston and H. L. K. Whitehouse.

H. J. B. BIRKS

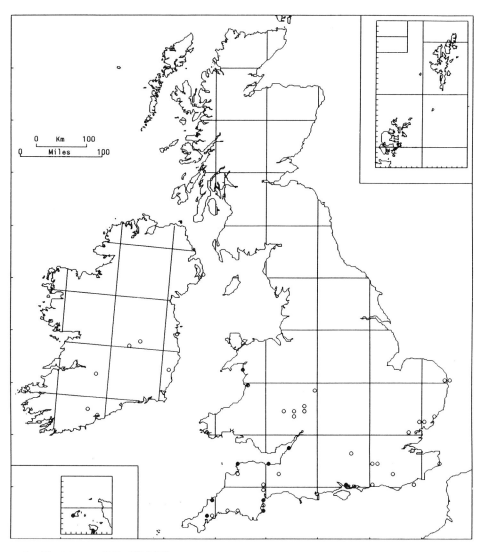

40/14. **Tortula cuneifolia** (With.) Turn.

A species of bare soil and rock crevices on wayside banks, in old quarries and on fallow land. It occurs on sandy and loamy soil, and also on thin soil over rock and on shale. Lowland. GB 9+30*, IR 10*.

Autoecious; capsules are common, maturing in spring.

Mediterranean and Atlantic Europe, north to Ireland, Germany and Hungary. Macaronesia, N. Africa, W. Asia.

T. L. Blockeel

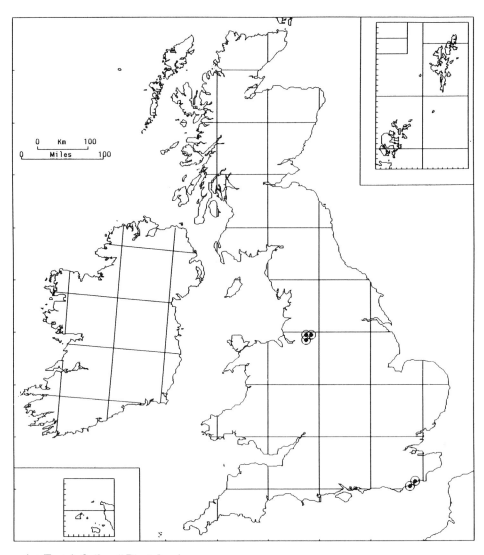

40/15. **Tortula freibergii** Dix. & Loeske

On sheltered or moderately exposed sandstone rocks and walls, sometimes in extensive patches. Near Manchester it occurs on the sandstone edging of a canal. The sites are on base-poor rock and associates include *T. marginata* and *T. muralis*. Lowland. GB 5.

Autoecious and perhaps sometimes dioecious; capsules common, maturing in spring.

Very rare endemic of W. Europe, also known from Portugal, Spain, France and Italy.

First collected in Britain in 1966, it occurs mainly in man-made or disturbed habitats, so there is slight doubt as to whether it is a British native (Crundwell & Nyholm, 1972). Further information about its occurrence near Manchester is given by Blockeel & Rumsey (1990).

T. L. Blockeel

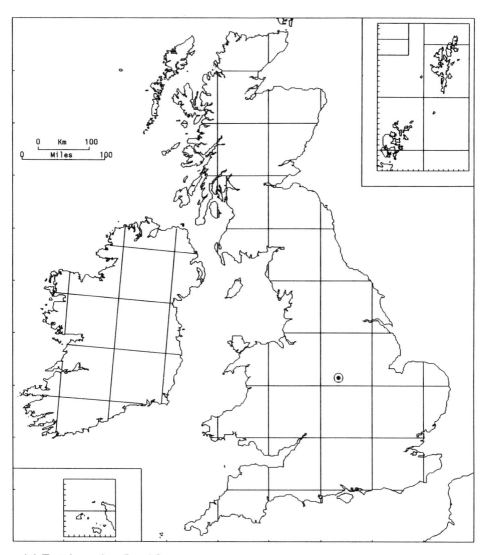

40/16. **Tortula amplexa** (Lesq.) Steere

Known only from a clay pit in Leicestershire, where it is a primary colonist of neutral or slightly acid clay-dumps. It most often occurs in nearly pure patches and appears unable to survive competition from other colonists such as *Ceratodon purpureus*. Lowland. GB 1.

Dioecious; only female plants are known in Britain. Rhizoidal tubers are common.

A native of western N. America, unknown elsewhere in Europe.

Presumably introduced. Discovered in 1973 in clay used for a pottery class. The clay was traced back to its origin in Leicestershire, where the plant was found growing in the wild (Side & Whitehouse, 1974).

T. L. BLOCKEEL

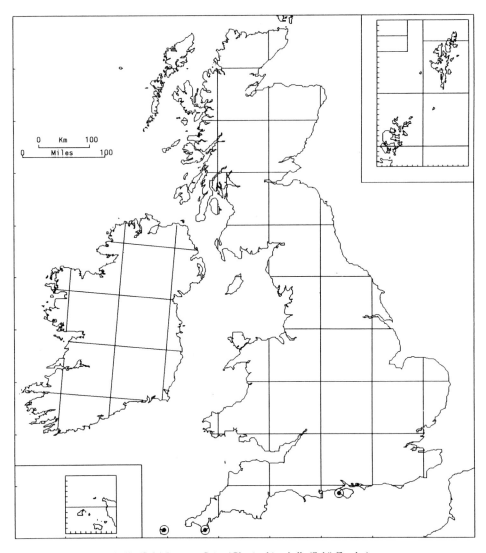

40/17. Tortula rhizophylla (Saki) Iwats. & Saito (*Chenia rhizophylla* (Saki) Zander)

A species of bare, neutral to slightly acidic soils. On the Isle of Wight it was found in a stubble-field. In Cornwall it has been found on gravelly soil on granite in the Isles of Scilly, and on a cliff-top path on the Lizard Peninsula. Lowland. GB 3.

Dioecious, female in Britain. Rhizoidal tubers are abundant.

In continental Europe known only from Spain and Italy. Outside Europe it has been found in the Canaries, Japan, Hawaii, N. America (Louisiana and Mexico) and Bolivia.

Almost certainly introduced. It was found on the Isle of Wight in 1964 and described as a new species, *Tortula vectensis* Warb. & Crundw. This name has now fallen into synonymy.

T. L. Blockeel

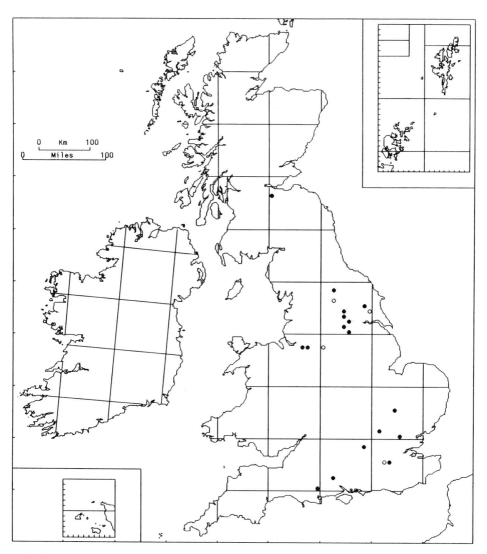

41/1. **Aloina brevirostris** (Hook. & Grev.) Kindb.

A pioneer colonist of bare and usually highly calcareous ground, most commonly on chalk and Magnesian limestone. It occurs on bare chalk on hillsides but is more common in temporary habitats in chalk pits, cuttings, quarries and lime-works, sometimes colonizing spoil-heaps and waste. In Scotland it is known from shale agglomerate on coal-bings. Lowland. GB 17+4*.

Synoecious; capsules abundant, maturing in autumn, winter and spring.

Circumpolar; a mainly boreal and Arctic species, reaching 82°N in Arctic Canada (Ellesmere Island). N., N.E. and C. Europe.

T. L. BLOCKEEL

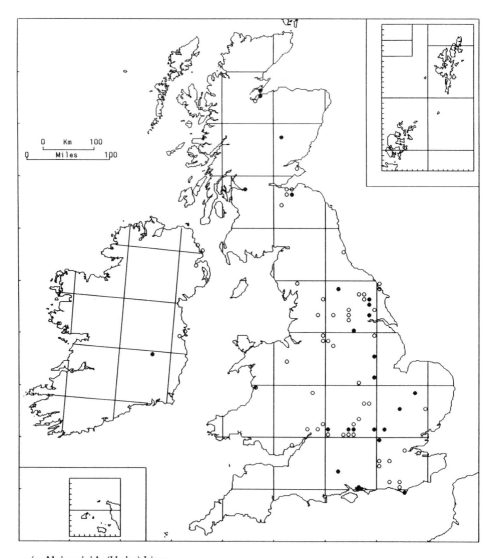

41/2. **Aloina rigida** (Hedw.) Limpr.

A plant of basic soil in a wide variety of habitats. It is found most commonly on bare ground on chalk and limestone, especially on banks and cliffs, and in chalk pits and limestone quarries and other excavated sites. More rarely it occurs in sand- and gravel-pits. It is also known, especially in Scotland, from soil and earthy rocks and crags among outcrops of base-rich rock, sometimes at considerable altitude. It was formerly common on old mud-capped walls and has become rare with the disappearance of this habitat. 0–500 m (Gleann Beag). GB 23+50*, IR 1+5*.

Dioecious; capsules abundant, maturing in autumn, winter and spring.

Widespread in the Northern Hemisphere, boreal-montane but apparently not reaching the Arctic. Most of Europe, but rare and montane in the south. In the tropics it extends south in the Andes to Bolivia and Peru.

T. L. BLOCKEEL

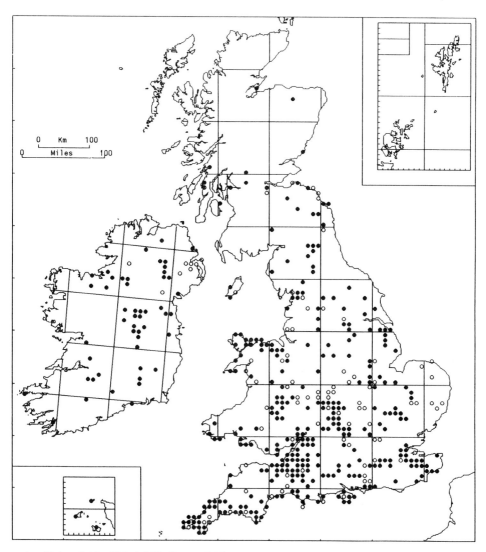

41/3a. Aloina aloides (Schultz) Kindb. var. **aloides**

A species of bare ground and soil in a variety of situations on base-rich substrata. The most characteristic habitat is the floor of chalk pits and limestone quarries, but it is also found on chalky and earthy banks, on sea-banks and sand-dunes, in gravel-pits, on soil and crumbling mortar on walls, and on base-rich sandstone. It occurred formerly with *A. rigida* on mud-capped walls. Mainly lowland, ascending to 460 m (Foel Fawr). GB 312+83*, IR 49+8*.

Dioecious; capsules abundant, maturing in late autumn to spring.

Most of Europe, north to S. Sweden. Macaronesia, N. Africa, W. Asia.

T. L. BLOCKEEL

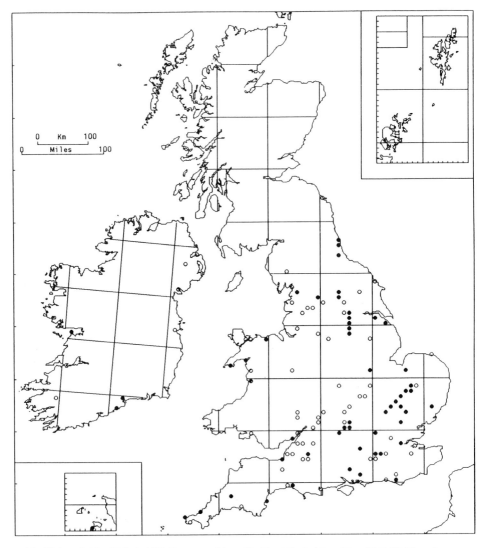

41/3b. **Aloina aloides** (Schultz) Kindb. var. **ambigua** (Br. Eur.) Craig (*A. ambigua* (B. & S.) Limpr.)

This variety is found in similar habitats to, and sometimes in association with, *A. aloides* var. *aloides*, but is apparently much less common in Britain than the type variety. Lowland. GB 50+56*, IR 3+5*.

Dioecious; capsules abundant, maturing in late autumn to spring.

Most of Europe north to Denmark. Macaronesia, N. Africa, W. and C. Asia, southwest U.S.A., Mexico, Australia.

Although correctly interpreted in the mid-19th century, it was misunderstood by H. N. Dixon and most subsequent British bryologists. Only records that have been confirmed microscopically since publication of Smith's (1978) flora are accepted here.

T. L. BLOCKEEL

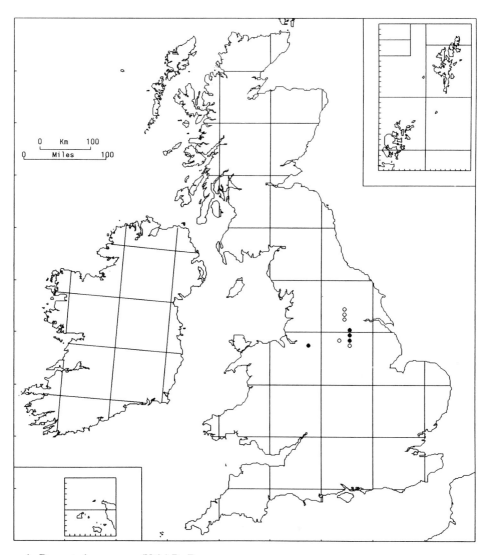

42/1. Desmatodon cernuus (Hüb.) Br. Eur.

A ruderal of highly calcareous soil, characteristically on lime-waste and quarry-spoil. It usually occurs at the foot of spoil-heaps, by paths and in quarry-hollows, in sites which provide some protection from desiccation. Most of the localities are in or near Magnesian limestone quarries, but at one former site it occurred on lime deposited from water pumped from a colliery. There is also an old record from Millstone Grit moorland, perhaps originating from imported lime. *Funaria hygrometrica* and *Leptobryum pyriforme* have often been noted as associates. Lowland. GB 4+5*.

Autoecious; capsules common, maturing in summer.

Circumpolar, reaching north to the High Arctic in Spitsbergen, Ellesmere Island and N. Greenland and south to the mountains of C. Asia and New Mexico (U.S.A.). N., C. and E. Europe.

The species is somewhat sporadic in occurrence and is no longer found in many of its former localities, but has been known continuously for over 70 years in the Don Valley in S. Yorkshire.

T. L. Blockeel

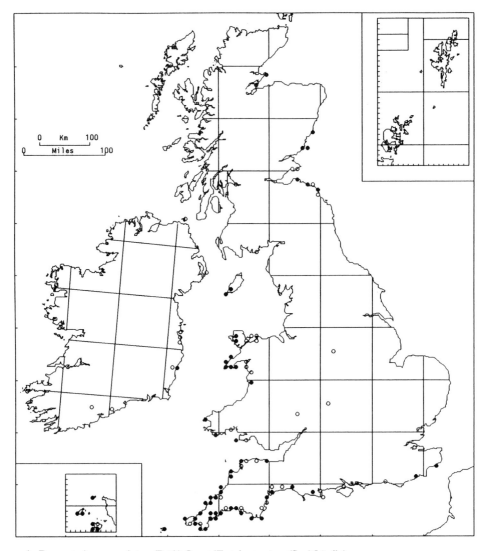

42/2. Desmatodon convolutus (Brid.) Grout (*Tortula atrovirens* (Sm.) Lindb.)

This species grows on sandy soil or on thin soil in crevices of rocks and walls, mostly on maritime banks and cliffs. It usually occurs in dry sunny situations, and is clearly salt-tolerant since it may be found close to the high-water mark. The substratum is commonly, but not exclusively, base-rich. Lowland. GB 54+33*, IR 1+9*.

Autoecious; capsules abundant, maturing in winter and spring.

Widespread in warm-temperate and subtropical regions of both Northern and Southern Hemispheres, especially in semi-deserts. S., W. and C. Europe, absent from Scandinavia and from much of E. Europe.

Flowers (1973) has observed that it tolerates mildly saline conditions in the deserts of N. America.

T. L. BLOCKEEL

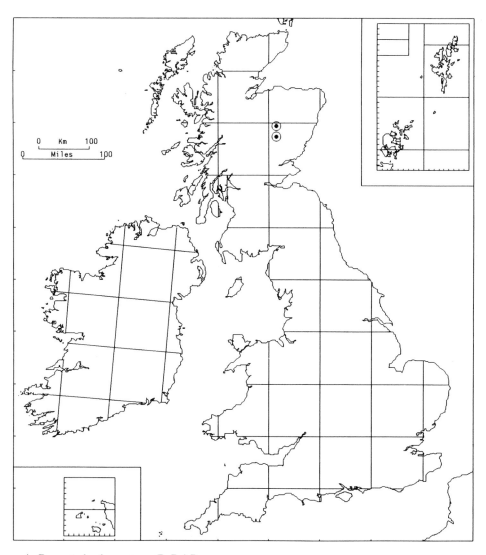

42/3. **Desmatodon leucostoma** (R. Br.) Berggr.

A montane species of soil among rocks and on ledges facing north or east. It is confined to calcareous outcrops and is commonly associated with *Stegonia latifolia*. 550 m (near Braemar). GB 2.

Autoecious; capsules common, maturing in summer.

A circumpolar arctic-alpine. N. Europe and mountains of C. Europe.

T. L. BLOCKEEL

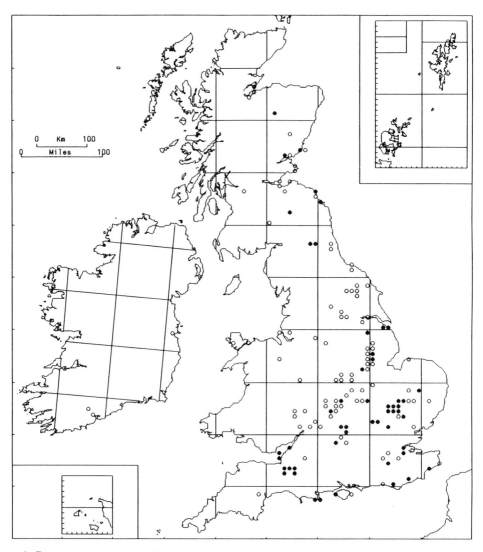

43/1. Pterygoneurum ovatum (Hedw.) Dix.

A plant of open habitats on chalk, limestone and other basic substrata, growing on banks, among rocks, in pits and quarries and on sea-cliffs. It also occurs on freshly disturbed soil on tracks and excavated ground, and has been recorded rarely from stubble-fields and pondside mud. It was formerly one of the most characteristic colonists of newly mud-capped calcareous walls, and has become increasingly rare with the disappearance of this habitat. Lowland. GB 52+88*, IR 3*.

Autoecious; capsules are common, maturing in late autumn and winter.

Much of Europe, north to S. Scandinavia; common in the Mediterranean region. Macaronesia, N. Africa, W. and C. Asia, N. America, Australia.

Chiefly a plant of semi-arid regions, it is one of the commonest mosses in Utah (U.S.A.), both on disturbed rocky ground and in open saline deserts (Flowers, 1973).

T. L. BLOCKEEL

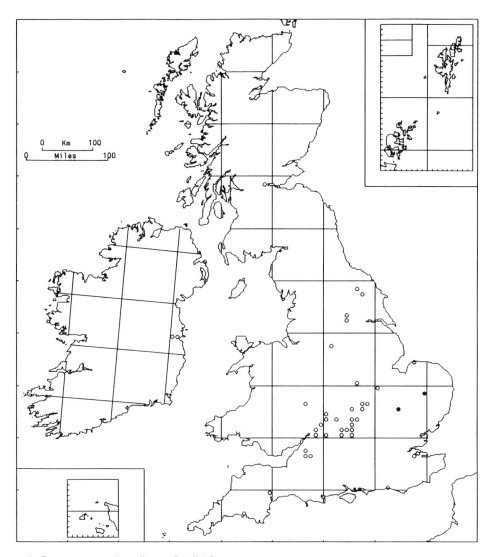

43/2. Pterygoneurum lamellatum (Lindb.) Jur.

This species occurs in similar habitats to *P. ovatum* but has apparently always been rarer and more restricted in distribution. Many of the old records are from calcareous mud-capped walls, a habitat that has now disappeared. The few recent records have been from chalk pits and calcareous roadside soil. Lowland. GB 2+36*, IR 3*.

Autoecious; capsules abundant, maturing in spring.

W., C. and E. Europe. C. Asia, Arctic Alaska, Arctic Canada north to Ellesmere Island, U.S.A. (Utah, Arizona).

P. lamellatum may conceivably now be extinct in Britain; it was last seen in 1970 at Cherry Hinton. It has a curiously disjunct world distribution including both the Arctic and hot semi-arid regions. In Alaska, its habitat is moist calcareous silt in frost-boils (Steere, 1978).

T. L. Blockeel

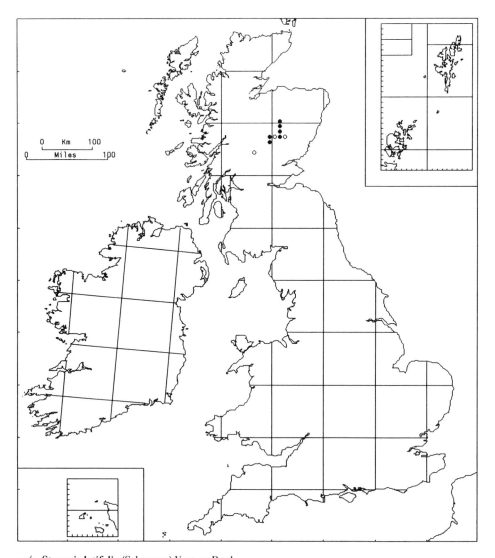

44/1. Stegonia latifolia (Schwaegr.) Vent. ex Broth.

A species of strongly base-rich soil on mountains, where it grows in open, well-drained situations among rocks, on ledges and in crevices. The underlying rock types include limestone and mica-schist. 550 m (near Braemar). GB 6+3*.

Autoecious; capsules are common, maturing in summer. Gemmae are unknown.

Circumpolar arctic-alpine. N. Europe; mountains of C. and W. Europe.

T. L. BLOCKEEL

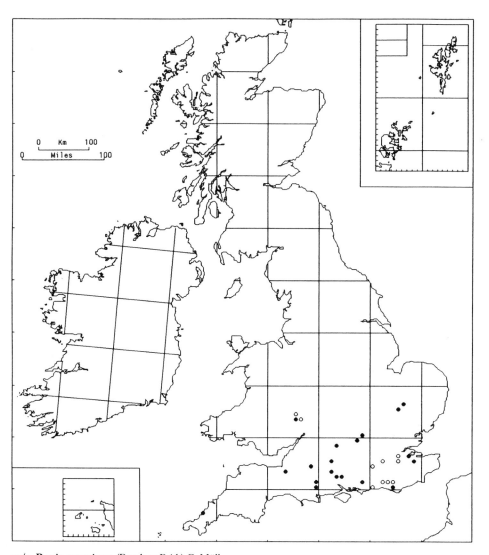

45/1. **Pottia caespitosa** (Bruch ex Brid.) C. Müll.

This ephemeral is primarily a plant of chalk grassland on downland and ancient earthworks, where it grows on bare soil (especially on steep banks), on trampled ground along trackways and on the edge of sea-cliffs; it is also found in disused chalk pits. *Phascum curvicolle* is recorded as an associate. At the western edge of its British range it occurs on calcareous sand on the coast of N. Cornwall and on bare Jurassic limestone banks in Herefordshire. Lowland. GB 19+10*.

Autoecious; capsules frequent, December to April.

Endemic to Europe, from Spain and the Balearics to Czechoslovakia and Romania; it reaches its northern limit in England.

Little detailed information is available about the ecology of this distinctive species, which would merit further study. It is apparently sporadic in its appearance, and at many of its sites it has been recorded only once or at long intervals. It was, for example, plentiful when discovered in Herefordshire on Common Hill, Fownhope, by the Rev. A. Ley in 1875, and again present in 1877; Ley was unable to refind it in subsequent years, but it was rediscovered there in 1932 and 1968. It is presumably able to persist as dormant spores.

D. F. Chamberlain & C. D. Preston

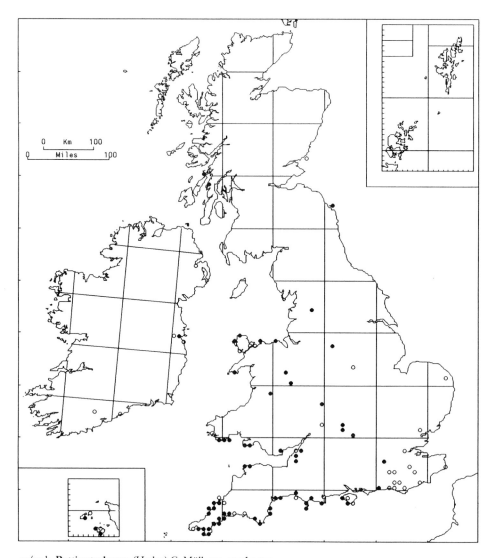

45/2a,b. **Pottia starkeana** (Hedw.) C. Müll. ssp. **starkeana**

An annual of disturbed or shallow, open soil. It grows on cliff-tops and rocky or earthy cliff-slopes, walls, banks, roadsides and pathsides, in quarries, chalk and limestone grassland, stubble-fields and (rarely) on woodland rides. At coastal sites it is most frequent on limestone but is also recorded from calcareous clay and sandstone as well as over non-calcareous rocks such as serpentine and slate. Inland populations are confined to calcareous ground or to lead-contaminated soil around disused mines. Associated species include *P. davalliana* and *P. lanceolata*. Lowland. GB 59+30*, IR 1+5*.

Autoecious; only identifiable with capsules, which mature from November to May.

Widespread in S., W. and C. Europe from the Mediterranean north to S. Sweden. Macaronesia, N. Africa, S.W. Asia, western N. America, Australia and New Zealand.

The concept of *P. starkeana* ssp. *starkeana* adopted here is that of D. F. Chamberlain (*in* Smith, 1978), including var. *starkeana* and var. *brachyodus* C. Müll. No ecological distinction between these varieties is apparent in the British Isles, and they are mapped together.

D. F. CHAMBERLAIN & C. D. PRESTON

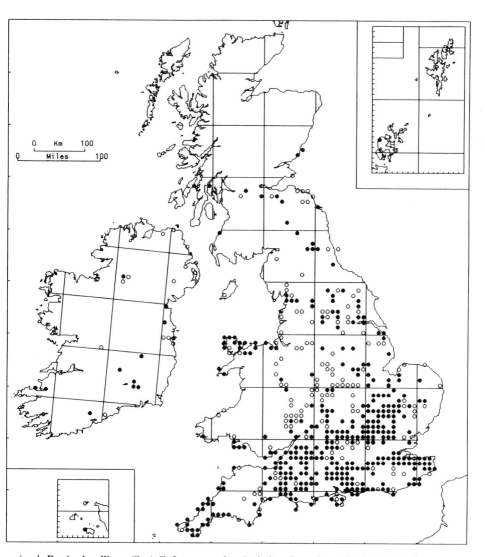

45/2c,d. **Pottia davalliana** (Sm.) C. Jens. *sensu lato* (including *P. starkeana* (Hedw.) C. Müll. ssp. *conica* (Schleich. ex Schwaegr.) Chamberlain and ssp. *minutula* (Schleich. ex Schwaegr.) Chamberlain)

An early colonist of disturbed calcareous soil, most frequently over chalk, limestone and calcareous clay. Characteristic habitats include stubble-fields, pathsides, quarries, thin soil around limestone outcrops, patches of bare soil in grassland, woodland rides and (formerly) earth-capped walls. Like some other Pottiaceae it is less strictly calcicolous in coastal sites, where it grows on earthy banks, thin soil over rocky cliffs, and sand-dunes. *Barbula unguiculata, Bryum rubens, Dicranella varia* and *Phascum cuspidatum* are frequent associates. Lowland. GB 354+146*, IR 12+16*.

Autoecious; capsules abundant, ripe winter and spring, or, in damp sites, sometimes summer. Rhizoid fragments can regenerate after more than one year's dry storage (Arts, 1987b).

Widespread in Europe north to S. Scandinavia. Macaronesia, Africa, S.W. Asia, N. America, Australia.

Pottia davalliana sensu lato was divided by Chamberlain (1969) into two taxa, treated as subspecies of *P. starkeana*. Most of the available records refer only to *P. davalliana sensu lato*, which is therefore mapped here. Ssp. *minutula* is less frequent than ssp. *conica*; it is apparently less maritime in Britain and more continental in Europe.

D. F. Chamberlain & C. D. Preston

243

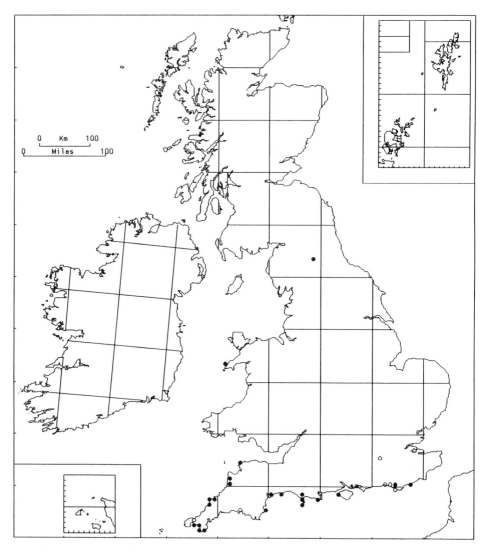

45/3. **Pottia commutata** Limpr.

A winter annual which usually grows on basic soil derived from chalk, limestone or calcareous sand. Its habitats include bare patches in short coastal turf, earthy cliffs and banks, crevices in Cornish 'hedges', pathsides, waste ground and fallow fields. At its outlying locality at Langdon Beck, Upper Teesdale, it was discovered on wet clay by a road. Usually lowland, but ascending to at least 350m in Teesdale. GB 22+6*.

Autoecious; capsules abundant, ripe January to April. Rhizoid fragments can regenerate after more than one year's dry storage (Arts, 1987b).

Mediterranean Europe, extending north along the Atlantic coast to the British Isles; Czechoslovakia, Hungary. Canaries, N. Africa, S.W. Asia.

P. commutata is a member of the critical *P. starkeana* aggregate. With the exception of the surprising record from Teesdale (the identity of which was confirmed by M. O. Hill), its British distribution reflects its Mediterranean affinities.

D. F. CHAMBERLAIN & C. D. PRESTON

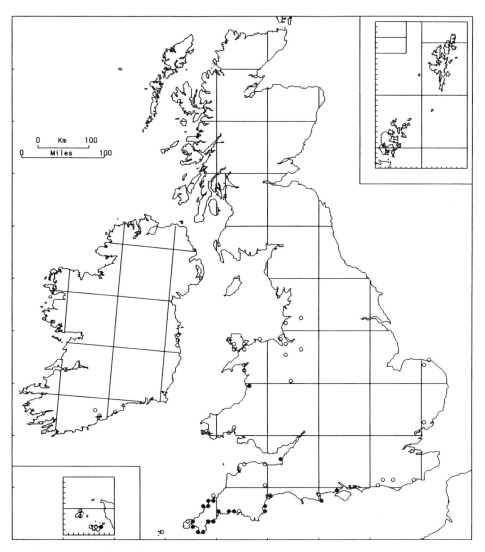

45/4. **Pottia wilsonii** (Hook.) Br. Eur.

An annual which grows on soil on rocky cliffs, Cornish 'hedges', pathsides and wall-tops, on sandy banks and on earthy hedgebanks. It shows no relationship to underlying geology, occurring over both acidic and basic rocks. Lowland. GB 20+34*, IR 6*.

Paroecious; capsules abundant, ripe January to May.

Mediterranean Europe from Spain to Greece, extending north in W. Europe to the British Isles; Romania. N. Africa, S.W. Asia, N. America (British Columbia).

P. wilsonii, long thought to be an exclusively Palaearctic species, has recently been discovered in semi-arid steppe in British Columbia (McIntosh, 1989). It has declined markedly in the British Isles since it was discovered by W. Wilson in Cheshire in 1828: it is no longer known in any inland sites and has apparently disappeared even from coastal localities towards the edge of its range.

D. F. Chamberlain & C. D. Preston

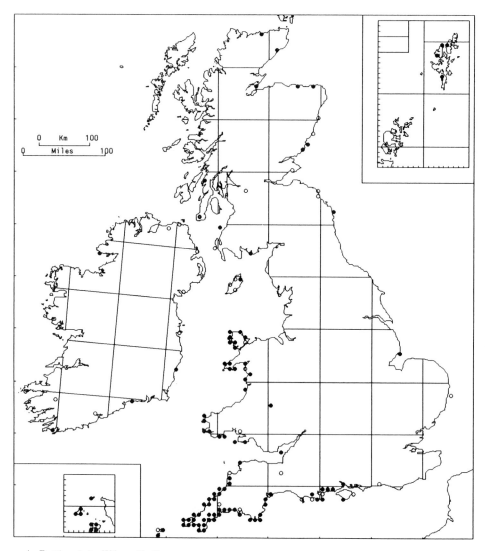

45/5. **Pottia crinita** Wils. ex Br. Eur.

A frequent annual in open habitats on exposed south-western coasts, growing on thin soil over rocky banks and cliffs, hedgebanks and streamsides, earthy crevices in rocks and Cornish 'hedges', sandy banks, trampled soil along paths and tracks and on ant-hills in coastal pasture. It occurs over a wide range of rock types including basalt, granite, limestone, sandstone, schist, serpentine and shale. *Desmatodon convolutus* and *Trichostomum brachydontium* are characteristic associates. It was, surprisingly, discovered inland on an old railway near Builth Wells in 1985. Lowland. GB 94+24*, IR 4+8*.

Autoecious; capsules frequent, maturing in winter and spring. Rhizoidal tubers are apparently absent (Arts, 1987b).

A Mediterranean-Atlantic species, reaching its northern limit in the Faeroes. Macaronesia, N. Africa, S.W. Asia.

D. F. CHAMBERLAIN & C. D. PRESTON

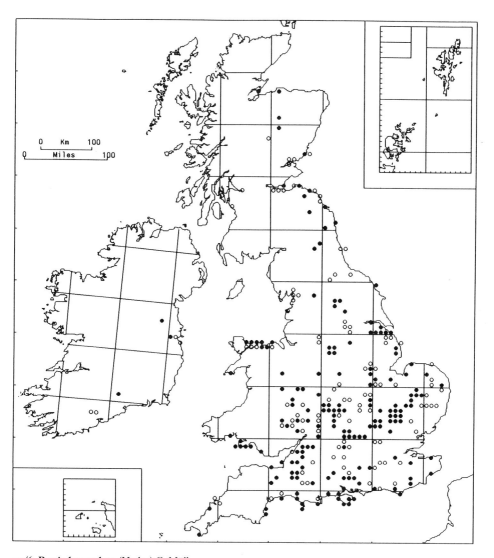

45/6. Pottia lanceolata (Hedw.) C. Müll.

A colonist of open or disturbed, usually calcareous, soil, most frequently found over chalk and limestone, but also recorded from calcareous clay and sand, dolerite, sandstone and serpentine. Typical habitats include thin rocky turf over limestone, open chalk downland (where it sometimes grows on ant-hills and molehills), quarries, soil on sea-cliffs and banks by the sea. It was formerly a component of the rich bryophyte communities that occurred on mud-capped walls, with associates such as *Aloina* spp., *Barbula* spp., *Phascum curvicolle* and *Pterygoneurum ovatum*. It only rarely occurs in stubble-fields. 0–550 m (Craig Leek). GB 174+105*, IR 3+4*.

Autoecious; capsules abundant, ripe in the early months of the year, rarely at other seasons. In nearby parts of Europe, rhizoidal tubers are frequent, and can remain viable even after five years' dry storage (Arts, 1987b). During & ter Horst (1983) demonstrated the presence of a bank of viable propagules in the soil of a Dutch chalk grassland.

Europe from the Mediterranean north to S. Scandinavia. Macaronesia, N. Africa, S.W., C. and E. Asia, N. America (Idaho, Texas).

D. F. Chamberlain & C. D. Preston

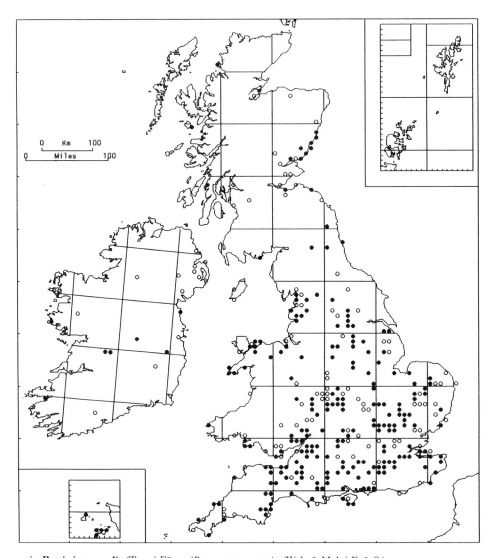

45/7. **Pottia intermedia** (Turn.) Fürnr. (*P. truncata* var. *major* (Web. & Mohr) B. & S.)

An ephemeral which is found both on calcareous substrates (sometimes growing with *P. davalliana*) and on acidic soils (with *P. truncata*), and also occurs in coastal habitats (with *P. heimii*). It grows on disturbed soil and in stubble-fields over chalk, limestone, calcareous clay, non-calcareous sandstone, sand and gravel; on roadside banks, earth-capped walls, garden paths, flower-beds, molehills and ant-hills; in quarries and coastal turf, on sandy soil on cliffs, mud by coastal dykes and on waste-ground, tracks and banks by the sea. Lowland. GB 218+101*, IR 7+13*.

Autoecious; capsules abundant, ripe November to March, rarely at other seasons. In continental Europe, rhizoidal tubers are frequent, remaining viable up to five years in dry storage (Risse, 1985a; Arts, 1987b).

Widespread in Europe from the Mediterranean north to Iceland and S. Scandinavia. N. Africa, S.W. Asia, China, Japan, N. America.

Possibly under-recorded; a rather critical taxon, treated by some recent authors as a variety of *P. truncata*. Chamberlain (*in* Smith, 1978) suggests that *P. intermedia* may be a hybrid derivative of *P. lanceolata* and *P. truncata*.

D. F. CHAMBERLAIN & C. D. PRESTON

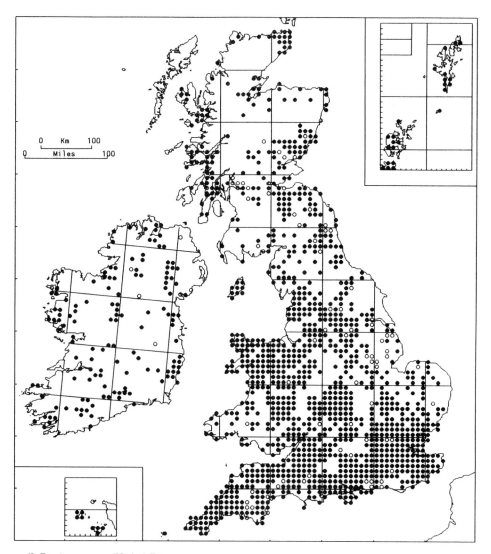

45/8. **Pottia truncata** (Hedw.) Fürnr.

An ephemeral of disturbed, fertile, non-calcareous mineral soil in both natural and man-made habitats. It is frequent in cultivated fields and gardens, in thin grass-leys, on disturbed ground in pastures (including molehills and ant-hills), on woodland rides, tracksides and roadsides, quarries, sand- and gravel-pits, ditch-, stream- and river-banks, and rocky cliff-slopes. Characteristic associates include *Bryum rubens*, *Ceratodon purpureus*, *Dicranella schreberana*, *D. staphylina*, *Ditrichum cylindricum* and *Phascum cuspidatum*. It avoids infertile peaty soil as well as highly calcareous substrates. Side (1977) found that *P. truncata* was replaced by *P. davalliana* in arable fields where the pH exceeded 8. Lowland. GB 1164+76*, IR 136+7*.

Autoecious; capsules frequent, maturing throughout the year but mainly in autumn and winter. Rhizoidal tubers have been recorded in Belgium, remaining viable even after four years' dry storage (Arts, 1987b).

Europe from the Mediterranean north to Iceland and S. Scandinavia. Macaronesia, N. Africa, Asia, N. & S. America, Australia, New Zealand.

D. F. Chamberlain & C. D. Preston

249

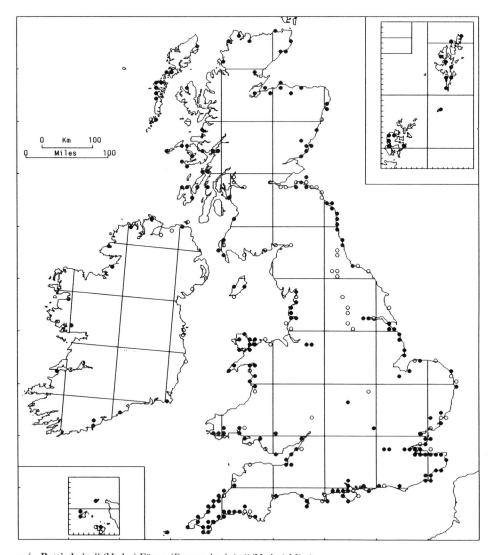

45/9. **Pottia heimii** (Hedw.) Fürnr. (*Desmatodon heimii* (Hedw.) Mitt.)

This halophyte is the most characteristic bryophyte of salt-marshes, where it grows in grazed turf, on footpaths and on disturbed ground on the upper marsh, descending to levels inundated by at least 100 tides a year (Adam, 1976). It is frequent in other coastal habitats, including sandy or muddy ground between boulders on beaches, soil at the foot of cliffs and sea-walls, the banks of dykes and tidal rivers, rock ledges and crevices, and short turf on cliff-slopes and cliff-tops. Although usually found within the spray zone, it extends up the R. Thames as far as Richmond (Surrey). Further inland it is recorded from the edge of pools by a salt spring in Warwickshire and on lime beds in Cheshire, as well as in habitats such as walls, paths and a canal-bank where its occurrence may have been only casual. Lowland. GB 210+67*, IR 11+9*.

Autoecious, rarely synoecious; capsules abundant, usually maturing from February to May but recorded throughout the year.

Atlantic coast of Europe north to Iceland and N. Norway; also at scattered sites inland in C. Europe. N. Africa, Asia, N. & S. America, Tasmania, New Zealand, Antarctica.

D. F. CHAMBERLAIN & C. D. PRESTON

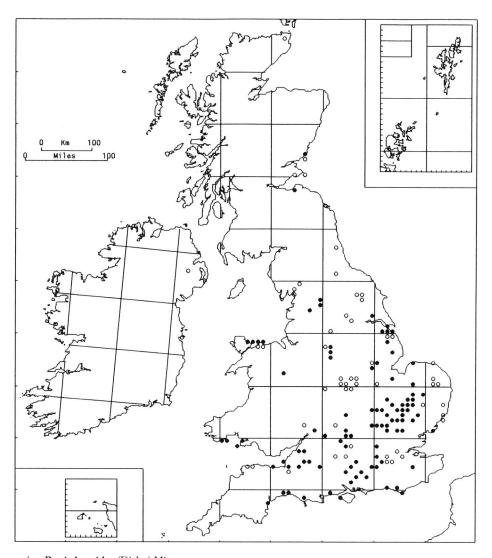

45/10. **Pottia bryoides** (Dicks.) Mitt.

A winter annual of open communities on freely draining basic soil, usually occurring in small quantity but sometimes locally abundant. It is often found in calcareous habitats, including thin soil and turf over limestone rocks, compacted ground in chalk and limestone quarries, chalky banks, tracks and paths, bare patches and ant-hills in chalk grassland, disturbed chalky soil by roads and (formerly) mud-capped walls. It also grows on gravel paths (many such records are from churchyards), on soil over concrete, and on sandy soil and sandstone cliffs by the sea. Associated species include *P. intermedia*, *P. lanceolata*, *P. recta* and *P. truncata*. Lowland. GB 98 + 53*, IR 2*.

Autoecious; capsules abundant, ripe winter and spring. Rhizoidal tubers have been found in a Belgian specimen, and were viable after two years' dry storage (Arts, 1987b).

W., C. & E. Europe, north to S. Sweden; rare in the Mediterranean region. Macaronesia, S.W. Asia, western N. America.

A design featuring *P. bryoides* has been adopted by the British Bryological Society as its logo.

D. F. Chamberlain & C. D. Preston

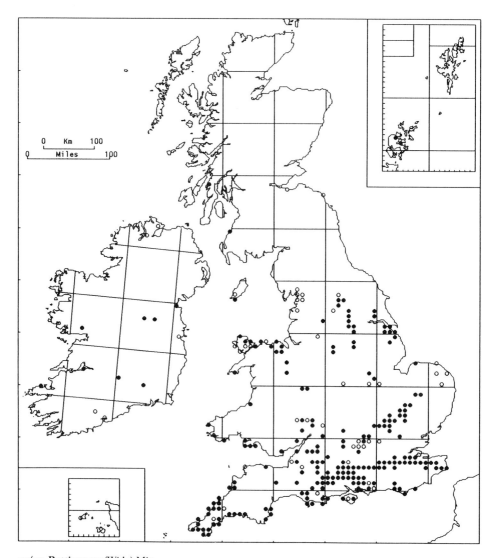

45/11. **Pottia recta** (With.) Mitt.

Commoner than *P. bryoides* in thin turf and on bare soil in areas of chalk and limestone, it occupies a similar range of habitats, including grassland, quarries, banks, tracks and roadsides. Unlike *P. bryoides* it is also frequent in calcareous stubble-fields, with associates such as *Barbula unguiculata, Bryum rubens, Dicranella varia, Phascum cuspidatum* and *Pottia davalliana*. It occurs more rarely on soils over sandstone and calcareous clay. In acidic areas it has been found by coastal lime-kilns. In Cornwall, it grows as a winter annual in cultivated fields (including bulb-fields) and on waste ground, paths, banks and crevices in 'hedges' near the coast. It is possibly under-recorded in such habitats elsewhere in S.W. England. Lowland. GB 205+43*, IR 8+5*.

Autoecious; capsules abundant, ripe September to April. Fragments of rhizoid can regenerate after more than one year's dry storage (Arts, 1987b).

C. & S. Europe from the Mediterranean north to the British Isles and Denmark. N. Africa, Caucasus, S.W. Asia.

Not dissimilar both morphologically and ecologically to *Phascum curvicolle*, with which it sometimes grows.

D. F. CHAMBERLAIN & C. D. PRESTON

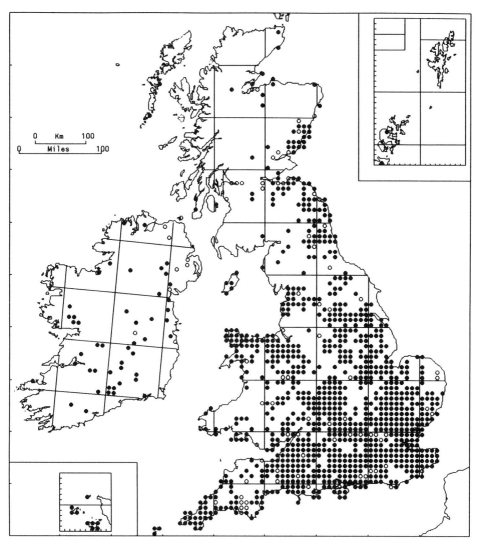

46/1. Phascum cuspidatum Hedw.

One of the commonest ruderal bryophytes in lowland England, likely to be found wherever soil has been disturbed. It is frequent in habitats such as cultivated ground (including stubble-fields and garden soil), waste ground, thin grassland, ant-hills in pasture, woodland rides, stream-banks, ditchsides, dumped soil, dredgings by rivers and banks by the sea. It favours mildly acidic to basic, fertile, mineral soil, avoiding highly acidic peaty substrates. *Barbula unguiculata*, *Bryum rubens*, *Dicranella staphylina*, *D. varia*, *Pottia davalliana* and *P. truncata* are frequent associates. Lowland. GB 938+75*, IR 47+9*.

Autoecious; capsules abundant, usually maturing from autumn to spring but also in summer in moist sites.

Throughout Europe to 64°N in Scandinavia. Macaronesia, N. Africa, Asia, N. America.

The map indicates all records of the species but in effect shows the distribution of var. *cuspidatum*, much the commonest variety. Var. *schreberanum* (Dicks.) Brid. occurs sporadically with var. *cuspidatum*; it is uncommon but probably under-recorded. Var. *piliferum* (Hedw.) Hook. & Tayl. is mapped separately.

D. F. Chamberlain & C. D. Preston

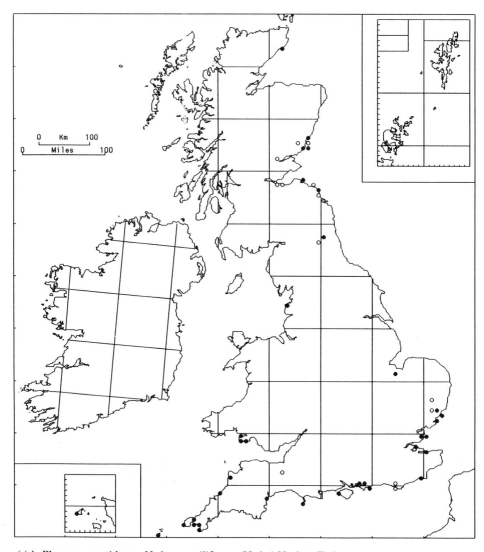

46/1b. **Phascum cuspidatum** Hedw. var. **piliferum** (Hedw.) Hook. & Tayl.

Like *P. cuspidatum* var. *cuspidatum*, var. *piliferum* is an ephemeral of open or disturbed, usually basic, soil over a range of rock types. It is most frequent on the coast, being found on tracks and banks, bare soil in trampled turf, waste sandy ground, dune-slacks, moist soil at the foot of sea-walls, wall-tops, crevices in Cornish 'hedges' and thin soil on rocky cliff-slopes. Inland, it is recorded from dry banks and outcrops of igneous rock. Lowland. GB 34+14*.

Autoecious; capsules abundant, usually maturing in winter and spring.

Widespread in Europe north to S. Scandinavia, becoming increasingly frequent towards the south. N. Africa, W. Asia, N. America.

Almost certainly under-recorded, being overlooked by bryologists who do not consider that *Phascum cuspidatum* merits a second glance.

D. F. CHAMBERLAIN & C. D. PRESTON

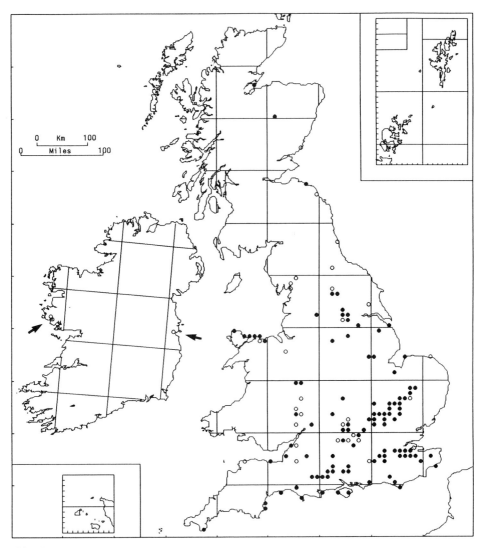

46/2. Phascum curvicolle Hedw.

A winter annual of calcareous ground, growing in sites kept open by drought or trampling, or colonizing soil after disturbance; it tends, however, to avoid regularly disturbed habitats. It occurs on shallow soil around limestone rock-outcrops and in thin turf on steep chalk slopes, on open ground in chalk pits and limestone quarries, by paths and tracks, and on disturbed soil in calcareous grassland (including areas from which turf has been cut). Although it sometimes grows in stubble-fields it is much less frequent in this habitat than *P. floerkeanum*. It was formerly a characteristic species of earth-capped limestone walls, growing in abundance with *Pottia lanceolata* in the later stages of the bryophyte succession on wall-tops (Jones, 1953). At its northern limit, in Scotland, it is confined to calcareous sea-cliffs. Lowland. GB 104+28*, IR 2*.

Autoecious; capsules abundant, maturing from October to April.

Widespread in S. and C. Europe, north to S. Scandinavia. N. Africa, S.W. Asia.

D. F. CHAMBERLAIN & C. D. PRESTON

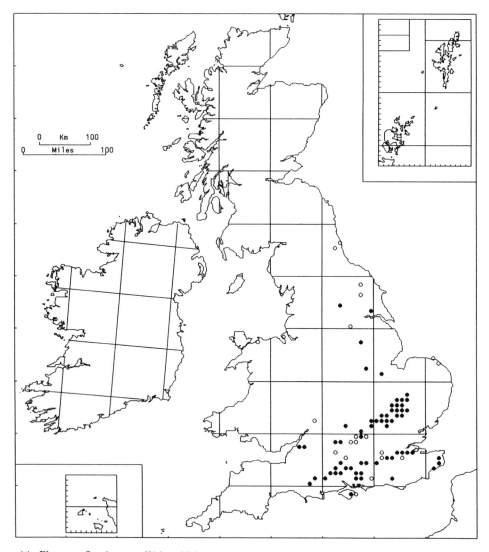

46/3. **Phascum floerkeanum** Web. & Mohr

A minute ephemeral of disturbed calcareous ground, mainly on soils derived from chalk or chalky clay but occasionally found over harder limestones. It is most frequent in stubble-fields, growing with species such as *Barbula unguiculata*, *Bryum klinggraeffii*, *B. rubens*, *Dicranella varia*, *Phascum cuspidatum* and *Pottia davalliana*. Other habitats include chalk pits, bare soil in chalk grassland, cultivated garden soil and woodland rides. It varies in quantity from year to year, being more abundant after a wet summer (Swinscow, 1959). Lowland. GB 62+21*.

Autoecious; capsules abundant, maturing from late August to October.

W., S. and C. Europe north to S. Sweden and S.W. Finland (Åland Islands); rare in the Mediterranean region. N. Africa, N. America.

Probably under-recorded, because of its small size and restricted season. Plants have usually disappeared by December, and are only rarely recorded in January and February.

D. F. CHAMBERLAIN & C. D. PRESTON

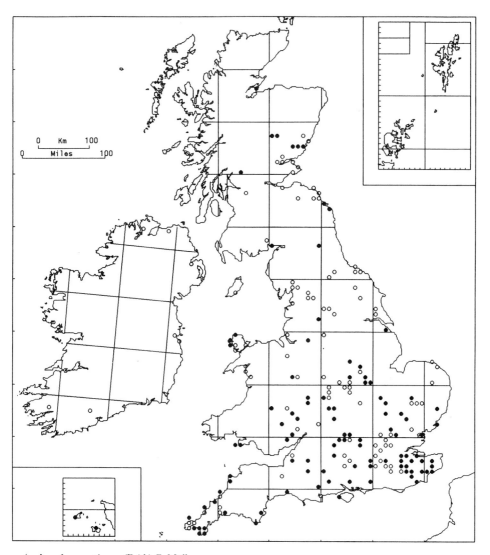

47/1. **Acaulon muticum** (Brid.) C. Müll.

An ephemeral of bare ground on neutral and base-deficient substrata. Habitats include stubble-fields, tracksides, woodland rides, gravel-pits, ant-hills and thin turf on banks. It is most characteristic of sandy and other loose-textured, well-drained soils, but sometimes also occurs on clay. Lowland. GB 102+94*, IR 7*.

Autoecious or dioecious; capsules abundant, maturing in winter and early spring, rarely in summer. Gemmae are unknown.

Widespread in Europe but rare in the north. Macaronesia, N. Africa, W. and C. Asia, N. America, Australia.

Most records are referable to var. *muticum*. Var. *mediterraneum* (Limpr.) Sérgio (sometimes treated as a distinct species, *Acaulon mediterraneum* Limpr.), with spiny spores, is known from two localities in S.W. England, but has not been seen since 1935. *A. minus* (Hook. & Tayl.) Jaeg. cannot be maintained as a taxon distinct from *A. muticum* (Hill, 1982).

T. L. Blockeel

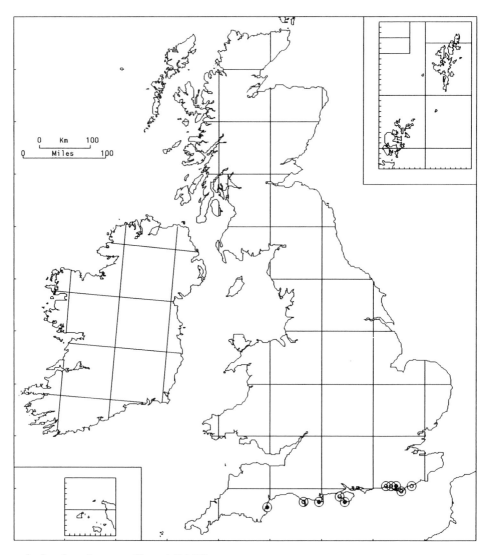

47/3. Acaulon triquetrum (Spruce) C. Müll.

An ephemeral of S.-facing slopes on coastal cliffs and banks, where it grows on soil and bare ground, usually calcareous. Rock types include limestones in Devon and Dorset, greensand on the Isle of Wight and chalk in Sussex. Lowland. GB 4+6*.

Autoecious; capsules abundant, maturing in winter and early spring. Gemmae are unknown.

Mediterranean region north to the Ukraine, C. Europe and England. Macaronesia, N. Africa, N. America, Australia.

It has always been rare in Britain, which is at the limit of its range. It has apparently declined in recent years as a result of pressures on its coastal habitat.

T. L. Blockeel

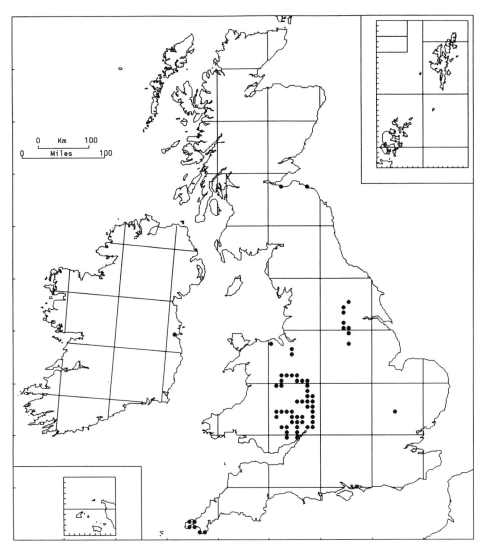

48/1. **Hennediella stanfordensis** (Steere) Blockeel (*Hyophila stanfordensis* (Steere) Smith & Whitehouse, *Tortula stanfordensis* Steere)

It occurs on shaded, trampled ground by footpaths and under trees on the banks of rivers. There are a few records from arable fields and one from a vertical roadside bank scraped by car-wheels near the R. Wye, at Ruckhall. In Yorkshire, it has been recorded from soil on rock ledges on Magnesian limestone cliffs. Lowland. GB 63, IR 1.

Apparently dioecious; most plants are female but young sporophytes have been seen at Hoarwithy (Whitehouse & Newton, 1988). Rhizoidal tubers are frequent.

There are isolated records from France and Greece. Outside Europe it is known from U.S.A. (California) and Australia.

First found in Britain in 1958 and in Ireland in 1978, this plant is probably an introduction, possibly from Australia. Blockeel (1990) has recently demonstrated that it belongs to the austral genus *Hennediella*. On footpaths it is clearly dispersed in soil on the footwear of pedestrians (Whitehouse, 1971); by rivers it may be water-dispersed. The rock-ledge habitat suggests other dispersal agents.

H. L. K. Whitehouse

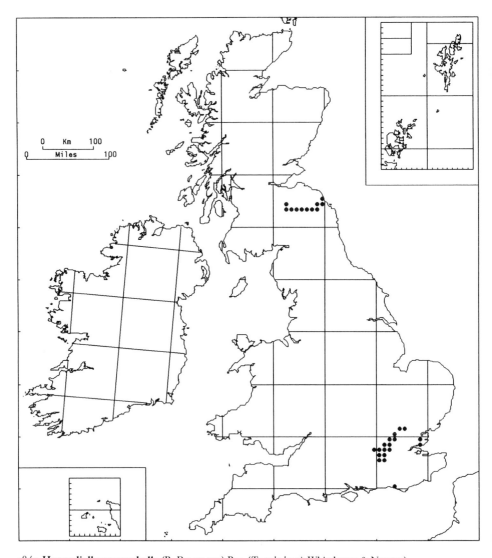

48/2. **Hennediella macrophylla** (R. Brown ter.) Par. (*Tortula brevis* Whitehouse & Newton)

This plant occurs on shaded trampled ground (often alongside exposed tree roots), in cultivated flower-beds, and under trees on river-banks. In Essex it grows along rabbit- and badger-tracks, and it has been found on fresh spoil outside a badger's sett, halfway up a cliff in a chalk quarry (Adams, 1990). Ascends to 170 m (R. Tweed near Edston). GB 29.

Monoecious; sporophytes occur regularly in spring but often fail to mature. Rhizoidal tubers are frequent and gemmae occur occasionally on the leaves; protonemal gemmae have been seen in culture.

In Europe known only from Britain. Tierra del Fuego, New Zealand.

First collected in Britain in 1965, it is probably an introduction from New Zealand. It was described as *Tortula brevis* by Whitehouse & Newton (1988), but this has now been recognized as a synonym of *H. macrophylla* (Blockeel, 1990). It is dispersed not only by badgers but also apparently by water, human footwear and other animals' feet.

H. L. K. WHITEHOUSE

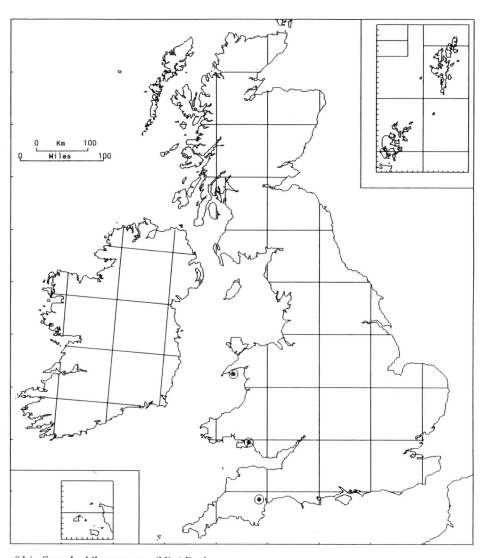

48A/1. **Scopelophila cataractae** (Mitt.) Broth.

This species, which is tolerant of heavy-metal contamination, is known from zinc-spoil heaps and waste in S. Wales, a disused lead-mine in N. Wales and an old metal-mine in Devon. Lowland. GB 3.

Dioecious; only male plants have been found in Britain. Rhizoidal tubers have been noted in specimens from N. Wales and continental Europe. Protonemal gemmae are produced copiously in culture, and have been observed in naturally growing European material (Arts, 1988a).

Spain, France, Holland, Belgium, Germany and Italy. Widely distributed in tropical Asia and America, north to Japan, Korea, N. Carolina and Arizona (U.S.A.).

First found in S. Wales in 1967 but not identified till 1982 (Corley & Perry, 1985). This locality and some of those in continental Europe are near smelting works; at all of these it was almost certainly introduced with ore. However, its occurrence on mine-waste in Devon and N. Wales means that it could possibly be a British native, though this is unlikely (Crundwell, 1986).

T. L. BLOCKEEL

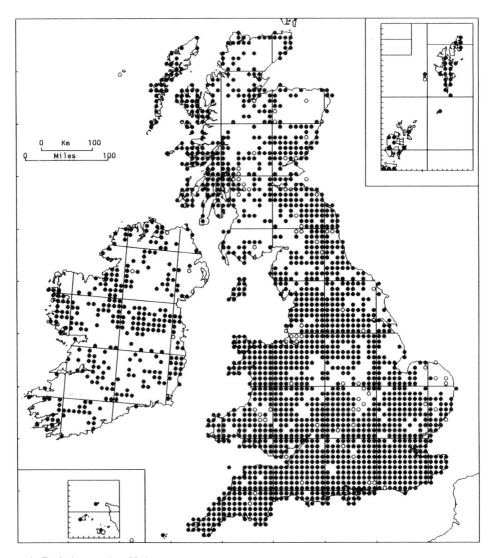

49/1. **Barbula convoluta** Hedw.

A very common species of disturbed neutral to base-rich soil in open situations. It occurs on paths and waste ground, in gardens, pastures and arable fields, in old pits and quarries, in chalk and limestone grassland, on cliff-tops and in soil-filled crevices on walls and rock ledges. It may also occur directly on crumbling wall-mortar and soft rock. It is primarily a lowland species but ascends to moderate altitudes in limestone districts and in man-made habitats. 0–490 m (Ben Lawers). GB 1763+82*, IR 290+6*.

Dioecious; capsules occasional to frequent, maturing in spring and summer. Rhizoidal tubers are known to occur frequently but are apparently not constantly present.

Almost throughout Europe. Widespread in the Northern Hemisphere; New Zealand.

Var. *commutata* (Jur.) Husn. is a luxuriant form of the species which is found in sheltered, nutrient-rich sites, for example on soil at the foot of walls. It has not been recorded consistently and is not mapped separately.

T. L. Blockeel

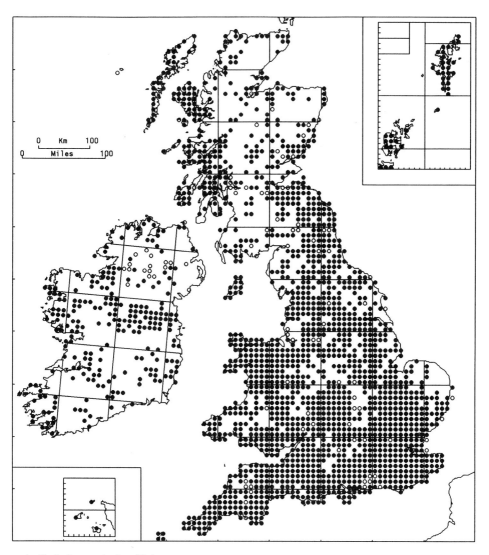

49/2. Barbula unguiculata Hedw.

Often a ruderal species occurring plentifully on neutral to base-rich soil on paths and waste ground, in gardens, arable fields, quarries and other disturbed habitats, and in the crevices of old walls. It is certainly most abundant in man-made situations but occurs also in more natural habitats such as soil on stream-banks, among calcareous rocks, and on stony ground and earthy ledges. It may occur in moderate shade provided the ground remains sufficiently open. 0–520 m (Gragareth). GB 1695+77*, IR 238+19*.

Dioecious; capsules are occasional to frequent, maturing in winter and spring. Protonemal gemmae are produced in culture (Whitehouse, 1987).

Circumboreal. Almost throughout Europe except the extreme north. Southern S. America, New Zealand.

T. L. BLOCKEEL

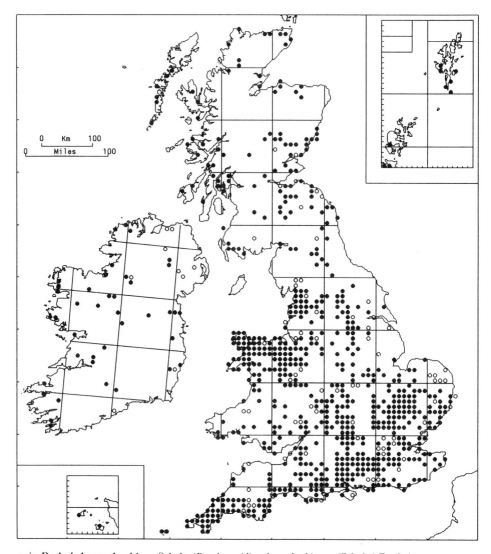

49/3. Barbula hornschuchiana Schultz (*Pseudocrossidium hornschuchianum* (Schultz) Zander)

Most common on well-drained calcareous soil, especially on compacted or stony ground. Typical habitats include paths, chalk and limestone banks, unsurfaced car-parks, old quarries and pits, and gravelly and sandy ground, including old dunes and sand-flats. It is found on bare ground kept open by trampling and other agencies rather than on regularly disturbed soil, and it requires open, well-illuminated situations. It may spread to areas where it is otherwise absent along tracks and paths which have been surfaced with limestone rubble and chippings. *Barbula convoluta* is an almost constant associate. 0–330 m (Berwyn Mts). GB 660+77*, IR 40+9*.

Dioecious; capsules are rare. Protonemal gemmae are produced in culture (Whitehouse, 1987).

Widespread in Europe, especially in the southern, central and western parts, becoming rare in the north. Macaronesia, N. Africa, Asia Minor, northwest N. America (probably introduced).

T. L. BLOCKEEL

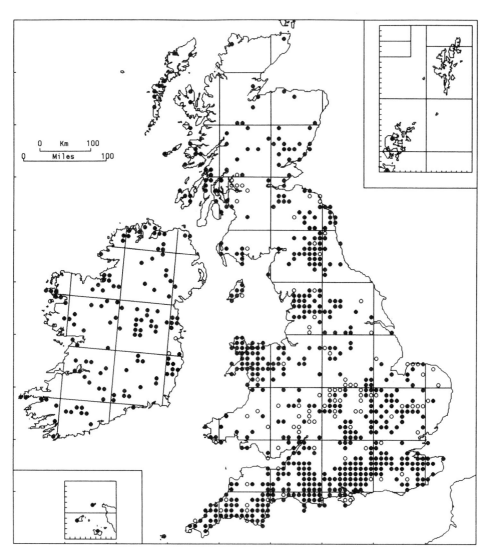

49/4. Barbula revoluta Brid. (*Pseudocrossidium revolutum* (Brid.) Zander)

A mainly lowland species, occurring in the crevices of calcareous rocks in well-lit situations. Natural habitats include limestone and other base-rich cliffs and crags, and the species may colonize old quarries and cuttings. It also occurs rather rarely on chalk banks and on hard stony ground. In general, however, it is commoner on old walls than natural rock, often on mortar where the building-stone is non-calcareous, or on churchyard monuments made of oolitic or other limestones. It is therefore widespread in areas where the underlying rock is acidic; indeed in some districts it is confined to man-made habitats. 0–600 m (Craig Leek). GB 590+106*, IR 125+4*.

Dioecious; capsules are rare, maturing in winter and spring. Leaf gemmae are rarely reported but Whitehouse (1964) found that they were frequent in Cambridgeshire. Rhizoidal tubers are apparently common in Belgium and France but have not yet been confirmed in the British Isles. Whitehouse (1987) has found protonemal gemmae in culture.

S., W. and C. Europe, north to S. Scandinavia. N. Africa, S.W. Asia, N. America (as var. *obtusula* (Lindb.) Mönk., which also occurs in continental Europe).

T. L. BLOCKEEL

265

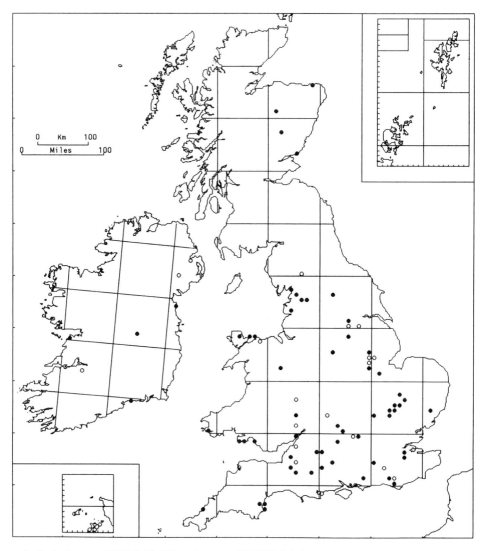

49/5. **Barbula acuta** (Brid.) Brid. (*Didymodon acutus* (Brid.) K. Saito)

This species is most often found on dry calcareous banks in southern regions, particularly in sunny chalk and limestone grassland and quarries, where it occurs in thin turf and on bare, open ground. It has also been reported from sea-banks, fixed dunes and basic clay. Much more rarely it grows in montane districts on earthy ledges and banks on base-rich rock. 0–600 m (Glen Fee). GB 57+17*, IR 4+3*.

Dioecious; capsules very rare, maturing in spring. Axillary gemmae have been reported in the species, but it is possible that such forms should be referred to *B. rigidula*.

Circumboreal. In Europe common in S., but widespread also in C. and N. where larger forms have been named *Didymodon validus* Limpr. South to C. America.

Often difficult to recognize in the field and thus overlooked, but certainly a rather rare plant in the British Isles, usually present in only small quantity and diminishing because of the destruction of chalk grassland. It is sometimes recorded in error, but the map should present a reasonably accurate overall picture.

T. L. BLOCKEEL

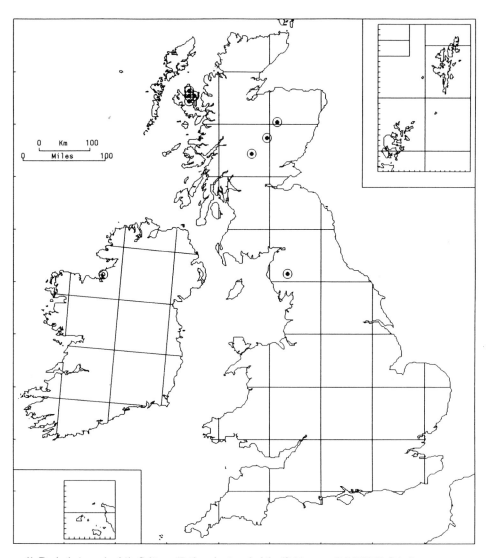

49/6. **Barbula icmadophila** Schimp. (*Didymodon icmadophilus* (Schimp. ex C. Müll.) K. Saito)

It grows in base-rich montane habitats, on cliffs and in gullies, usually on crumbling or rotten rock. In Skye it grows in dry habitats on basalt on ledges and in short turf, and also on the sides of an artificial cutting. The Irish locality is on limestone. 60 m (Skye) to 600 m (Cumbria). GB 8, IR 1*.

Dioecious; capsules and gemmae are unknown in the British Isles.

Circumboreal. In Europe mainly in C. and N. mountainous regions.

In view of the geographically scattered localities in which this species occurs, its restriction to a small number of isolated stations is difficult to explain.

T. L. BLOCKEEL

267

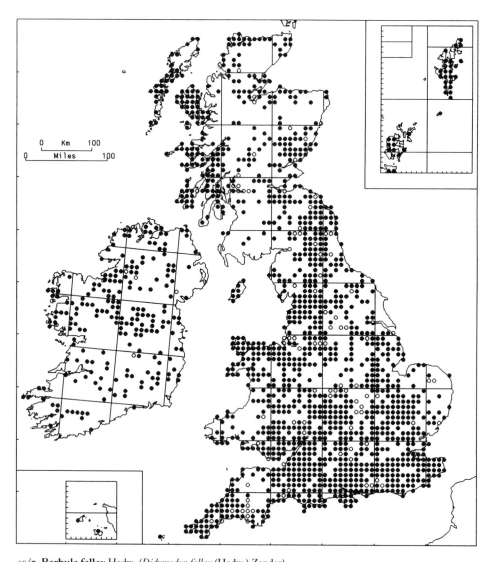

49/7. **Barbula fallax** Hedw. (*Didymodon fallax* (Hedw.) Zander)

This is a moss of open ground with a neutral to base-rich substratum. It occurs on soil in calcareous grassland, on roadside embankments, on tracks and paths, on the banks of streams and rivers, and in quarries and pits. It is also characteristic of fixed dunes and sand-flats, and bare clay by ditches, ponds, streams and on landslips. On limestone and other basic rock it is found on earthy banks and ledges, and on quarry-spoil. Mainly lowland, but recorded at 490 m as an associate of *Aongstroemia longipes* on Ben Lawers, and probably ascending higher. GB 1323+93*, IR 211+5*.

Dioecious; capsules are occasional, maturing in winter and spring. Gemmae occur on the protonema in culture (Whitehouse, 1987).

Circumboreal. Very common over much of Europe, becoming rare in the far north.

T. L. BLOCKEEL

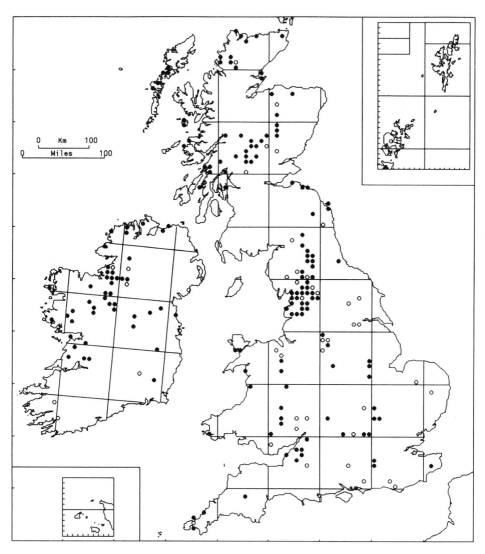

49/8. Barbula reflexa (Brid.) Brid. (*Didymodon ferrugineus* (Schimp. ex Besch.) M. Hill)

Characteristically occurring in well-drained situations among calcareous rocks, especially limestone. It may be found on thin soil on rocks and in scree, sometimes also on deeper soil. Though often in well-lit situations it is able to grow in light shade, for example in open ash-woods on limestone. It is also found on calcareous dunes and in machair, where it is locally abundant in some W. Highland and Hebridean sites. It occurs, albeit rarely, in thin turf in chalk grassland and in chalk pits. 0–785 m (Coire-cheathaich). GB 124+39*, IR 45+8*.

Dioecious; capsules are unknown in Britain and appear to be rare throughout its range.

Widespread in parts of C. and N. Europe, especially in upland districts. N. Africa, Asia, eastern and central N. America.

T. L. BLOCKEEL

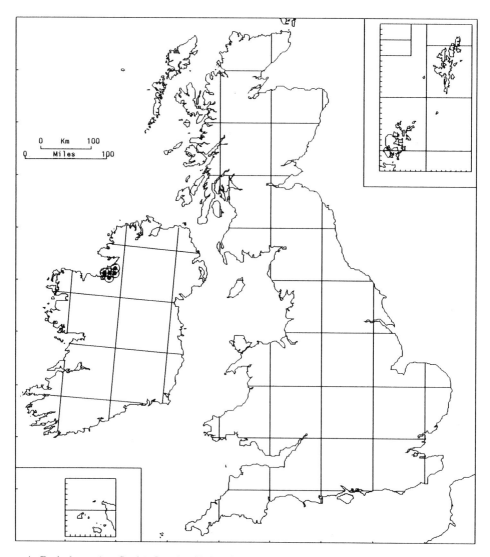

49/9. **Barbula maxima** Syed & Crundw. (*Didymodon maximus* (Syed & Crundw.) M. Hill)

This species is restricted to W.- to N.-facing limestone cliffs in the Benbulbin range in W. Ireland. It occurs on permanently wet ledges or on damp soil at the foot of rocks, often in large, deep tufts. *Gymnostomum insigne* is frequently in close proximity. 300 m. IR 5.

Sterile; related species are dioecious. Gemmae are unknown.

Northwest Territories of Canada.

This plant was formerly regarded as a variety of *B. reflexa* but Syed & Crundwell (1973) demonstrated that it is specifically distinct. Until recently it was thought to be an Irish endemic. In the most recent Canadian checklist (Ireland *et al.*, 1987) the Canadian record is signified as in need of confirmation, but the species is accepted for N. America by Anderson *et al.* (1990).

T. L. BLOCKEEL

270

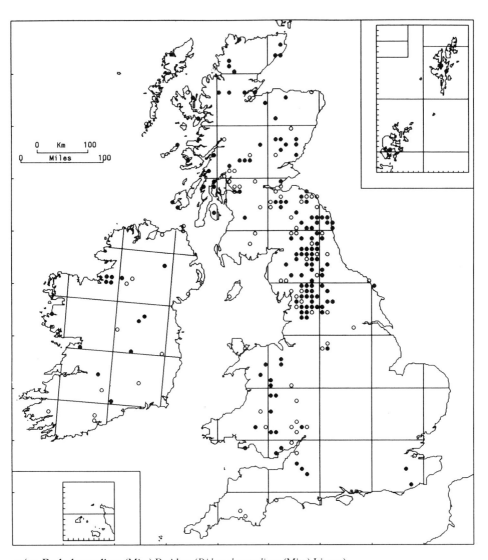

49/10. **Barbula spadicea** (Mitt.) Braithw. (*Didymodon spadiceus* (Mitt.) Limpr.)

Most commonly growing on stones and tree roots by clear, fast-flowing streams and rivers, especially in limestone districts. The plants are often deeply embedded in alluvial sand and detritus, and may be associated with *Dichodontium flavescens* and, more rarely, *Barbula nicholsonii*. In some areas, as in the Yorkshire Dales, it is characteristically found by the larger rivers, but in mountainous areas it may also occur by streams and seepages in calcareous gullies and ravines. 0–550 m (Whernside). GB 138+67*, IR 13+14*.

Dioecious; capsules are occasional to frequent, maturing in late spring and summer. Gemmae occur on the protonema in culture (Whitehouse, 1987).

Widespread in Europe; absent in the far north and most of the Mediterranean region. Asia.

T. L. BLOCKEEL

271

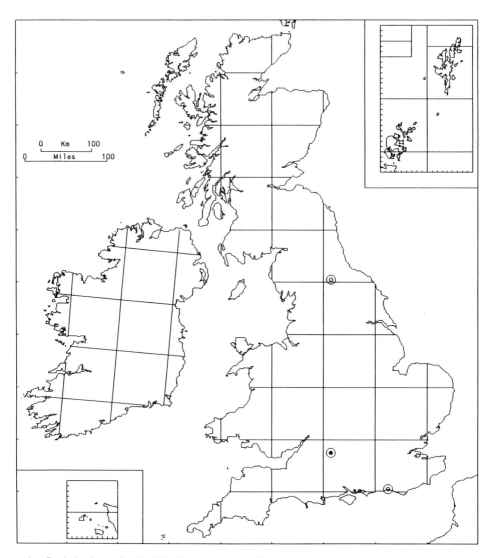

49/11. **Barbula glauca** (Ryan) Möll. (*Didymodon glaucus* Ryan)

This species has been recorded from chalk in Sussex and Wiltshire and Carboniferous limestone in Yorkshire, but is known to persist only in Wiltshire, and there precariously. The habitat is in chalk crevices and on soil at the foot of chalk-faces. No voucher is known for the Yorkshire record, which was made about 1914 on rocks near Richmond. In Sussex, the locality was a chalk pit near Shoreham but the plant has not been seen there since 1915. Lowland. GB 1+2*.

Dioecious; British plants are female or lack gametangia. Axillary gemmae are produced in chains on modified rhizoids in the leaf axils, and have been reported to occur also on the protonema.

A European endemic, scattered through C. Europe south to the Italian Lakes; very rare in Scandinavia.

T. L. BLOCKEEL

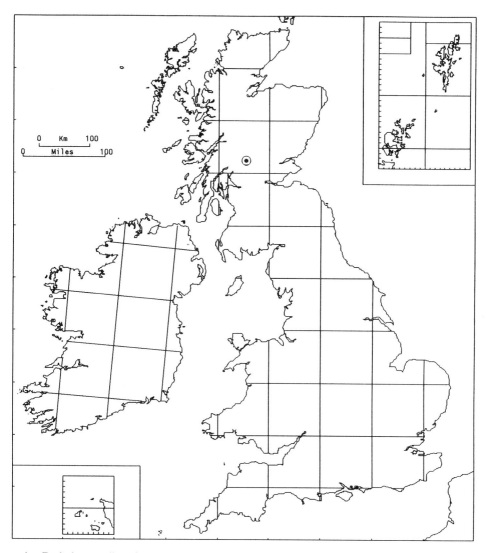

49/13. **Barbula mamillosa** Crundw. (*Didymodon mamillosus* (Crundw.) M. Hill)

This species is known in Britain only from the type locality in Perthshire, where it occurs in compact tufts on montane calcareous crags. 600 m. GB 1.

Dioecious; the type-gathering is male. Axillary gemmae are abundant.

Occurs in Germany (specimen confirmed by A.C. Crundwell) and reported from Iceland. Unknown elsewhere.

First described by Crundwell (1976a), it is an inconspicuous plant and may well be more widespread than the records suggest.

T. L. BLOCKEEL

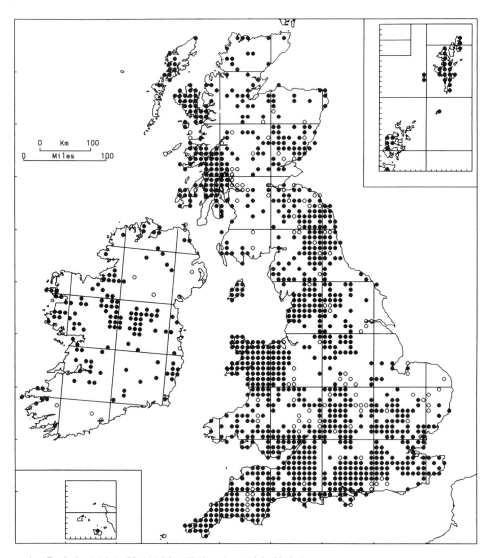

49/14. **Barbula rigidula** (Hedw.) Mitt. (*Didymodon rigidulus* Hedw.)

A common species in the lowlands on a wide variety of calcareous and base-rich rocks. It occurs widely on natural rock-outcrops in open to lightly shaded situations, but is equally plentiful on man-made structures. These include concrete, brick, roof-tiles and paving-stones in addition to walls and buildings constructed of natural materials. Although it may occur in open situations, it is not tolerant of intense insolation and is usually found where there is some shelter or shade. 0–750 m (Knock Ore Gill). GB 1078+116*, IR 133+11*.

Dioecious; capsules occasional, maturing in autumn, winter and spring. Gemmae are usually abundant on protonema in the leaf axils.

Circumboreal. Almost throughout Europe, but rare in the Mediterranean region.

T. L. BLOCKEEL

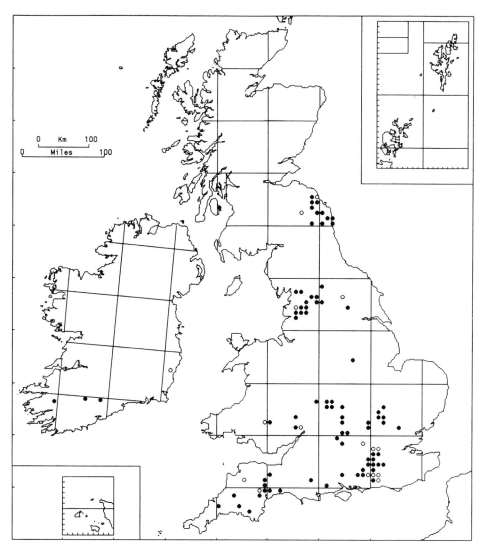

49/15. Barbula nicholsonii Culm. (*Didymodon nicholsonii* Culm.)

The natural habitat of this species is on stones and tree roots on the banks of streams and rivers where subject to periodic flooding. In such situations it may be embedded in alluvial sand and detritus, and may be associated with *Barbula cylindrica*, *B. spadicea* and *Bryoerythrophyllum recurvirostrum*. In parts of lowland Britain it is most common on retaining walls and bridge-supports but in some western and northern districts it occurs widely on natural substrata. In S. England it is occasional on tarmac, concrete and gravel paths. Lowland. GB 78+15*, IR 3+1*.

Dioecious; populations are usually female and sporophytes have been reported only once, in Oxfordshire (Corley *et al.*, 1987). Whitehouse (1987) has found protonemal gemmae in culture, but the occurrence of axillary gemmae requires confirmation. Rhizoidal tubers have been reported from Belgium.

Long thought to be confined to the British Isles but now known from Spain, France, Belgium and Germany. Reported from Turkey and N. America.

An under-recorded species, being often overlooked or mistaken for other species of the genus, especially when growing on unremarkable substrates such as tarmac or gravel paths.

T. L. Blockeel

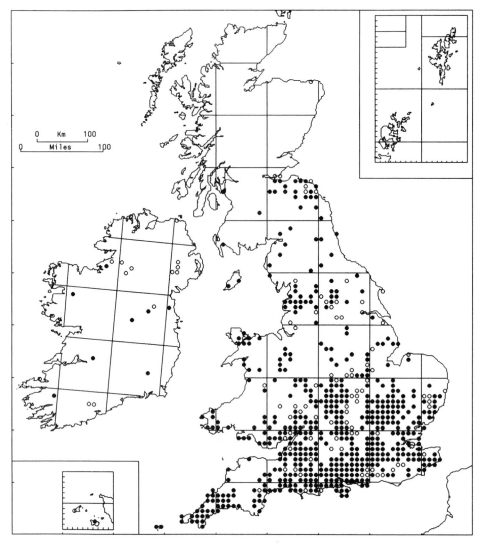

49/16. **Barbula trifaria** (Hedw.) Mitt. (*Didymodon luridus* Hornsch. ex Spreng.)

Commonly on hard calcareous ground, stones and walls, occurring directly on the rock or on thin soil. Although sometimes in very dry situations, as on wall-tops, the species is often found in more moist situations, as on stones in woods, at the base of walls and in gutters. It is also widespread on stones and walls, more rarely tree-roots, on the banks of rivers in places subject to flooding, sometimes associated with *B. nicholsonii*. Lowland. GB 523+84*, IR 8+14*.

Dioecious; capsules are very rare in Britain, maturing in spring. Gemmae sometimes occur on the protonema but may be produced only when the plant grows in deep shade (Whitehouse, 1980).

Widespread in Europe, especially in the south and west; absent from most of Scandinavia. Macaronesia, N. Africa, W. Asia, western N. America, C. America.

T. L. BLOCKEEL

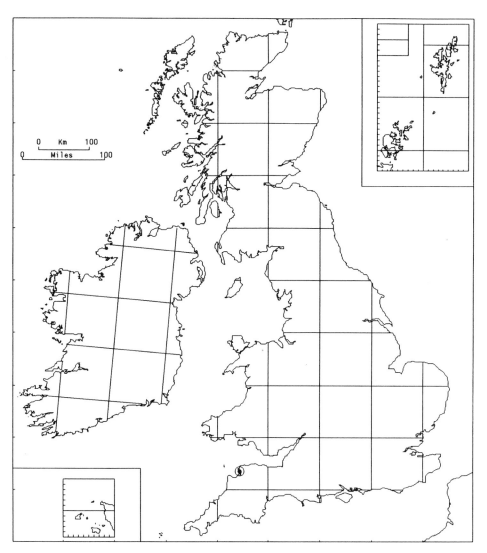

49/17. **Barbula cordata** (Jur.) Braithw. (*Didymodon cordatus* Jur.)

Found as small tufts, growing on soil derived from soft sandstone rock on a S.-facing sea-cliff and in smaller quantity where the same rock outcrops on a nearby roadside bank. The soil on which it grows is partially stabilized by crustose lichens, including *Collema tenax, Fulgensia fulgens, Placynthium nigrum, Squamaria crassa* and *Toninia caeruleonigricans*. Annual vascular plant and bryophyte associates include *Cardamine hirsuta, Cerastium diffusum, Erodium cicutarium, Myosotis ramosissima, Aloina aloides, Barbula convoluta, B. revoluta, B. trifaria, Bryum bicolor, Ceratodon purpureus, Desmatodon convolutus, Pottia starkeana* ssp. *starkeana, Trichostomum brachydontium, T. crispulum* and *Zygodon viridissimus*. Lowland. GB 1.

Dioecious; only female plants are known in Britain. Axillary gemmae are abundant in the wild, and gemmae are produced on the protonema in culture.

C. and S. Europe east to the Crimea and Caucasus; absent from Scandinavia.

B. cordata was first discovered at its single British locality in 1903, growing on an earth-covered wall-top, but it has not been seen on walls in recent years and its occurrence in this habitat was presumably casual.

R. D. PORLEY

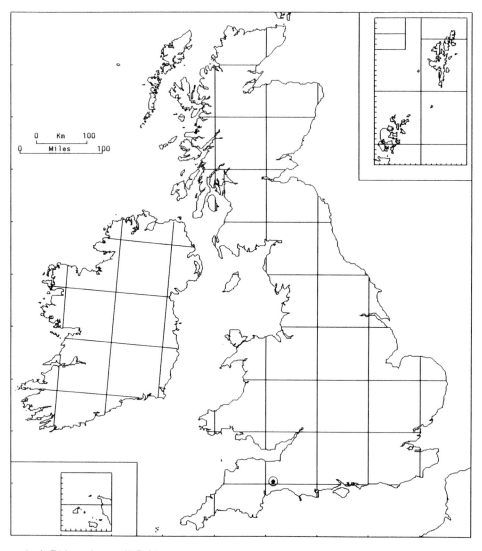

49/17A. Didymodon reedii Robins.

This species has been reported from a single locality in Devon, in the crevices of a mortared stone-wall of acid flinty sandstone by a stream (Appleyard, 1985). Lowland. GB 1.

The English plant appears to be dioecious and has produced protonemal gemmae in culture. In N. America, *D. reedii* is sterile but produces plentiful axillary gemmae.

Eastern N. America.

The identification with the rare N. American species *D. reedii* is extremely dubious, since the English plant has pluripapillose lamina cells and lacks axillary gemmae. It is probably a form of *Barbula vinealis*.

T. L. BLOCKEEL

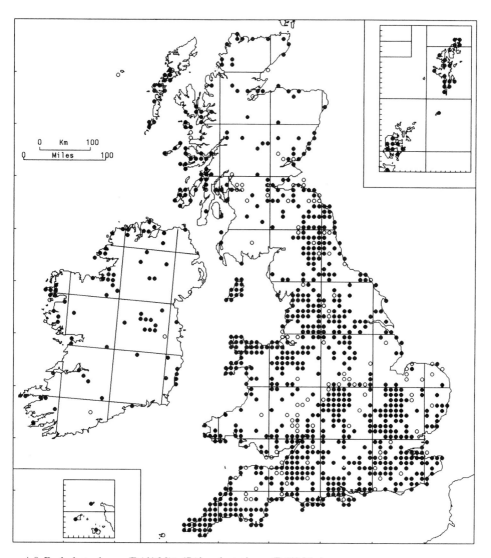

49/18. **Barbula tophacea** (Brid.) Mitt. (*Didymodon tophaceus* (Brid.) Lisa)

This species occurs in a wide range of base-rich habitats. These include tufa and wet calcareous rocks on stream-banks in woodland, in ravines and on cliffs and coastal banks. It also occurs on stones by streams, in base-rich flushes and seepages, and on moist walls, especially at the base. It is often on a thin layer of soil but also grows directly on rock where this is sufficiently soft or porous. It is by no means confined to rupestral habitats, frequently colonizing gravelly tracks, bare clay and soil, and bare ground and spoil-heaps in quarries and pits. It is especially plentiful on the coast, forming extensive sheets in association with *Dicranella varia* on wet undercliffs and by streams. 0–440 m (Horseshoe Pass, Llangollen). GB 850+115*, IR 74+4*.

Dioecious; capsules are frequent, maturing in winter and spring. Rhizoidal tubers occur (Side, 1983) but appear to be rare.

Almost throughout Europe, but rare in Scandinavia. N. Africa, Asia, N., C. and S. America.

T. L. BLOCKEEL

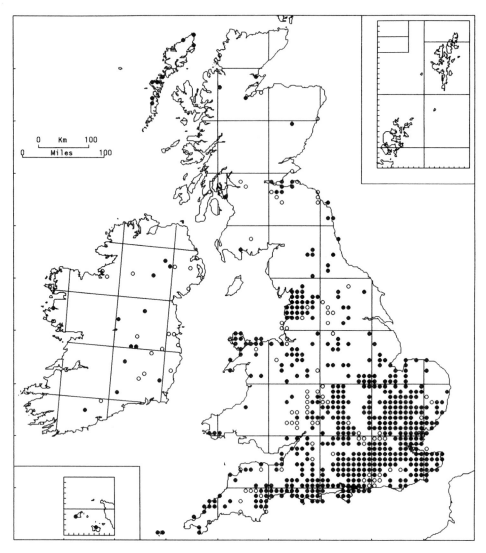

49/19. **Barbula vinealis** Brid. (*Didymodon vinealis* (Brid.) Zander)

This species occurs in dry habitats on base-rich rocks and walls. Although it is widespread on well-illuminated, natural rock-outcrops, such as limestone, in most districts it is more frequent on walls, often where there is some accumulation of soil or detritus, but also on bare stone. It may colonize artificial substrata, including concrete and roof-tiles. It is also frequent on hard or stony calcareous ground and in short dry turf. Lowland. GB 539+88*, IR 12+18*.

Dioecious; capsules are rare, maturing in spring and summer.

Common in Europe, except N. Scandinavia, and abundant in Mediterranean countries. Widespread in the Northern Hemisphere, south to C. America.

Separation from *B. cylindrica* may be difficult and errors in the details of the distribution are likely.

T. L. BLOCKEEL

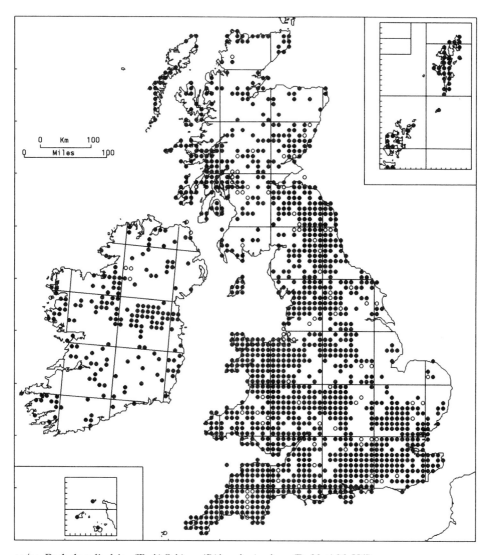

49/20. **Barbula cylindrica** (Tayl.) Schimp. (*Didymodon insulanus* (De Not.) M. Hill)

This species occurs in moister habitats than the closely related *B. vinealis*. It is widespread in base-rich woodland but does not require shade in sufficiently moist habitats. It is common on the sheltered side of walls and at their bases, including those of artificial materials such as brick and concrete, and may also grow on pavements and tarmac. Equally characteristic are stones, cliffs, tree roots and tree-bases by streams and rivers. The plants usually grow where there is some accumulation of silt or sand and often occur on soil in sheltered western hedgebanks. 0–430 m (Skye). GB 1368+85*, IR 186+6*.

Dioecious; capsules are very rare, maturing in summer. Rhizoidal tubers have been reported (Ellis & Smith, 1983) but they appear to be exceptional.

Most of Europe, commoner in northern parts than *B. vinealis* but more rare in the south. Widespread in the Northern Hemisphere.

T. L. BLOCKEEL

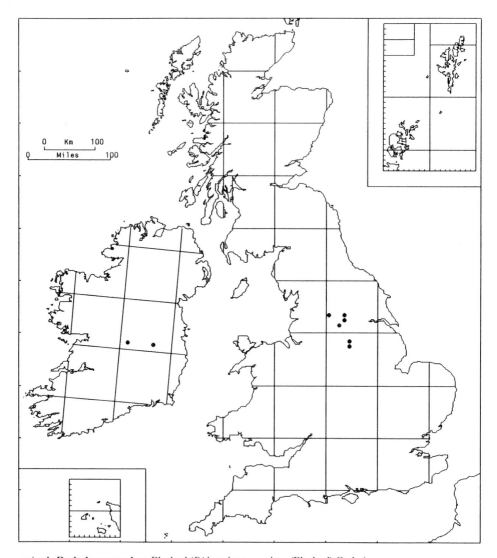

49/20A. Barbula tomaculosa Blockeel (*Didymodon tomaculosus* (Blockeel) Corley)

Usually found as scattered stems among other bryophytes on moist disturbed soil in open situations. Most of the known localities are on heavy clay in arable fields on the Coal Measures in Yorkshire and Derbyshire, but the species has also been collected on ground trampled by cattle at the edge of a pasture on clay soil overlying Millstone Grit. The Irish collections consist of very poor material detected under the microscope, and appear to have come from a more friable soil than the English plants. Associates include *Bryum* spp., *Dicranella schreberana*, *D. staphylina*, *Pottia truncata* and *Pseudephemerum nitidum*. Lowland. GB 6, IR 2.

Dioecious; all collections to date are female. Abundant rhizoidal tubers are present in all gatherings.

Not yet recorded outside the British Isles.

This species has only recently been described (Blockeel, 1981). Although it is easily overlooked and probably under-recorded, it appears to be genuinely rare.

T. L. Blockeel

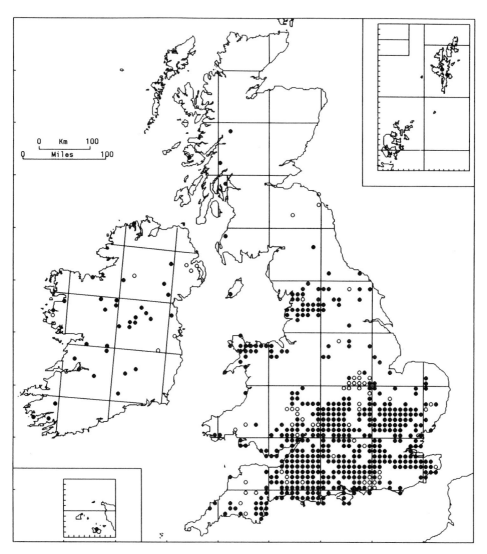

49/20B. **Barbula sinuosa** (Mitt.) Garov. (*Didymodon sinuosus* (Mitt.) Delogne, *Oxystegus sinuosus* (Mitt.) Hilp.)

This species is most characteristic of moist calcareous rocks on the banks of streams and rivers, where it forms lax but sometimes extensive patches. It may occur within the flood zone and is then often embedded in alluvial sand. In this habitat it occasionally colonizes tree roots. It is also widespread on calcareous substrates remote from water, but always where shaded and well sheltered, e.g. on stones and about tree roots in base-rich woodland, on old walls, and more rarely on concrete and paving, especially below N.-facing church-walls. Lowland. GB 477+58*, IR 33+8*.

Dioecious; capsules are unknown, all populations apparently being female. The leaf apices are fragile and presumably serve as a means of propagation; gemmae occur on the protonema in culture (Whitehouse, 1987).

Widespread in Europe, except the extreme north. Turkey. Reported from N. America.

T. L. BLOCKEEL

283

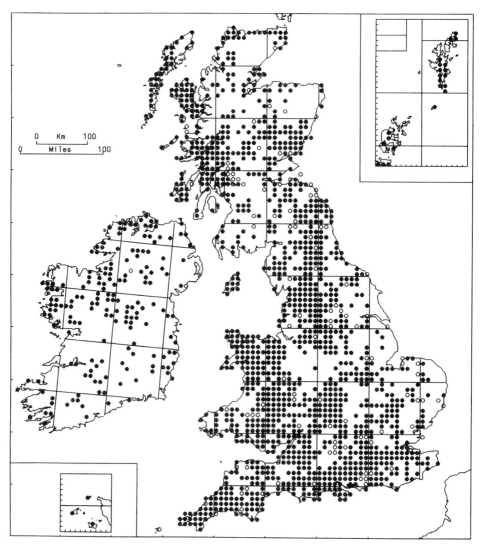

49/21. **Bryoerythrophyllum recurvirostrum** (Hedw.) Chen (*Barbula recurvirostra* (Hedw.) Dix.)

Usually occurring in crevices and on thin soil on and among base-rich rocks in sheltered to moderately shaded places. Thus it is found in woodland, on stream-banks, on sheltered walls, in old quarries, and on rock ledges and crags. It is also frequently present in the bryophyte communities which develop at the base of trees in the flood zone of streams and rivers, with *Barbula cylindrica*, *Homalia trichomanoides*, *Tortula subulata* etc. More rarely it occurs on deeper soil, but then usually among rocks and tree roots. 0–1180 m (Ben Lawers). GB 1306+109*, IR 176+4*.

Synoecious; capsules are common. Rhizoidal tubers are unknown in Europe but occur in Japan (Saito, 1975).

Very widespread in the Northern Hemisphere, including the Arctic; scattered occurrences in temperate parts of the Southern Hemisphere.

T. L. BLOCKEEL

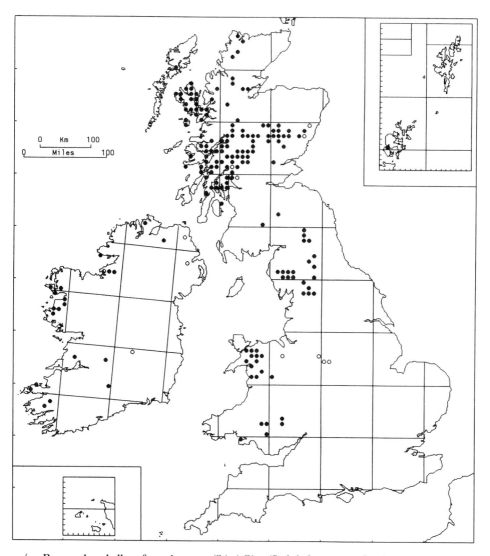

49/22. Bryoerythrophyllum ferruginascens (Stirt.) Giac. (*Barbula ferruginascens* Stirt.)

A montane species occurring as scattered stems or pure tufts on and among base-rich rocks, on cliffs, in gullies and ravines and on stream-banks. It is often on wet rock-ledges mixed with other bryophytes, but also grows as pure tufts in rock crevices and on soil among rocks. It appears to be indifferent to rock type, provided that there is an adequate supply of bases, either directly or by seepage. Thus it grows on limestone, schists, basalt and igneous rocks. In some districts it is a frequent colonist of basic tracksides, e.g. in Kielder Forest, Northumberland. Ascends to 1075 m (Aonach Beag), and occurs down to sea-level in W. Scotland (by Crinan Canal); rarely below 200 m in England and Wales. GB 153+9*, IR 20+4*.

Dioecious; capsules unknown in Europe but reported from N. America. Rhizoid tubers normally abundant.

Incompletely circumpolar. Widespread in N. Europe and mountains of C. Europe. Apparently absent from the Far East, but present in western and eastern N. America, south to the mountains of Colorado, Virginia and Mexico.

T. L. Blockeel

285

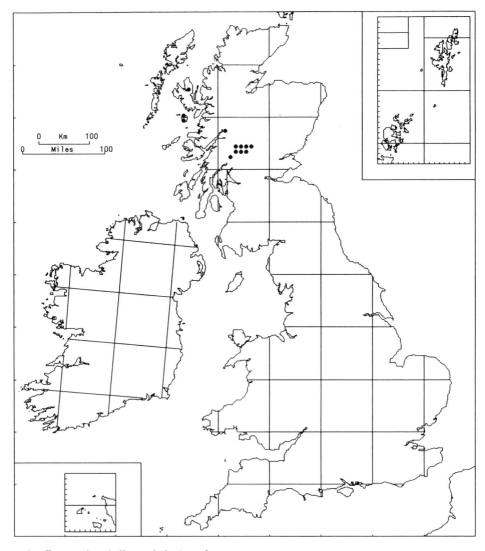

49/23. Bryoerythrophyllum caledonicum Long

This species occurs as lax patches or scattered stems on permanently wet calcareous rocks in the mountains. In mainland Scotland it is found on sloping schist rocks where there is a constant seepage of water and some accumulation of soil. Thus it grows on ledges, on inclined rock and at the edges of streams. Localities are usually N.- or E.-facing, and are kept open by the soft unstable nature of the substratum. The species is absent from sites which are intermittently dry. *Blindia acuta* and *Bryoerythrophyllum ferruginascens* are the most frequent associates. 380 m (Rhum) to 1060 m (Ben Lawers). GB 12.

All collections are sterile and gemmae are unknown.

Endemic to Scotland.

B. caledonicum was first collected at the end of the 19th century but was then and subsequently referred to a form of *Leptodontium flexifolium*. Zander (1978) recognized it as a *Bryoerythrophyllum* and treated it as a form of *B. jamesonii* (Tayl.) Crum. Its correct status was elucidated by Long (1982).

T. L. BLOCKEEL

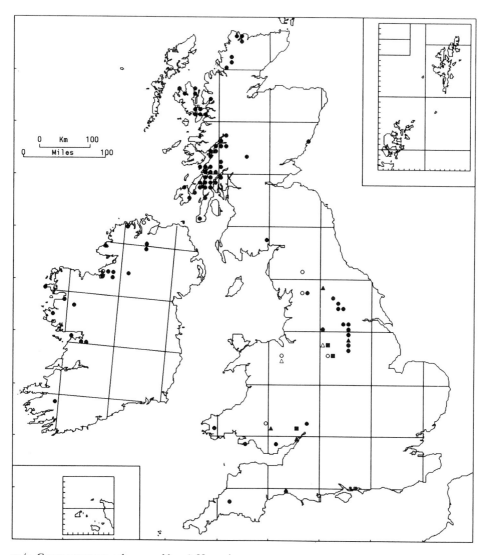

50/1. **Gymnostomum calcareum** Nees & Hornsch.

On limestone rocks or calcareous basalt, particularly where soft, slightly moist and shaded, very rarely on the mortar of walls. It is usually in low-lying, sheltered situations. 0–550 m (Croagh Patrick). GB 75+7*, IR 17+3*.

Dioecious; mostly female in the British Isles, male plants (shown as triangles on map) rare in Britain and not recorded from Ireland, capsules (shown as squares) very rare. Gemmae occur on the protonema in culture.

Widely distributed in Europe, particularly in the Mediterranean region. Macaronesia, N. and S. Africa, Asia, S. America, Australasia.

This species has been much confused with related taxa. Mapped records are mainly based on herbarium specimens confirmed for a recent revision (Whitehouse & Crundwell, 1991). The first record from Ireland was from Benbulbin in 1871 and the first from Britain was from Miller's Dale, Derbyshire in 1874.

H. L. K. WHITEHOUSE

287

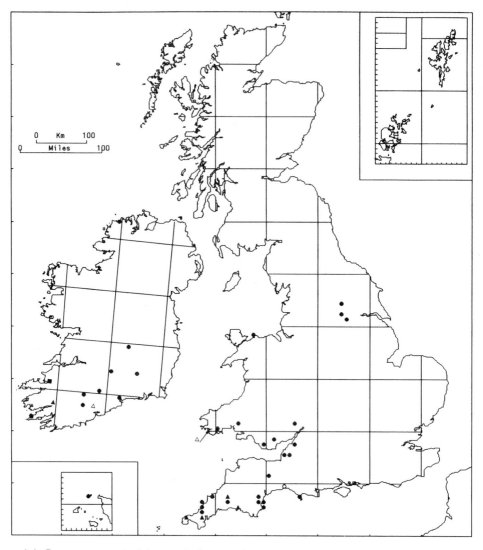

50/1A. **Gymnostomum viridulum** Brid. (*G. luisieri* (Sérgio) Sérgio ex Crundw.)

On earth in crevices in limestone in shady or sunny situations, on well-drained calcareous soil, on crumbling mortar such as of derelict mine-buildings, on limestone wall-tops, and on the underside of bridges over streams. Lowland. GB 27+2*, IR 11+1*.

Dioecious; mostly female, male plants (shown as triangles on map) confined to the extreme south-west of Britain and Ireland, sporophytes (shown as squares) known only from one locality in Ireland. Gemmae are frequent on protonema and in upper leaf axils.

A Mediterranean plant, extending to Czechoslovakia, Switzerland, Belgium and W. France. Madeira, Canaries, E. Africa, W. Asia.

Although first collected near Cork, Ireland, in 1850 and near Tadcaster, England, in 1900, it was formerly confused with *G. calcareum* and has only recently been distinguished. Mapped records are all based on herbarium specimens confirmed by H. L. K. W. or A. C. Crundwell.

H. L. K. Whitehouse

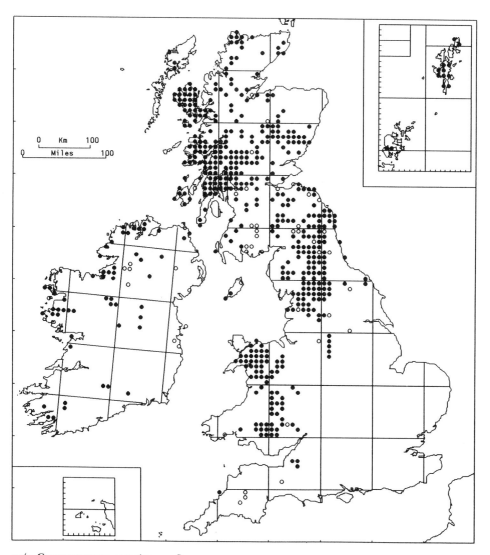

50/2. Gymnostomum aeruginosum Sm.

A species of damp to wet, base-rich rocks, occurring in a wide range of habitats. On limestone it occurs on irrigated rock and in flushes, and in crevices kept moist by seepage. It is not, however, a strong calcicole, and may occur on slightly base-enriched substrata such as shale and sandstone. It occurs widely in upland and montane districts where habitats include seepages, gullies and ravines, cliffs and crags, and more rarely old walls. It is less common in the lowlands where it is found on stream-banks, especially in deep valleys, and on coastal banks and cliffs. 0–1175 m (Ben Lawers). GB 465+38*, IR 54+11*.

Dioecious; capsules occasional, maturing in late summer and autumn. Protonemal gemmae are produced in culture (Whitehouse, 1987).

Almost throughout Europe. Widespread in the Northern Hemisphere but the precise distribution is difficult to define because of disagreement among authors about the taxonomic boundaries of the species.

T. L. BLOCKEEL

289

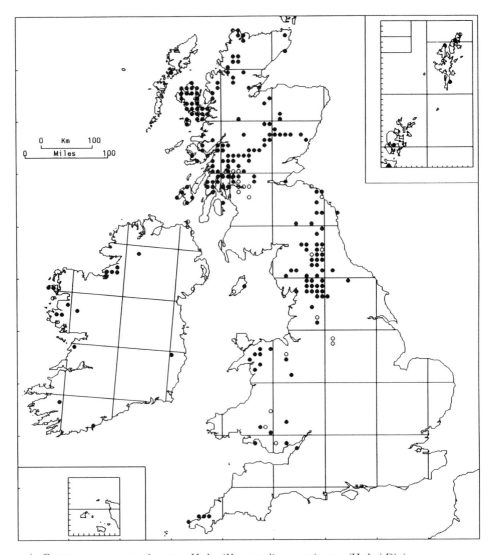

50/3. **Gymnostomum recurvirostrum** Hedw. (*Hymenostylium recurvirostrum* (Hedw.) Dix.)

A strong calcicole of damp rock crevices and rock-faces, especially in ravines. It also occurs on boulders and limestone pavement, and on hummocks and gravelly ground in calcareous flushes. More rarely is is found on old damp walls and on metalliferous mine-waste. Though not quite confined to upland and montane districts, it is certainly most common in them. 0–1070 m (Aonach Beag). GB 194+22*, IR 16+5*.

Dioecious; capsules rare, maturing in autumn.

Nearly cosmopolitan, occurring in all continents. Most of Europe but rare in the Mediterranean region.

T. L. BLOCKEEL

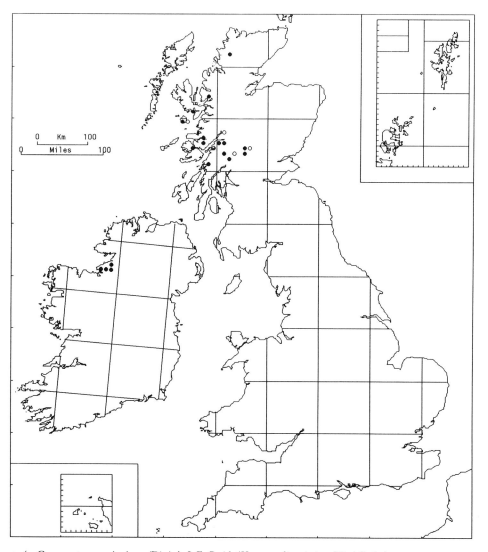

50/4. Gymnostomum insigne (Dix.) A. J. E. Smith (*Hymenostylium insigne* (Dix.) Podp.)

In similar habitats to, and sometimes associated with, *G. recurvirostrum*, occurring on wet calcareous rock-ledges in montane areas with a high rainfall. It also grows in sufficiently humid places in the lowlands, for example in ravines. 0–680 m (Creag an Lochain). GB 14+5*, IR 4.

Dioecious; capsules very rare.

In continental Europe known only from Spain. Disjunct in Pacific N. America (British Columbia).

Most other Atlantic species are calcifuges; calcicoles with this distribution pattern are exceedingly few.

T. L. Blockeel

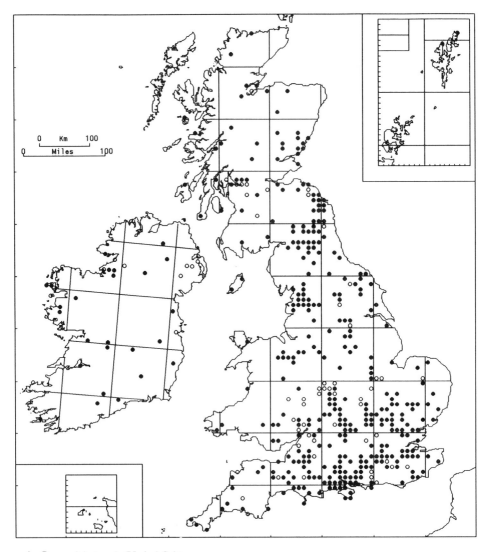

51/1. **Gyroweisia tenuis** (Hedw.) Schimp.

On damp, shady limestone or sandstone rocks, particularly by water, and on the mortar of walls, usually where water drips from above. When on sandstone, the site is usually, perhaps always, base-enriched. The species is very shade-tolerant, and even grows around electric lights in tunnels and caves. 0–400 m (above Delnabo, Banff). GB 345+52*, IR 25+11*.

Dioecious; male plants much rarer than female, sporophytes frequent in the north and west but rare in S.E. England. Vegetative propagation is by gemmae produced on the protonema.

Widespread in C. Europe, rare in Scandinavia and the Mediterranean region. Isolated records from Madeira, Iran and a few places in N. America.

H. L. K. Whitehouse

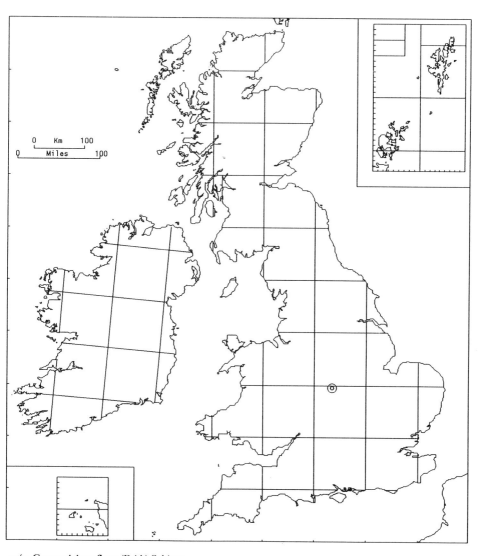

51/2. **Gyroweisia reflexa** (Brid.) Schimp.

On sandstone rocks in a quarry in Lutterworth Road, Nuneaton. Lowland. GB 1*.

Dioecious; sporophytes were frequent at Nuneaton. Gemmae, produced on the protonema, were also present.

Mediterranean Europe. Macaronesia, N. Africa, W. Asia; reported from N. America.

Found by Miss C. A. Cooper in September 1933. A bungalow had been built in the quarry about 1927 and the quarry was turned into a rock garden (H. H. Knight, *in litt.* to W. E. Nicholson, 1934). *G. reflexa* was seen in the quarry for several years, the last record being in 1938 by Miss K. E. Smith, who lived there (Anon., 1939). The quarry was visited by C. D. Preston and H. L. K. W. in June 1984. Little rock remained exposed, as houses were being built on the site, and the plant was not found.

H. L. K. WHITEHOUSE

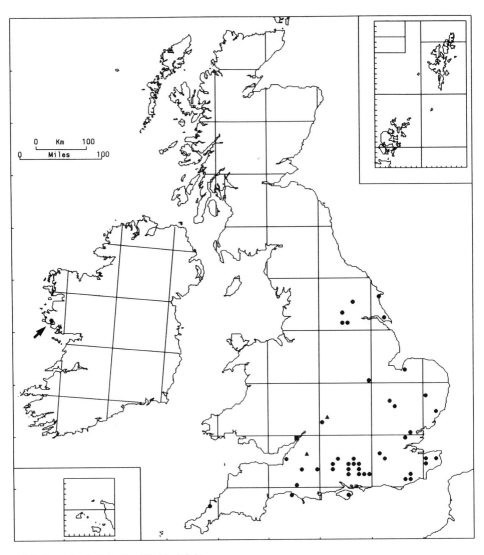

51A/1. **Leptobarbula berica** (De Not.) Schimp.

On shaded limestone, both large rocks and detached stones, chiefly in woodland, and on shaded brickwork or stonework and the mortar of walls or conduits, such as at the base of the north wall of churches. On walls it is often associated with *Tortula marginata*. Associates on limestone include *Barbula sinuosa*, *Fissidens pusillus*, *Tortella inflexa* and *T. tortuosa*. Lowland. GB 46, IR 1.

Dioecious; males (shown as triangles on map) much rarer than females, capsules (shown as squares) very rare. Gemmae occur on the protonema.

A Mediterranean species, occurring also in W. and N. France, W. Germany, Belgium and the Netherlands. Madeira, N. Africa, W. Asia.

The earliest British record so far traced is from the bowl of a font in Winestead churchyard, S.E. Yorkshire in 1948, so it is possible that the plant has spread recently. Formerly confused with *Gymnostomum calcareum* (Appleyard *et al.*, 1985) and *Gyroweisia tenuis* (Whitehouse & During, 1986).

H. L. K. WHITEHOUSE

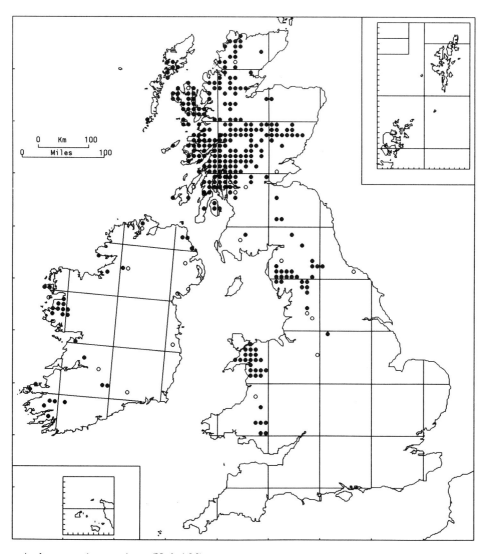

52/1. Anoectangium aestivum (Hedw.) Mitt.

A locally abundant species of rock crevices in montane habitats, often forming large cushions. It requires some base-enrichment but is by no means confined to strongly calcareous sites. Characteristically it occurs in gullies, on cliffs and ravine-walls, and on stream-banks. It is often near waterfalls and on wet rocks, and although not confined to irrigated sites requires some degree of moisture or shelter. 0–1180 m (Ben Lawers), mostly above 300 m but descending to sea-level in the north and west. GB 278+15*, IR 30+10*.

Dioecious; capsules rare to occasional, maturing in spring and summer. Gemmae are unknown.

Circumboreal. A northern and montane species, widespread in C. and N. Europe. Scattered localities also in equatorial countries and the Southern Hemisphere.

T. L. BLOCKEEL

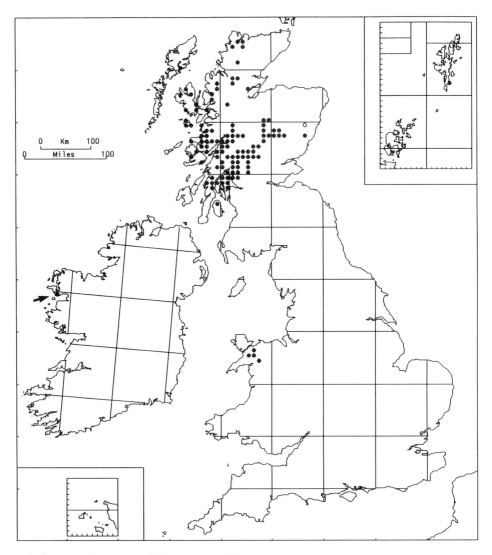

52/2. Anoectangium warburgii Crundw. & M. Hill

A montane species growing in loose patches or as scattered stems on wet rocks. It commonly occurs at the edge of streams and by waterfalls, but also on irrigated surfaces and on rocks kept moist by seepage. There is a tendency to grow on soft or crumbling rock. The substratum is usually base-rich, and rock types on which it has been recorded include mica-schist, limestone and basalt in Scotland, and volcanic tuff in Wales. Base-poor rock may also be colonized when enriched by seepage. Although normally in natural habitats, it has been found on wet slabs in a W. Highland quarry. 0–1070 m (Aonach Beag). GB 112+3*, IR 1.

Dioecious, most Scottish plants are female, but the Welsh population is entirely male; capsules are very rare, maturing in spring and early summer. Gemmae occur in the axils of the older leaves.

Apparently endemic.

The species is easily overlooked and for a long time was confused with *Gymnostomum calcareum*. It had been collected at least as early as 1871, but was not described until a century later (Crundwell & Hill, 1977).

T. L. Blockeel

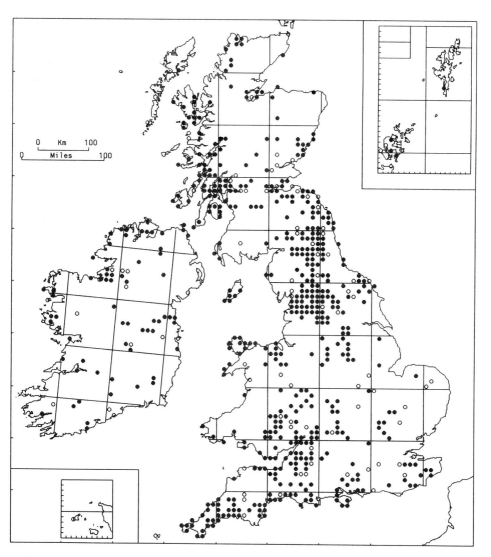

53/1. Eucladium verticillatum (Brid.) Br. Eur.

Locally plentiful on a wide variety of moist to wet base-rich rocks, but especially on limestone. It commonly occurs on dripping cliffs and may become encrusted with lime deposits, being well known as a tufa-forming moss. The most characteristic situations are on the banks of streams and in wooded valleys and ravines, but the species also occurs on moist cliffs, by waterfalls, in calcareous flushes, and on drier, shaded rocks in woodland. It sometimes colonizes man-made habitats such as wet walls. Common associates are *Barbula tophacea*, *Cratoneuron commutatum* and *Gymnostomum aeruginosum*. Lowland. GB 476+75*, IR 63+13*.

Dioecious; capsules are rather rare, maturing in spring. Protonemal gemmae are regularly produced when the plant grows in deep shade, but are apparently absent when it is well illuminated (Whitehouse, 1980).

Much of Europe, especially common in the south and west. Widespread in other temperate and sub-tropical regions of the Northern Hemisphere.

The gemmiferous form that grows in deep shade may lack leafy shoots; it has been found in caves and in a rabbit-burrow and could be quite widespread.

T. L. BLOCKEEL

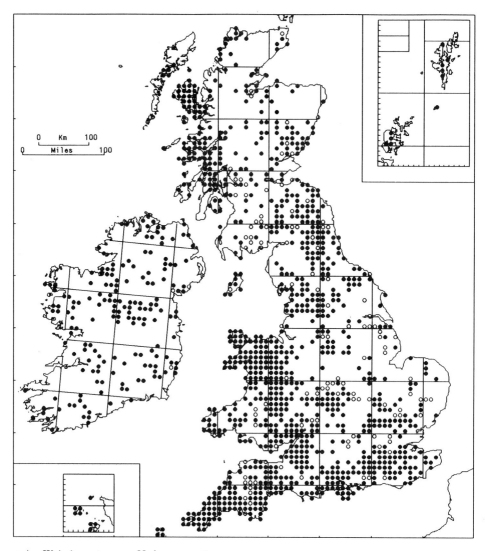

54/1a. **Weissia controversa** Hedw. var. **controversa**

This plant occurs in a wide variety of habitats on bare soils which are relatively stable and not frequently disturbed. Thus it occurs on banks in pastures, by streams and by paths, on soil on both calcareous and acidic rock-ledges on crags, cliffs and in gullies, on coastal banks and cliffs, and on open calcareous ground. Although often in quite open habitats, it also occurs away from direct light, as on soil in rock crevices and under turfy overhangs. It probably requires at least slight mineral enrichment. 0–550 m (Skye). GB 982+121*, IR 172+4*.

Autoecious; capsules abundant, maturing in spring and summer.

Almost cosmopolitan. Nearly throughout Europe.

T. L. Blockeel

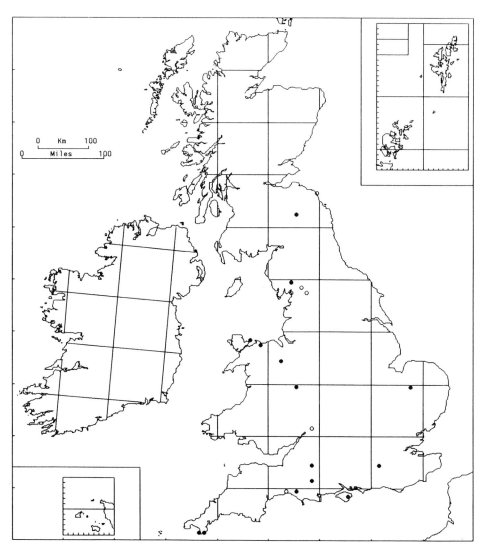

54/1b. **Weissia controversa** Hedw. var. **crispata** (Nees & Hornsch.) Nyholm

This variety occurs in similar habitats to var. *controversa* but is a fairly strict calcicole. It occurs in chalk grassland, on cliffs, in quarries, and among limestone and other calcareous rocks, including limestone pavement. Lowland. GB 14+6*.

Autoecious; capsules abundant, maturing in spring.

S., W. and C. Europe, north to S. Scandinavia. Widespread in the Mediterranean area and adjacent regions. Reported from N. America.

T. L. Blockeel

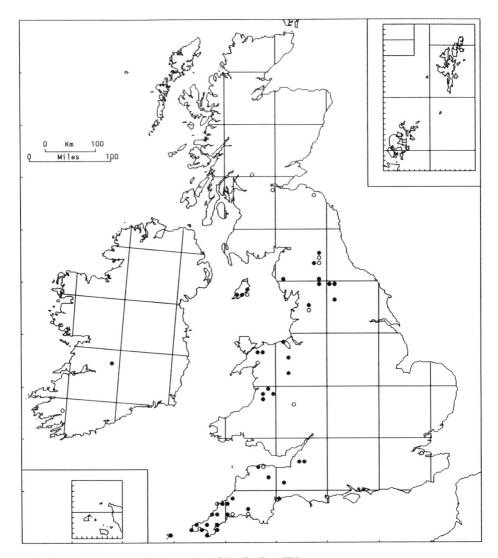

54/1c. **Weissia controversa** Hedw. var. **densifolia** (Br. Eur.) Wils.

In coarser tufts and patches than var. *controversa*, and characteristic of open ground rich in heavy metals. It is most plentiful on spoil-heaps and diggings about old lead-mines, but also occurs in natural habitats, as on rotten rock-ledges in the Lake District and river gravels with high levels of zinc in Northumberland. Lowland. GB 42 + 13*, IR 1 + 1*.

Autoecious; capsules abundant.

Widely scattered in C. Europe, south to Italy and Yugoslavia.

T. L. BLOCKEEL

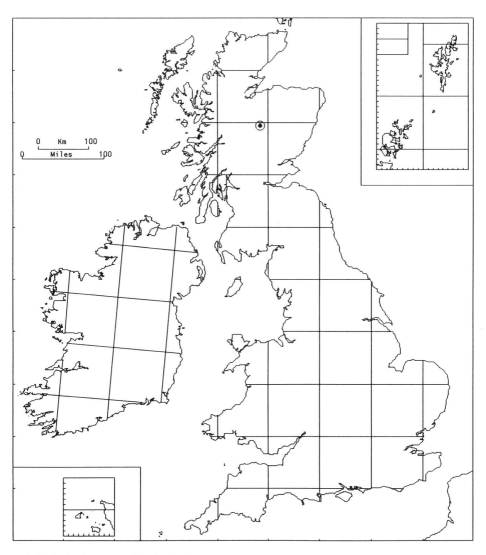

54/2. Weissia wimmerana (Sendtn.) Br. Eur.

Known only from earthy crevices among calcareous rocks in a steep S.W.-facing gully in Glen Feshie. 500 m. GB 1.

Paroecious; capsules present.

Montane parts of C. and N. Europe, south to the Pyrenees. Asia.

Very similar to *W. controversa*, of which it should probably be treated as a variety (Smith, 1978). It was first discovered in Britain in 1956 (Warburg, 1957) and has possibly been overlooked elsewhere in Scotland.

T. L. BLOCKEEL

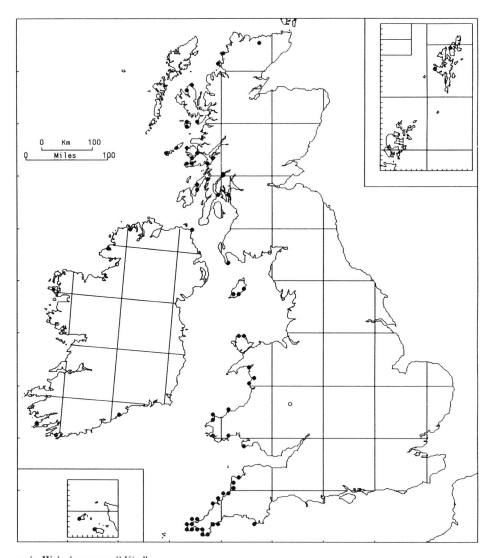

54/3. **Weissia perssonii** Kindb.

Almost confined to coastal districts, this species occurs on banks, sea-cliffs and cliff-tops, both on soil and in crevices with a thin layer of humus. In coastal habitats the substratum may be non-calcareous, even peaty, but inland the species is probably a calcicole. Common associates near the sea include *Schistidium maritimum* and *Tortella flavovirens*. Lowland. GB 60+2*, IR 7+1*.

Autoecious; capsules abundant, maturing in late winter and spring.

Faeroes, Norway, Sweden. Endemic to Europe.

Although collected in Wales in 1908, it was not distinguished by British bryologists until much later (Crundwell, 1971). It is probably still under-recorded.

T. L. BLOCKEEL

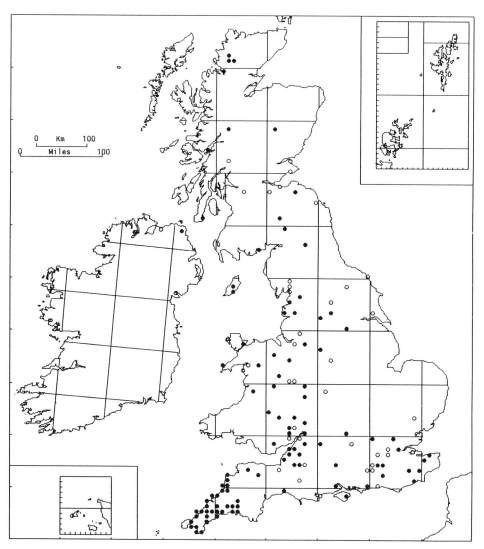

54/4. **Weissia rutilans** (Hedw.) Lindb.

It occurs in damp pastures and arable fields, on woodland rides, on damp earthy banks, on moist sand and clay, and on the banks of streams, ditches and reservoirs. Although mainly a lowland species, it is also recorded in flushes and on soil among calcareous rocks in the uplands. Its habitats are similar to those of *W. controversa* but it is nearly confined to moist non-calcareous soil. 0–450 m (Plynlimon). GB 102+39*, IR 1+3*.

Autoecious; capsules abundant, maturing in winter and spring.

Widespread in C., W. and N. Europe. N. Africa.

T. L. BLOCKEEL

303

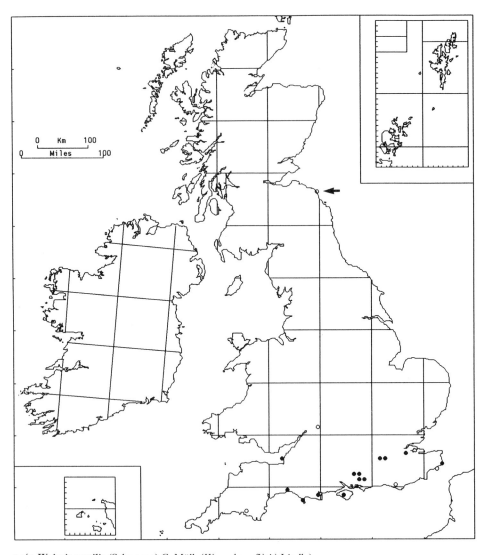

54/5. Weissia tortilis (Schwaegr.) C. Müll. (*W. condensa* (Voit) Lindb.)

A rare species of open, dry calcareous habitats, occurring as small tufts which sometimes become detached from the ground. It is recorded from calcareous turf, broken ground and scree overlying chalk and limestone, and soil-filled crevices among limestone boulders. Lowland. GB 13+5*.

Autoecious; capsules common, maturing in spring.

S., W. and C. Europe. Macaronesia, N. Africa, S.W. Asia.

Often confused in the past with forms of *W. controversa*. Only confirmed records have been mapped.

T. L. BLOCKEEL

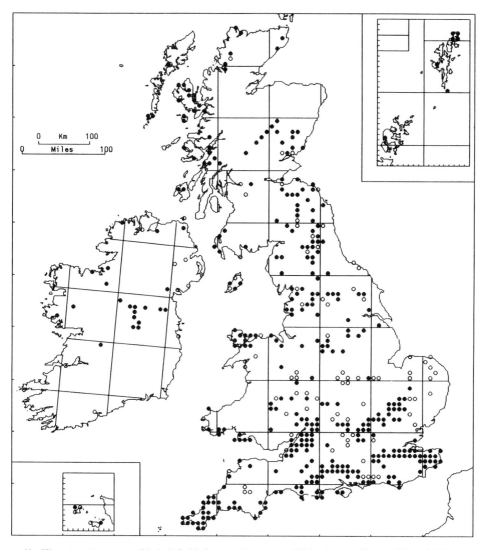

54/6a. **Weissia microstoma** (Hedw.) C. Müll. var. **microstoma** (*W. brachycarpa* (Nees & Hornsch.) Jur. var. *obliqua* (Nees & Hornsch.) M. Hill, *W. hedwigii* Crum)

It grows on earthy banks, on broken ground on chalk, limestone and other basic substrata, and in thin turf in grassland. Much more rarely, it occurs on basic soil among rocks in ravines and on montane crags. Overall, it occurs in a similar range of habitats to *W. controversa*, but is less common, more confined to open and unstable habitats, and more calcicolous. 0–430 m (Skye). GB 348+84*, IR 21+9*.

Autoecious; capules abundant, maturing in winter and spring.

Widespread in Europe, except much of Scandinavia. N. Africa, Caucasus. The species is widely scattered in eastern N. America, but the illustration in Crum & Anderson (1981) shows a form that should be ascribed to var. *brachycarpa*.

T. L. BLOCKEEL

305

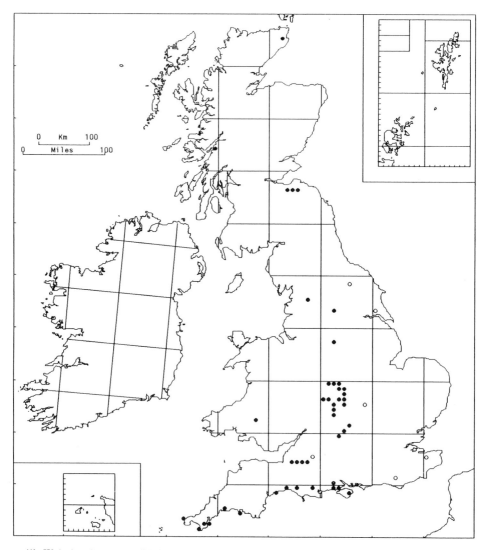

54/6b. **Weissia microstoma** (Hedw.) C. Müll. var. **brachycarpa** (Nees & Hornsch.) C. Müll. (*W. brachycarpa*
(Nees & Hornsch.) Jur. var. *brachycarpa*)

In wetter habitats than var. *microstoma*, and usually on non-calcareous clay, loam or marl. It is known from fields,
woodland rides and glades, and ditch-banks, pits and reservoir-margins. Lowland. GB 44+6*.
    Capsules abundant, maturing in spring.
    Widespread in Europe but distribution not well known. Eastern N. America (illustrated by Crum & Anderson,
1981).

T. L. BLOCKEEL

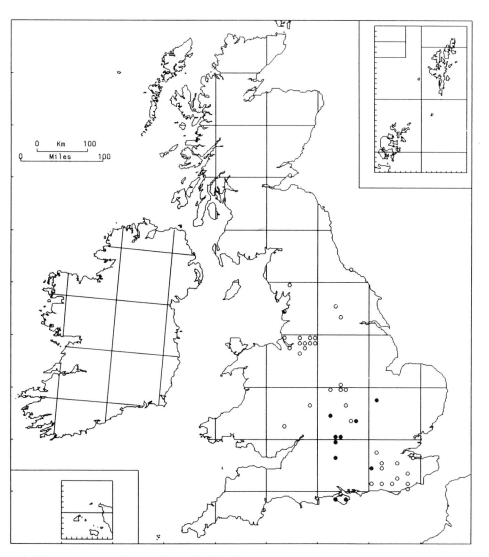

54/7. Weissia squarrosa (Nees & Hornsch.) C. Müll.

A rare species of moist, bare non-calcareous soil in fields, on fallow ground, by ditches and ponds, and in open ground in woods. It has been recorded from clayey and loamy soil, and muddy ground. Lowland. GB 10+39*.

Autoecious; capsules abundant.

Scattered in C. Europe and Scandinavia.

T. L. BLOCKEEL

307

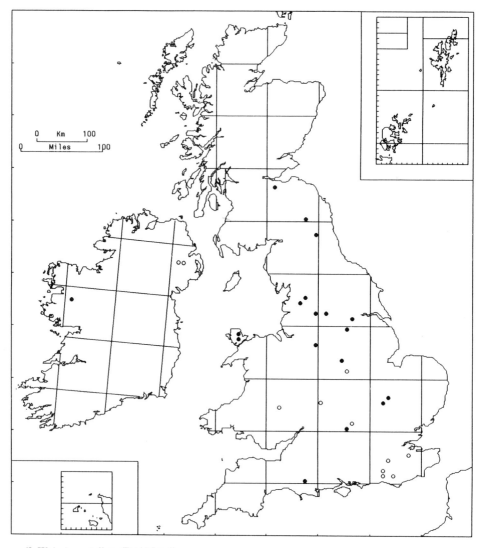

54/8. **Weissia rostellata** (Brid.) Lindb.

An ephemeral species which is an early colonist of moist bare ground. It is most frequent on gravelly ground and mud exposed in late summer and autumn on the margins of reservoirs and occasionally rivers, but has been recorded also in a number of other situations, including the banks of ditches, woodland rides, and bare patches and turfy hollows in fields and pastures. It is usually on clay or rich organic soil which is either water-retentive or kept moist by its physical situation. *Pseudephemerum nitidum* is a common associate, and by reservoirs it may occur in the communities characterized by *Physcomitrium sphaericum* and bulbiliferous *Pohlia* spp. Lowland. GB 17+9*, IR 1+2*.

Autoecious; capsules abundant.

A rather rare species of N.W., C. and E. Europe.

Some of the records are old, and although the species is ephemeral and mobile in its occurrence, there may have been a real decline in some districts.

T. L. BLOCKEEL

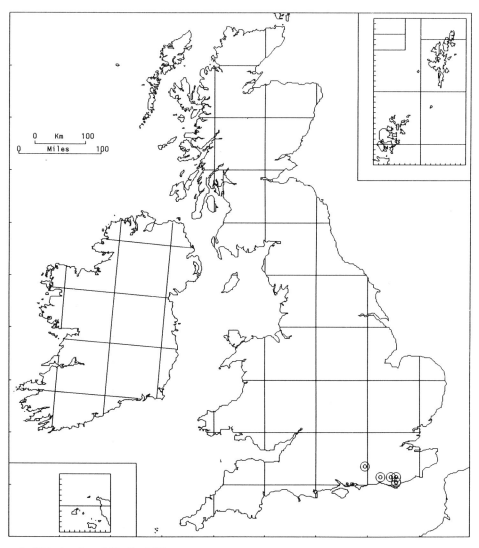

54/9. Weissia mittenii (Br. Eur.) Mitt.

This plant was recorded from fallow and stubble-fields, bare clay, and damp soil on woodland rides. Its habitats were similar to those of *W. multicapsularis*, with which it sometimes grew. Nicholson (1908) noted that both species required a recently exposed soil surface, and could not always be refound in successive seasons. Lowland. GB 5*.

Autoecious; capsules mature in spring; they are present in all gatherings but are often malformed.

Not recorded outside southern England.

Extinct; it was found in only five localities and has not been seen since 1920. Its taxonomic status is uncertain: it may be of hybrid origin.

T. L. BLOCKEEL

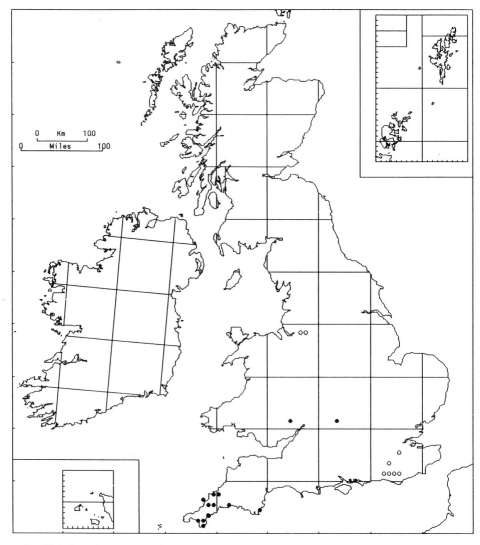

54/10. **Weissia multicapsularis** (Sm.) Mitt.

A rare species of moist bare ground in non-calcareous habitats. It occurs most commonly on clay, and has been recorded from woodland rides, fields, brick-pits, earthy banks and sea-cliffs. Lowland. GB 13+8*.

    Autoecious; capsules abundant, maturing in spring.

    Outside Britain known only from France.

T. L. BLOCKEEL

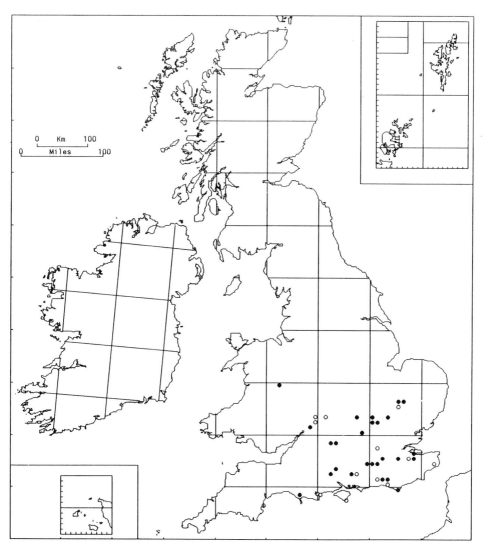

54/11. **Weissia sterilis** Nicholson

Occurring as scattered stems or patches on broken ground and in bare patches in turf. Unlike *W. rostellata* and *W. multicapsularis*, it is confined to dry calcareous ground on chalk and limestone, particularly on S.-facing slopes, with an outlying occurrence on dolerite in Wales. Lowland. GB 26+13*.

   Autoecious; capsules common, maturing in winter and spring.

   Elsewhere known only in N.E. France.

T. L. BLOCKEEL

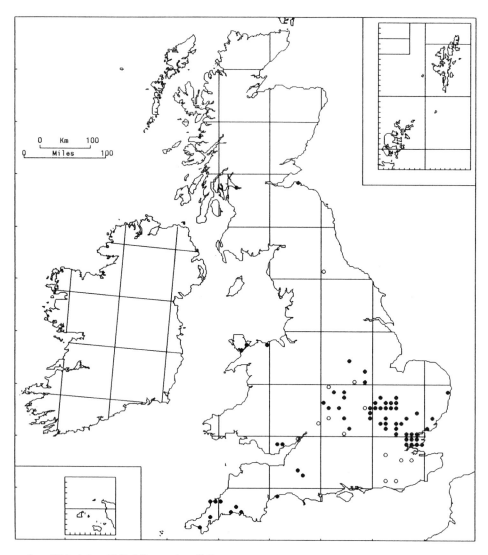

54/12a. **Weissia longifolia** Mitt. var. **longifolia**

A species of bare open habitats on non-calcareous loamy and sandy soil. It has been recorded from banks, ditches, stubble-fields, pastures, paths and gardens. In calcareous habitats it is replaced by var. *angustifolia*. Lowland. GB 66+13*.

Autoecious; capsules abundant, maturing in winter and spring.

Widespread in Europe, and adjacent parts of Africa and S.W. Asia.

T. L. BLOCKEEL

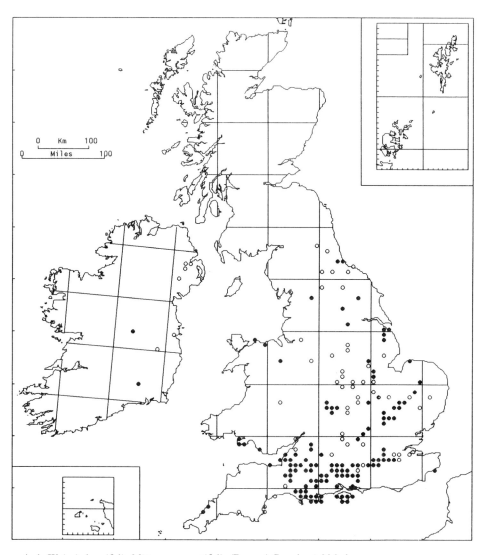

54/12b. **Weissia longifolia** Mitt. var. **angustifolia** (Baumg.) Crundw. & Nyholm

A rather rare plant of bare or disturbed calcareous soil. It is most common on chalk, occurring in turf and broken ground, in old pits and quarries, on banks and at the edge of paths. It is also found widely over limestones and other basic rocks. Lowland. GB 119+53*, IR 2+7*.

Autoecious; capsules abundant, maturing in late winter and spring.

Widespread in Europe, especially the south, and in adjacent parts of Africa and S.W. Asia. Continental plants are less clearly distinct from var. *longifolia* than British plants.

T. L. BLOCKEEL

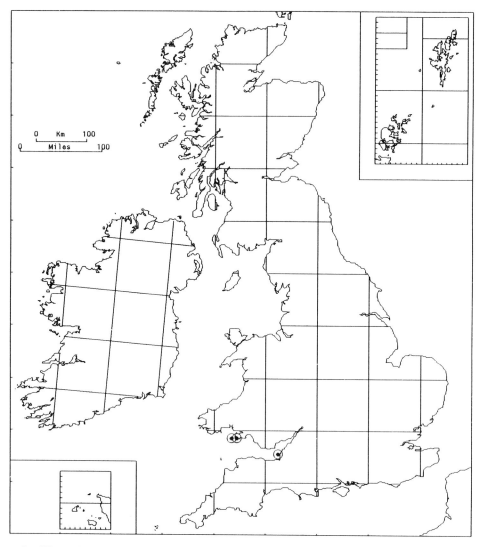

54/13. **Weissia levieri** (Limpr.) Kindb.

A very rare species, restricted to sunny S.-facing habitats on Carboniferous limestone. In the Gower Peninsula, S. Wales, it grows on soil in turf on cliff-tops and has also been recorded from a lane bank. On Brean Down in Somerset the habitat is on soil in rock crevices. Lowland. GB 3.

Autoecious; capsules common.

A rare but widely scattered species in S. and C. Europe, also known from adjacent parts of the Mediterranean region.

T. L. BLOCKEEL

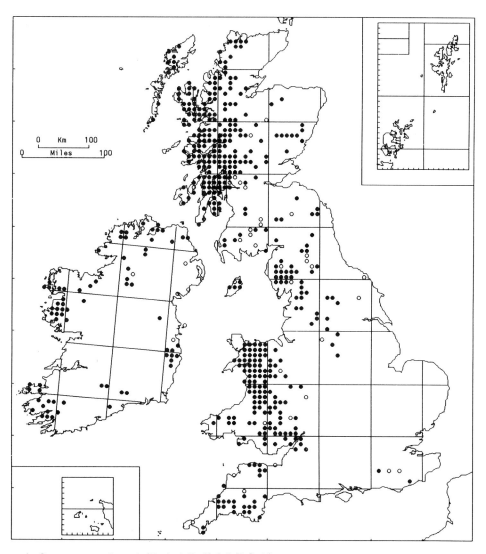

55/2. Oxystegus tenuirostris (Hook. & Tayl.) A. J. E. Smith

Unusual among Pottiaceae in occurring in base-poor situations. It nearly always grows on rocks (rarely tree roots) close to water, by streams, springs and oligotrophic upland pools and lakes. In southern and eastern districts it is characteristic of wooded streams, but in the uplands it also occurs on open moorland and mountain-slopes. In these areas it may grow on rock ledges on cliffs and in gullies. Although not found in strongly calcareous habitats, the species probably requires some slight mineral enrichment. 0–610 m (Ben More, Mull). GB 408+34*, IR 71+7*.

Dioecious; capsules are very rare, maturing in spring and summer. Rhizoidal tubers have been reported from continental material but have apparently not been confirmed in the British Isles.

Circumboreal, and occurring widely on tropical mountains. Widespread in S., W. and C. Europe, becoming rare in the Mediterranean region.

Var. *holtii* (Braithw.) A. J. E. Smith is of doubtful value (Smith, 1978) and is not distinguished from var. *tenuirostris* on the map.

T. L. Blockeel

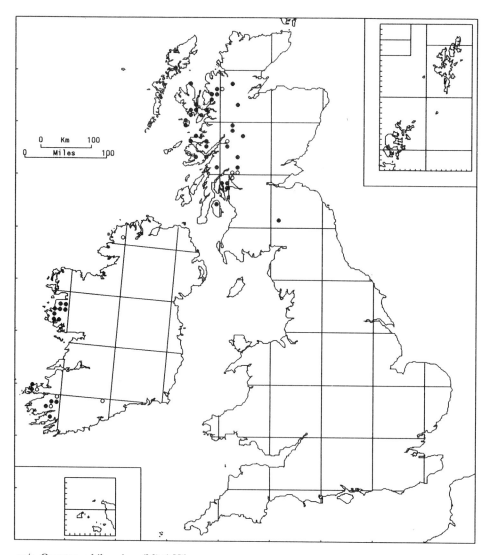

55/3. **Oxystegus hibernicus** (Mitt.) Hilp.

Occurs in two main types of habitat. In the lowlands it grows as scattered stems or small tufts among other bryophytes on wet ledges in humid ravines, among block-litter and, rarely, in sea-caves. It is most frequent in the vicinity of waterfalls, and may be associated with *O. tenuirostris*. At higher altitudes, it occurs on damp, often dripping base-rich ledges on N.- or E.-facing crags and in gullies. Here, *Leptodontium flexifolium* is a common associate. A less frequent habitat is intermittently flushed turf. 0–800 m (Beinn Dorain). GB 34+6*, IR 14+5*.

Dioecious; capsules are very rare.

Believed to be endemic to Britain and Ireland.

T. L. Blockeel

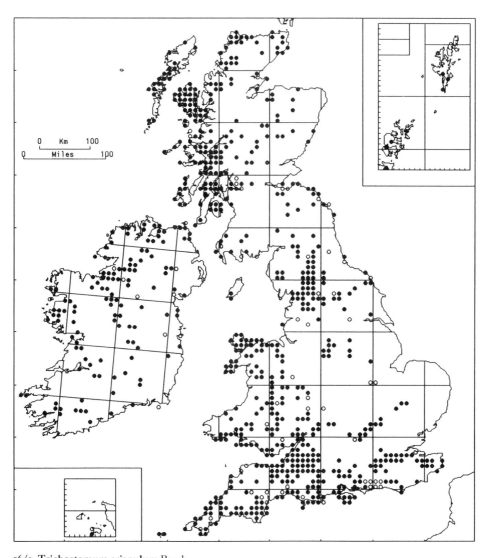

56/1. **Trichostomum crispulum** Bruch

A common species of base-rich rocks. It occurs frequently in the crevices of limestone outcrops, both in shade and on exposed crags, but also on a wide range of other substrata, sometimes only slightly basic. It grows in base-rich turf, including chalk grassland, on rocky banks, on sand-hills, and rarely on clay in woodland and on mine-waste. Although frequently in dry habitats it is sometimes found on wet rocks in gullies and on stream-banks. 0–1175 m (Ben Lawers). GB 601+53*, IR 121+10*.

Dioecious; capsules are rare, maturing in spring.

Widespread in Europe, especially in S., W. and C. regions. N. Africa, Asia, E. Canada.

T. L. BLOCKEEL

317

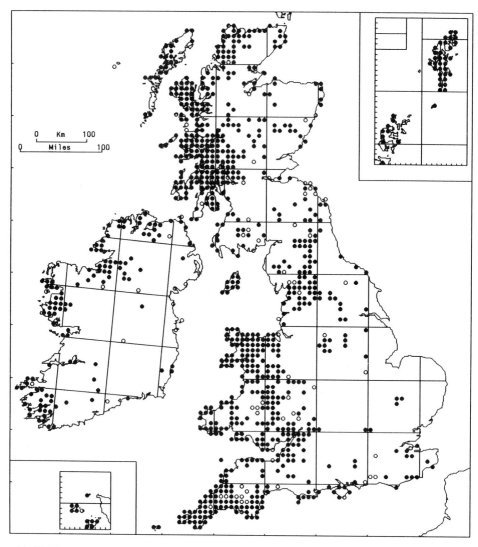

56/2. **Trichostomum brachydontium** Bruch

In coastal districts, especially in the west, this salt-tolerant species is indifferent to substratum, occurring widely both on acidic and base-rich rocks, in turf and on dunes, in exposed situations and in shade. Inland, however, it is calcicolous, growing typically on dry but shaded base-rich rocks in woodland and ravines, or, more rarely, on mountains or in calcareous turf, as on chalk banks. Mainly lowland, 0–700 m (Aonach Beag). GB 777+65*, IR 115+13*.

Dioecious; capsules are rather rare, maturing in spring.

Common in S. and W. Europe, becoming rare to the N. and E. and absent from most of Scandinavia. Widespread in Africa, Asia and S. America, but absent from N. America.

T. L. Blockeel

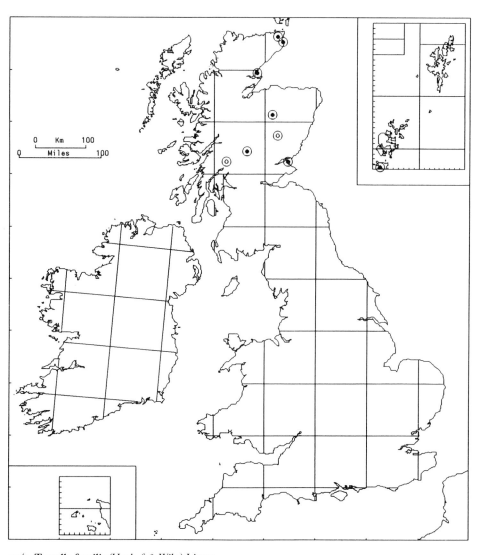

57/1. **Tortella fragilis** (Hook. f. & Wils.) Limpr.

This species is found in two quite separate habitats. Inland, it occurs on limestone and other basic rock-ledges in the mountains, but is apparently very rare there. It is somewhat commoner on the coast of N. and E. Scotland, where it occurs on damp sand in dune-slacks and on damp flats. 0–1060 m (Ben Lawers). GB 6+2*.

Dioecious; capsules are unknown in Britain. Gemmae are unknown: vegetative propagation occurs by means of the deciduous fragile leaf apices.

Circumboreal, frequent in the Arctic and extending to the High Arctic. Northern and montane parts of Europe. Lesotho and New Zealand (Campbell Is.).

This species has been confused by British bryologists with forms of *T. tortuosa*. The map is based on specimens determined by A. C. Crundwell or M. O. Hill.

T. L. BLOCKEEL

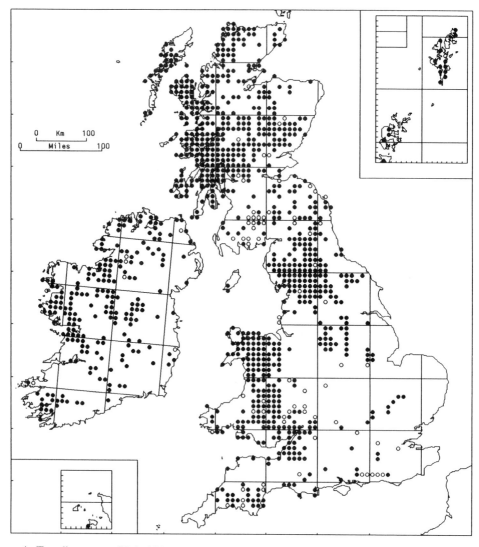

57/2. **Tortella tortuosa** (Hedw.) Limpr.

A locally abundant species growing in robust tufts and patches among sheltered base-rich rocks. Although most plentiful on upland limestone, for example on Carboniferous limestone in the Pennines, it is not a strong calcicole and may also occur on siliceous and other non-calcareous rocks with only slight base-enrichment. Habitats include wooded cliffs and boulders, rock ledges and crevices in ravines and gullies and on exposed crags and cliffs, recesses in scree and block-litter on mountain and moorland slopes, and old sheltered walls. In W. Scotland it grows in machair; in S. and E. England it is a rare component of grassland on moist, often N.-facing banks on chalk and limestone hills. 0–1200 m (Ben Lawers). GB 793+75*, IR 207+8*.

Dioecious; capsules rare, maturing in summer.

Circumboreal. Almost throughout Europe.

T. L. Blockeel

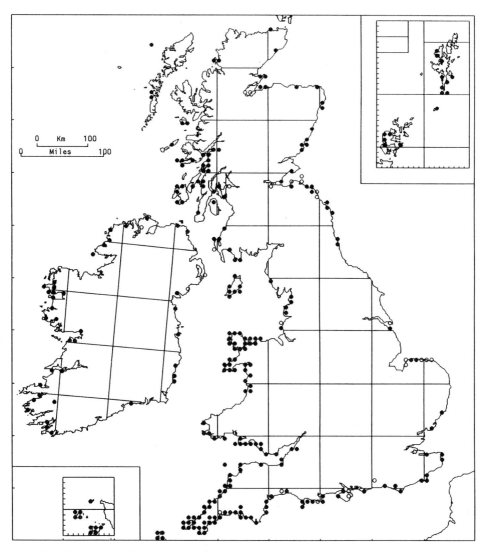

57/3. **Tortella flavovirens** (Bruch) Broth.

An exclusively maritime species in Britain, occurring on calcareous sand-dunes and sand-flats and on coastal banks. It may occur on stabilized sand-banks and shingle, and in soil-pockets and crevices among rocks, especially where the soil is light and sandy. In W. Scotland it occurs frequently with *Schistidium maritimum* on rocks regularly splashed by salt-spray. Lowland. GB 221+26*, IR 38+3*.

Dioecious; capsules are very rare.

A Mediterranean-Atlantic species in Europe, extending north to S. Scandinavia. Macaronesia, N. Africa, S.W. Asia, eastern N. America.

Var. *glareicola* (Christens.) Crundw. & Nyholm does not differ ecologically from the type and is much less common.

T. L. BLOCKEEL

321

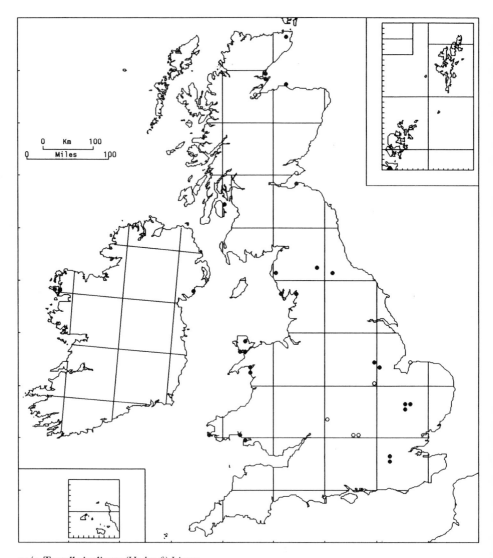

57/4. **Tortella inclinata** (Hedw. f.) Limpr.

In maritime districts this species usually grows on calcareous dunes. Inland it occurs on chalk banks, on thin soil on limestone rocks, and in open turf over limestone in long-established grassland. 0–500 m (Upper Teesdale). GB 26+7*, IR 3.

Dioecious; capsules are very rare. Gemmae are unknown, but short deciduous shoots are sometimes present. Most of Europe but rare in the Mediterranean region. N. Africa, Asia Minor, N. America.

T. L. BLOCKEEL

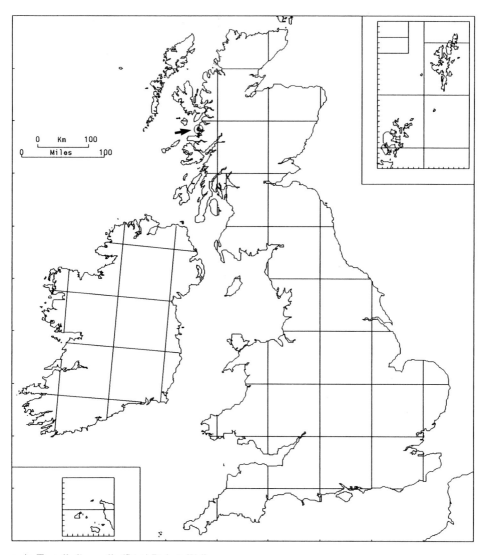

57/5. Tortella limosella (Stirt.) Rich. & Wall.

This species is known only from the type locality, a sandy seashore at Arisaig. GB 1*.

Sterile; related species are dioecious.

Unknown outside the British Isles.

It has not been collected recently. Its status in uncertain and it might possibly be an aberrant form of *T. flavovirens*.

T. L. Blockeel

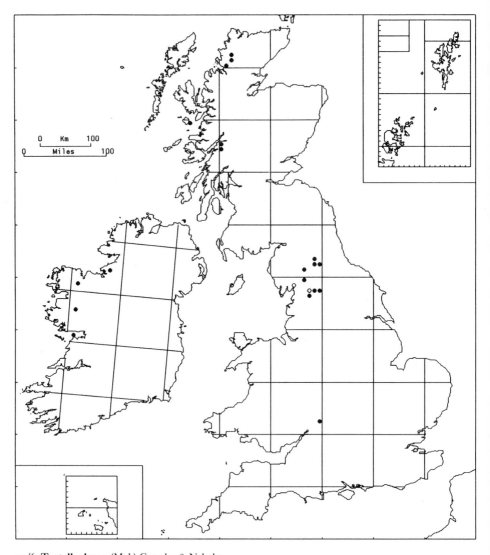

57/6. Tortella densa (Mol.) Crundw. & Nyholm

A rather rare species growing in the crevices of limestone rocks and more rarely on bare soil among rocks and in grassy flushes. It is a strong calcicole occurring on various limestones, but especially on the Carboniferous. Most stations are at moderate altitudes in exposed situations. It has recently been found on heavy-metal mine-waste in N.W. England. 0–580 m (Beinn Sgulaird). GB 15+3*, IR 4.

Dioecious; capsules are unknown.

N., W. and C. Europe. Caucasus.

T. L. Blockeel

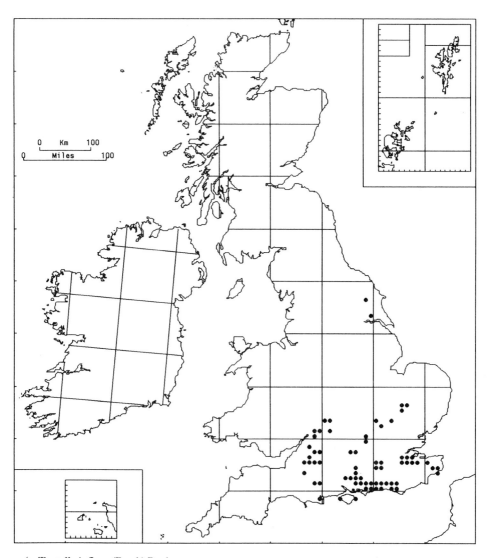

57/7. **Tortella inflexa** (Bruch) Broth.

Strictly confined to chalk and soft oolitic limestone, this species nearly always occurs on pebbles and detached fragments of rock, but sometimes also on larger stones. It may occur in pure patches, often covering the whole upper surface of small pebbles, or may be mixed with *Fissidens pusillus*, *Rhynchostegiella tenella* and *Seligeria paucifolia*. It has been found in woodland, on scrubby banks, in stubble-fields and in chalk and limestone grassland. Lowland. GB 68.

Dioecious; capsules are very rare.

A Mediterranean-Atlantic species, reaching its northern limit in Britain.

An inconspicuous species, not recognized in Britain until 1957 although the earliest herbarium specimen dates from 1904. It is probably native to Britain but may have extended its range during the present century.

T. L. BLOCKEEL

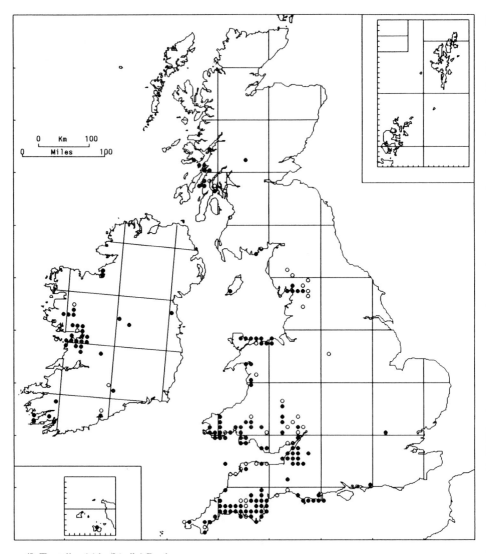

57/8. **Tortella nitida** (Lindb.) Broth.

A species of dry sunny rocks. It occurs widely on natural rock, often on limestone, but also on a range of other basic substrata. In some districts, where natural rock-outcrops are infrequent, the most usual habitat is on old wall-tops, especially those of limestone and other base-rich materials, including concrete. Lowland. GB 124+40*, IR 33+5*.

Dioecious; capsules unknown in Britain. Vegetative propagation by fragile leaves with deciduous tips.

A Mediterranean-Atlantic species, north to C. Europe and Scotland. Macaronesia, N. Africa, S.W. Asia.

The species has been confused in the past with forms of *T. tortuosa* and *T. flavovirens* and there are probably some erroneous records.

T. L. Blockeel

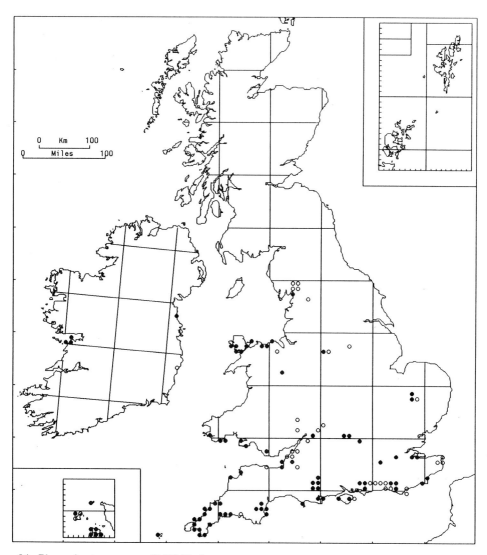

58/1. Pleurochaete squarrosa (Brid.) Lindb.

This species is most plentiful in coastal districts, on sandy and calcareous ground. It occurs on sand-dunes, in chalk and limestone grassland, in turf on sea-cliffs, and on stony limestone ground, rarely on limestone pavement. It may grow in scrubby ground and in light shade, but is most common in open habitats. 0–300 m (Peak District). GB 73+33*, IR 4+2*.

Dioecious; capsules and gemmae are unknown in Britain.

An abundant species in the Mediterranean region, extending north in W. Europe to the British Isles and the island of Gotland; also widespread in C. Europe. It occurs widely across the Northern Hemisphere, especially in regions with Mediterranean and sub-continental climates, south to E. Africa and Peru.

T. L. BLOCKEEL

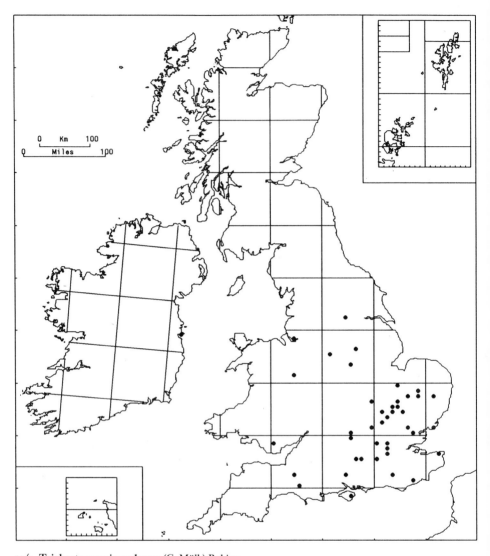

59/1. Trichostomopsis umbrosa (C. Müll.) Robins.

A plant of damp, shady calcareous places. A favourite habitat is on mortar at the base of walls that are backed by earth, such as railway arches. It is also recorded from shaded soil in old chalk or limestone quarries and on banks near chalk springs. Lowland. GB 41, IR 1.

Dioecious; only female plants have been found in Britain. Tubers occur on the rhizoids.

Spain, Portugal. N., C. and S. America.

Probably an introduction from America, first discovered in Britain near Winchester in 1958 but not published till twenty years later (Crundwell & Whitehouse, 1978). The European population is entirely female; a report of sporophytes in the Canaries (Long *et. al.*, 1981) is now thought to be erroneous (D. G. Long, pers. comm.). *T. umbrosa* is under-recorded because of its small size and bryologists' lack of attention to wall-bases.

H. L. K. WHITEHOUSE

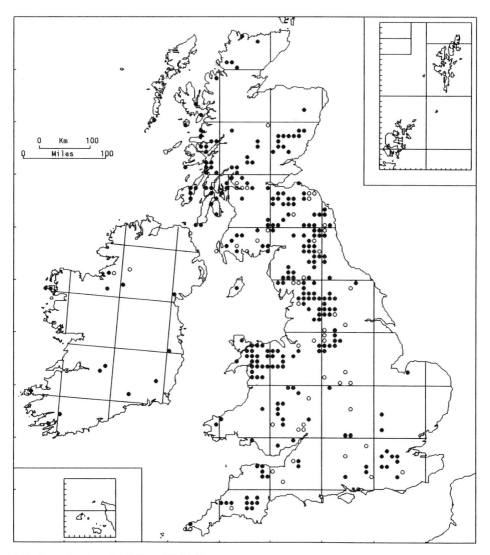

60/1. Leptodontium flexifolium (With.) Hampe

A species of well-drained raw humus or peaty sand on heath and moorland, in acid grassland and in open woodland, a characteristic habitat being thin soil on top of boulders. It occurs also on other decaying organic substrata, including old thatched roofs and wood. Localities are usually well illuminated and the species does not occur in deep shade. 0–750 m (Pen yr Ole Wen, Snowdonia). GB 257+51*, IR 14+4*.

Dioecious; capsules occasional, maturing in spring. Vegetative propagation is by deciduous bulbiform branchlets; narrowly flask-shaped gemmae are sometimes produced on protonema in leaf axils, but are apparently rare.

W. Europe, from the Pyrenees north to Denmark. Widespread in the tropics and sub-tropics, north to C. Asia and southern U.S.A.

Gemmae of different shape are produced in leaf axils of some continental and exotic populations. The taxonomic implications of this variation require further elucidation (Newton & Boyce, 1987).

T. L. BLOCKEEL

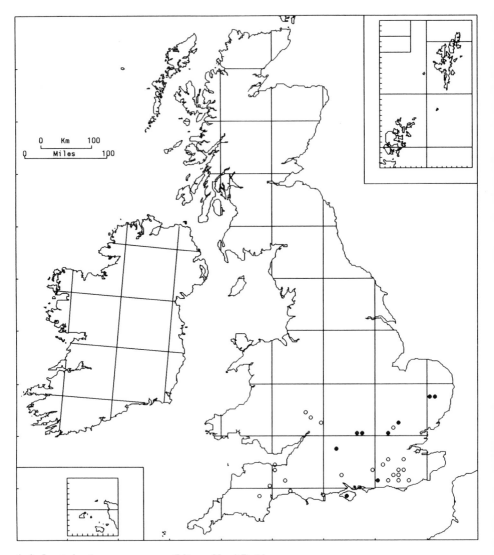

60/2. **Leptodontium gemmascens** (Mitt. ex Hunt) Braithw.

A species of decaying vegetation, characteristically on old thatched roofs but also on wood. It has recently been found in acid grassland and heath, where it occurs at the base of grassy tussocks, on decaying *Juncus*, and even on rabbit-droppings. In these habitats associated species include *Ceratodon purpureus*, *Dicranum scoparium*, *Hypnum jutlandicum*, *Pohlia nutans* and *Polytrichum juniperinum*. Lowland. GB 9+22*.

Dioecious; capsules unknown; no data on gametangia. Gemmae abundant, produced in leaf axils as well as at leaf apices (Frahm, 1973).

W. Europe, from Pyrenees to Denmark, very rare or overlooked. Outside Europe known only from Marion Island in the sub-Antarctic.

It is apparently decreasing because of the scarcity of suitably decayed thatch.

T. L. Blockeel

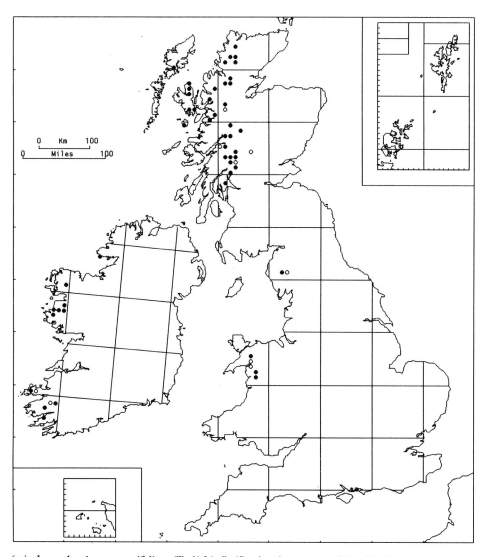

60/3. Leptodontium recurvifolium (Tayl.) Lindb. (*Paraleptodontium recurvifolium* (Tayl.) Long)

A montane species, growing in more or less pure tufts or scattered among other bryophytes and higher plants, on and at the base of cliffs, on rock ledges and banks, and in rock crevices, often where water drips from above or within the spray zone of a waterfall. The aspect is usually between north and east, and the substratum is probably always base-enriched. 100–720 m (Aonach Beag). GB 37+7*, IR 11+3*.

Dioecious; capsules unknown; according to Dixon (1924), abortive archegonia have been found in Britain. Gemmae are absent but the leaves are fragile and serve as a means of vegetative propagation.

Outside Britain known only from western N. America (British Columbia and Alaska).

T. L. BLOCKEEL

331

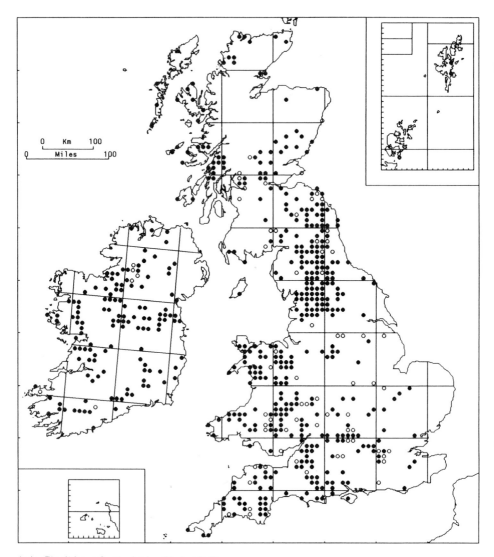

61/1. **Cinclidotus fontinaloides** (Hedw.) P. Beauv.

This species occurs most plentifully on limestone rocks in and by streams and rivers, especially in upland districts. In such places it may smother stones and boulders near water-level, often in association with *Orthotrichum cupulatum* and *Schistidium alpicola*. It also occurs on other basic rocks and on wood, always close to running water or on irrigated surfaces. Localities are usually well illuminated, and the species is most plentiful in periodically submerged sites, rarely if ever growing permanently under water. It is notably abundant in the turloughs (seasonal lakes) on the limestone of W. Ireland. In lowland Britain, where streams are slow flowing, it grows mainly by weirs and locks, and sometimes also on embankment-walls and tree roots. 0–300 m (near Malham Tarn). GB 398+66*, IR 128+6*.

Dioecious; capsules are frequent in upland districts, occasional in the lowlands, maturing in spring and summer.

Widespread in Europe, north to C. Scandinavia. Macaronesia, N. and E. Africa, W. Asia.

T. L. Blockeel

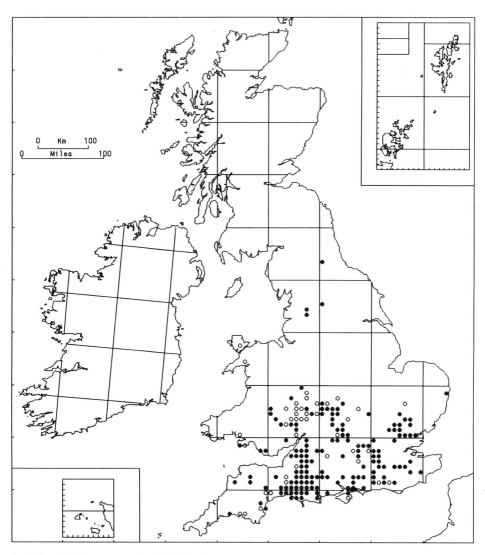

61/3. Cinclidotus mucronatus (Brid.) Mach.

Usually occurring on tree roots and bases on the banks of streams and rivers that are subject to periodic flooding, and often embedded in alluvial sand or silt. It is less common on stones and rocks, although at The Strid, Bolton Abbey, it grows downwards from the under-surface of grit rocks overhanging the R. Wharfe. *Tortula latifolia* is a common associate. Rarely, it is also found far removed from running water, on walls, stones and tree-bases. Lowland. GB 152+42*.

Dioecious; capsules rare in Britain, maturing in spring and summer.

S., W. and S.E. Europe, not reaching Scandinavia. Macaronesia, N. Africa, S.W. Asia.

T. L. BLOCKEEL

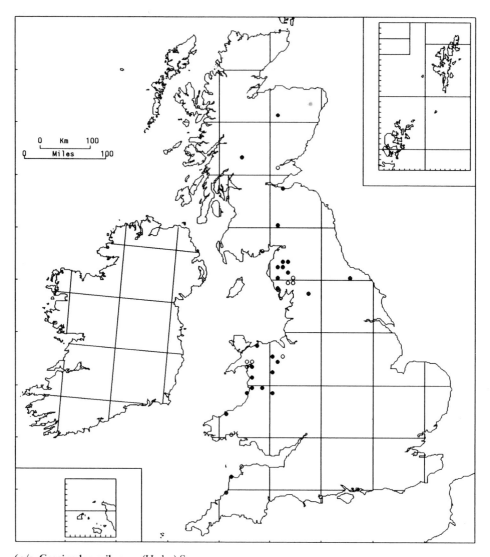

62/1. **Coscinodon cribrosus** (Hedw.) Spruce

C. cribrosus is found on dry, very acidic, frequently friable slate or shale, in outcrops or on walls. It also sometimes occurs on harder rocks such as granite. 0–500 m (Selcoth Burn). GB 26+11*.

Dioecious; capsules rare, spring.

In Europe it reaches from about 69°N in Fennoscandia south to Sardinia and Sicily. Widespread in the Northern Hemisphere but scattered and rare through much of its range, extending from the High Arctic in N. Ellesmere Island (Canada), south to N. Africa, Turkey, Himalaya and Arizona (U.S.A.).

A. J. E. SMITH

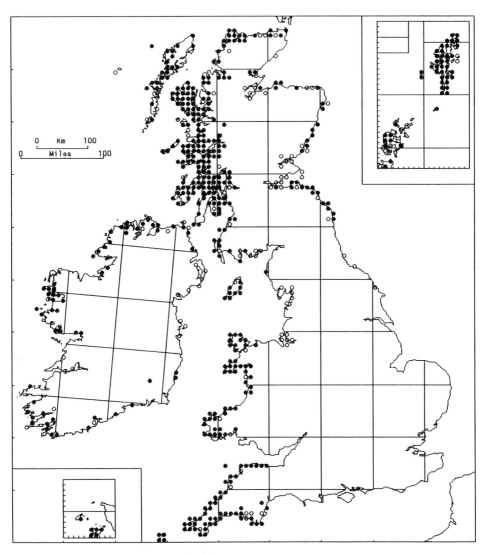

63/1. **Schistidium maritimum** (Turn.) Br. Eur.

This is the most markedly maritime species in the bryophyte flora of the British Isles. It is common and locally abundant in dry or periodically moist crevices and ledges of boulders, rocks and sea-cliffs above the main *Xanthoria parietina* zone but well within the normal extent of sea-spray. The zone occupied by *S. maritimum* varies according to the degree of exposure of the site and can extend from sea-level to 240 m on the exposed basalt sea-cliffs of western Skye. Common associates in exposed situations include *Armeria maritima*, *Sedum anglicum*, *Ulota phyllantha* and *Ramalina* spp., whereas in sheltered areas and crevices they are *Asplenium marinum* and *Trichostomum brachydontium*. It also occurs as a low-growing sterile form in closed but grazed upper salt-marsh turf, with *Carex extensa*, *Festuca rubra*, *Glaux maritima*, *Juncus gerardii* and *Plantago maritima*. The only inland record mapped is from a wall by the R. Barrow near Graiguenamanagh in Co. Carlow. Seemingly indifferent to rock type except that it is very rare on limestone. Lowland. GB 388+84*, IR 56+19*.

Autoecious; sporophytes common.

Coasts of N. and W. Europe south to France and east to the Baltic and N. European Russia. Japan, Aleutian Islands, eastern and western N. America. Its world distribution is mapped by Bremer (1980).

H. J. B. BIRKS

335

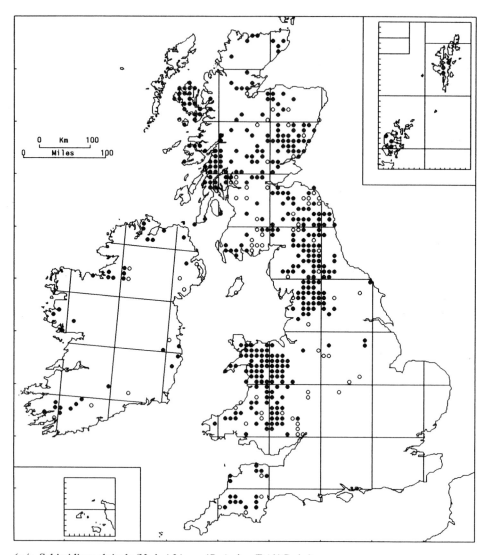

63/2. **Schistidium alpicola** (Hedw.) Limpr. (*S. rivulare* (Brid.) Podp.)

Occurs as large (up to 10 cm) often pendent tufts at or slightly above water-level on rocks, boulders, large stones, and tree roots in fast-flowing rivers and large streams, sometimes also in small streams and rocky ravines. In base-rich areas it is commonly associated with *Brachythecium plumosum*, *Cinclidotus fontinaloides* and *Fontinalis antipyretica*, whereas in more acid upland areas it grows with *Amblystegium fluviatile*, *Fontinalis squamosa*, and *Racomitrium aciculare*. Also occurs on reservoir-walls, on boulders by lakes, and, more rarely, on dripping basic cliffs. 0–760 m (Glen Callater). GB 386+69*, IR 31+9*.

Autoecious; sporophytes common, ripe spring.

N. and C. Europe. Azores, C. Africa, W. Asia, Siberia, Himalaya, Japan, N. America, Greenland, southern S. America.

Small tufted forms on limestone or other base-rich rocks are often referred to var. *alpicola*, while longer trailing forms are referred to var. *rivulare* (Brid.) Limpr. In a recent revision, Bremer (1980) recognized two infraspecific taxa, *S. rivulare* ssp. *rivulare* and *S. rivulare* ssp. *latifolium* (Zett.) B. Bremer, treating *S. alpicola* as a *nomen confusum*. The common plant in Britain and Ireland is ssp. *rivulare*, but ssp. *latifolium* is known from several localities in Britain.

H. J. B. BIRKS

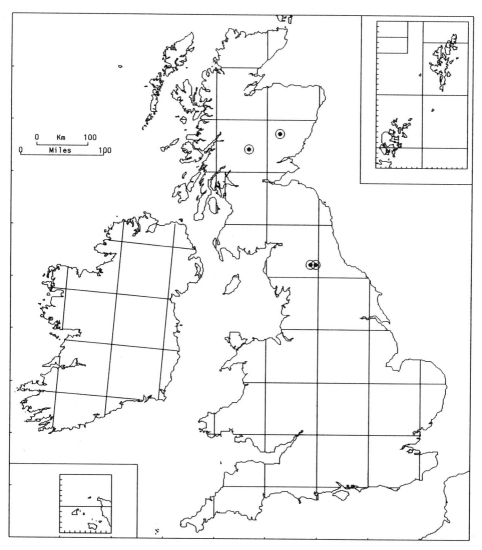

63/3. **Schistidium agassizii** Sull. & Lesq.

This taxon has two distinct habitats in Britain. On Ben Lawers, it occurs as pure tufts on mica-schist boulders in and around high-altitude hollows that are seasonally flooded by snow melt-water. Associates include *Carex saxatilis*, *Marsupella emarginata*, *Scapania undulata* and *Calliergon sarmentosum*. Elsewhere in Scotland and N. England it occurs in its more typical Scandinavian habitat, namely on rocks and stones at or near the flood-zone of fast-flowing rivers and large streams, growing with *Brachythecium plumosum*, *Fontinalis antipyretica*, *Schistidium alpicola* and the lichen *Dermatocarpon aquaticum* (cf. Holmes, 1976). 300–1100 m (Ben Lawers). GB 4.

Autoecious; sporophytes common.

N., W. and C. Europe, from Iceland and Spain east to Estonia and Karelia. Western and eastern N. America, Greenland.

A morphologically variable plant, its form possibly depending on rate of water flow and degree of submergence. The first British record was from Ben Lawers (Birks & Birks, 1967).

H. J. B. BIRKS

337

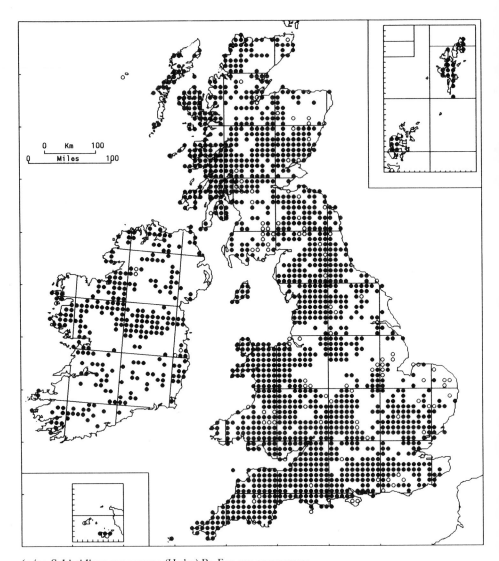

63/4a. **Schistidium apocarpum** (Hedw.) Br. Eur. var. **apocarpum**

It occurs in both natural and man-made habitats in the lowlands and uplands, growing on a wide range of calcareous or mildly basic rocks, including limestone, gneiss, schist, sandstone, gabbro, basalt and other igneous rocks. It is commonly found on dry, often sun-exposed, or periodically moist, even flushed, sloping rock-outcrops and boulders, often with *Homalothecium sericeum*, *Orthotrichum anomalum* and *Tortula subulata*. It also occurs on dry basic rocks and boulders in ravines, wooded valleys and, less commonly, streams, and in crevices and hollows of dry rocks and clints in bare limestone pavement with *Bryoerythrophyllum recurvirostrum*, *Encalypta streptocarpa* and *Tortella tortuosa*. It is common on dry limestone walls with *Bryum capillare* and *Homalothecium sericeum*, and on tombstones, bridges, concrete slabs, wall-tops, derelict masonry and other stonework, often with *Grimmia pulvinata* and *Tortula muralis* on mortar. It is sometimes found on rocks in quarries and old mines; more rarely it occurs on trees and tree roots along river banks. Most abundant in limestone districts but common in many non-calcareous areas. 0–1050 m (Ben Lawers). GB 1524+108*, IR 283+5*.

Autoecious; sporophytes abundant.

Cosmopolitan.

H. J. B. BIRKS

338

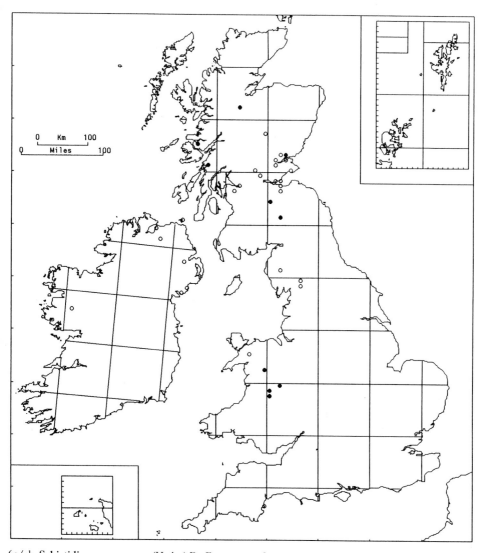

63/4b. **Schistidium apocarpum** (Hedw.) Br. Eur. var. **confertum** (Funck) Möll.

On dry basic igneous rocks in the north and west; also recorded from calcareous sandstone cliffs and roadside boulders, all at low to moderate elevations. GB 10+19*, IR 7*.

Autoecious; sporophytes abundant.

Europe, Iceland. N. Africa, Caucasus, N. and C. Asia, Japan, N. America, S. America (Patagonia), Falkland Islands.

A poorly understood taxon which has been inadequately recorded.

H. J. B. Birks

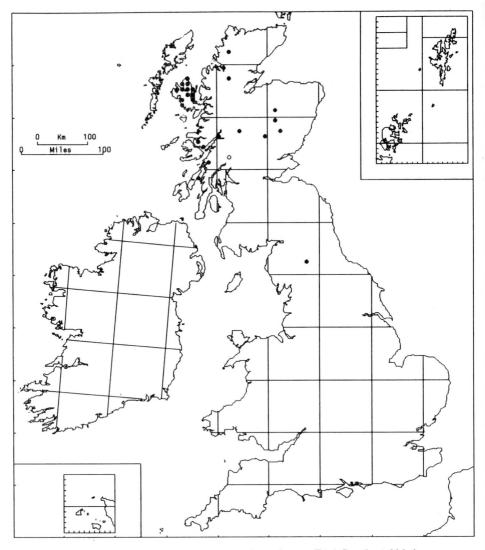

63/4c. **Schistidium apocarpum** (Hedw.) Br. Eur. var. **homodictyon** (Dix.) Crundw. & Nyholm

On dry, sun-exposed S.- or W.-facing boulders or earthy, often crumbling ledges and rock-faces of limestone, or on cliffs of calcareous schist, basalt or andesite, with *Amphidium lapponicum*, *Anomobryum concinnatum* and *Grimmia funalis*. It also occurs on dry calcareous basalt walls of low-lying ravines in northern Skye with *Orthotrichum rupestre* and *Tortula subulata*. 0–750 m (Clova). GB 26+1*.

Autoecious; sporophytes common.

Sweden.

A very variable and difficult taxon of uncertain status.

H. J. B. Birks

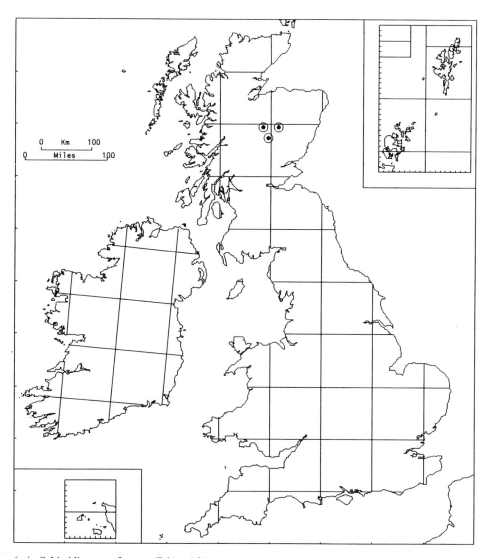

63/5. Schistidium atrofuscum (Schimp.) Limpr.

It occurs as large, dark, readily disintegrating tufts frequently mixed with silt and small stones on dry, crumbling, W.- or S.-facing, strongly calcareous schists or metamorphosed limestone cliffs. Associates include *Desmatodon leucostoma, Myurella julacea, Rhytidium rugosum, Schistidium apocarpum, S. strictum* and *Stegonia latifolia*. 550–650 m. GB 3.

Autoecious; sporophytes occasional.

Spain, Norway, Alps, Tatra Mts, Balkans. Turkey, N. America.

A local and curiously restricted plant throughout its known range. It was first recorded in the British Isles by Warburg (1957) and is possibly overlooked.

H. J. B. Birks

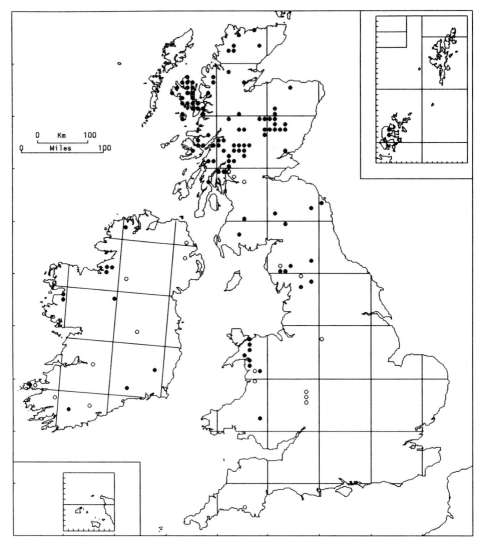

63/6. Schistidium strictum (Turn.) Loeske

Occurs on dry or periodically moist, basic inland cliffs of limestone, sandstone, calcareous schist, basalt or related igneous rocks, often N.- or E.-facing, and frequently growing with *Amphidium lapponicum*, *Neckera crispa* and *Tortula tortuosa*. More rarely it occurs on shaded basic rock-walls of low-lying ravines or on sheltered basic sea-cliffs. 0–1000 m (Snowdon). GB 102+13*, IR 11+8*.

Autoecious; sporophytes frequent.

Europe. Caucasus, Siberia, Himalaya, China, Japan, N. America.

A variable plant in size, colour and growth-form.

H. J. B. Birks

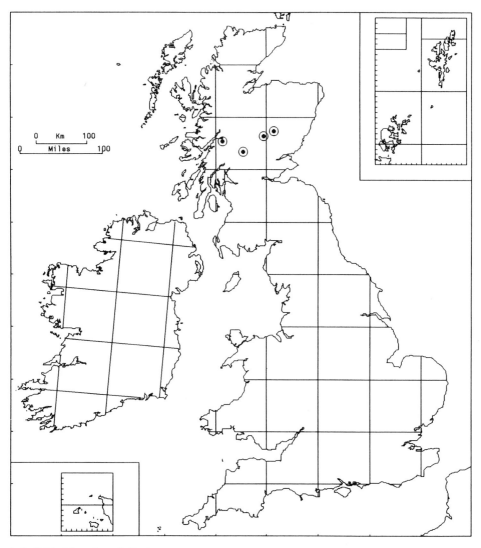

63/7. **Schistidium boreale** Poelt

On dry or seasonally moist, calcareous schist cliffs and boulders, growing with other montane calcicoles such as *Polystichum lonchitis*, *Saxifraga aizoides*, *Anomobryum concinnatum*, *Hypnum bambergeri*, *Pseudoleskeella catenulata* and *Schistidium trichodon*. 400–850 m (Caenlochan Glen). GB 4.

Autoecious; sporophytes frequent.

Fennoscandia, N. Russia, Czechoslovakia. Novaya Zemlya.

For details of its discovery in the British Isles, see Warburg (1965). In view of its habitat requirements, it is likely to be commoner in the Scottish Highlands than the map suggests.

H. J. B. BIRKS

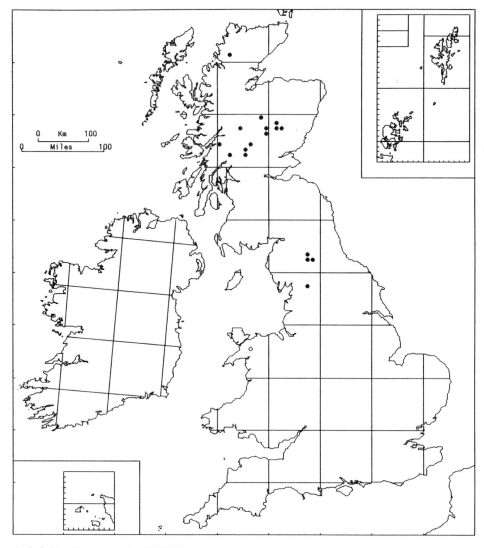

63/8. **Schistidium trichodon** (Brid.) Poelt

A very local plant of dry, often S.- or W.-facing, or seasonally damp limestone, calcareous schist or volcanic tuff cliffs, growing with calcicoles such as *Asplenium trichomanoides*, *Scapania aspera*, *Ctenidium molluscum*, *Ditrichum flexicaule*, *Pseudoleskeella catenulata*, *Schistidium apocarpum* (including var. *homodictyon*) and *Tortella tortuosa*. 100–1190 m (Ben Lawers). GB 17+1*.

Autoecious; sporophytes common.

Fennoscandia and C. Europe. E. Canada.

Probably overlooked in parts of the C. Highlands, although curiously absent from many suitable-looking localities. See Crundwell (1959) for a discussion of its taxonomic status.

H. J. B. Birks

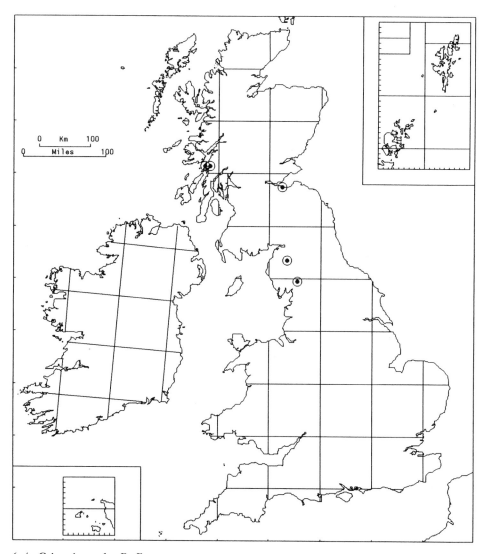

64/1. Grimmia anodon Br. Eur.

Although a very rare species, where *G. anodon* does occur it may do so in quantity. It grows as very small tufts or hemispherical cushions in crevices on exposed, more or less vertical basic rocks on cliffs, in quarries and, in one locality, on a road-bridge across a railway. It is evidently able to withstand pollution, occurring not only in the railway locality but also on cliffs near the centre of Edinburgh. 150–250 m (Arthur's Seat). GB 5.

Autoecious; capsules common, spring.

Throughout Europe, north to Spitsbergen. Very widely distributed in the Northern Hemisphere, from the High Arctic south to Tenerife, N. Africa, Iran, Himalaya and northern Mexico.

In the northern part of its range it is mostly scattered and rare, but it is common in semi-deserts further south. According to Flowers (1973), it is the commonest lowland moss in Utah, where it occurs abundantly in deserts and on mountain slopes up to 2000 m. Likewise it is common in mountainous parts of C. Asia.

A. J. E. SMITH

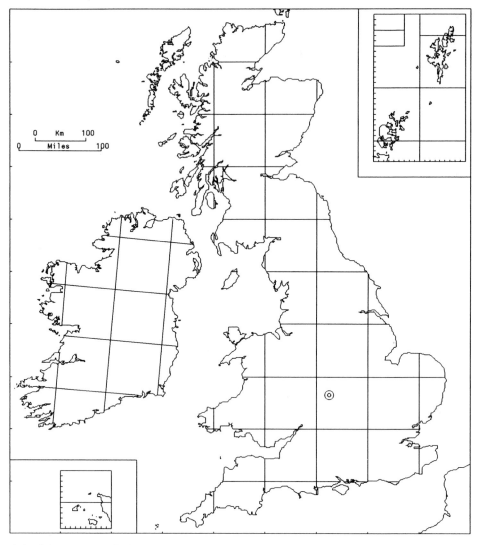

64/2. Grimmia crinita Brid.

Forming low grey patches (like mouse-fur) on limestone walls or mortar, *G. crinita* was almost certainly introduced at its only known British locality, where it grew on the dry mortared wall of a canal-bridge. Lowland. GB 1*.

Autoecious; capsules common, spring.

Mediterranean Europe north to Poland and Baltic Russia. Tenerife, N. Africa, Sinai, Palestine, Iran, Caucasus.

It is now extinct in Britain, records extending from 1872 to 1882. In 1874 it was described as occurring 'very sparingly and only in one spot'.

A. J. E. Smith

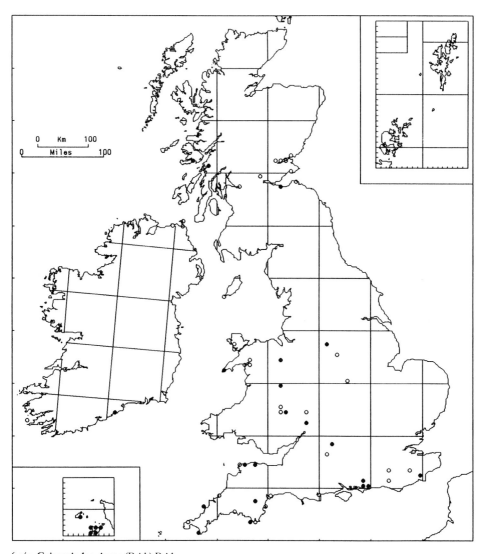

64/3. **Grimmia laevigata** (Brid.) Brid.

On exposed acidic to slightly basic rock, on cliffs, on outcrops, by rivers, on sarsen-stones and on roofing slates, particularly near the coast. 0–350 m (Roundton). GB 20+30*, IR 1+4*.

Dioecious; capsules rare, spring.

Europe north to S. Fennoscandia. Macaronesia, N., E. and S. Africa, Asia, Hawaii, N. America, Mexico, Brazil, Australasia.

G. laevigata is a decreasing species. It may be adversely affected by acid precipitation, as well as by eutrophication of rock surfaces by nitrates in agricultural chemicals and rainwater.

A. J. E. SMITH

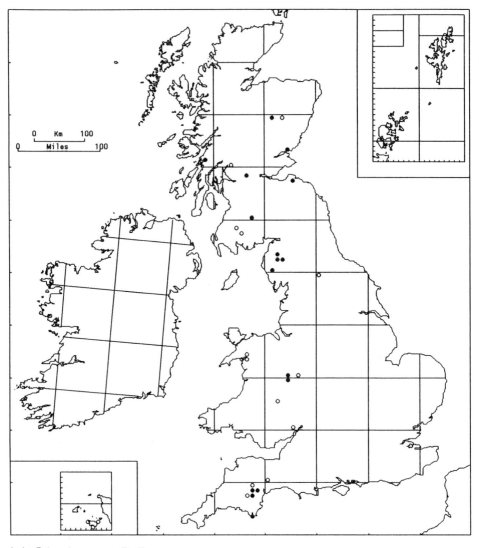

64/4. **Grimmia montana** Br. Eur.

On hard, exposed rocks ranging in type from serpentine to acidic siliceous. 0–600 m (Clogwyn du'r Arddu). GB 16+14*.

Dioecious; capsules rare, spring.

Europe north to Fennoscandia. Azores, Morocco, N. and C. Asia, western N. America, Greenland.

Like several other *Grimmia* species, *G. montana* is decreasing.

A. J. E. SMITH

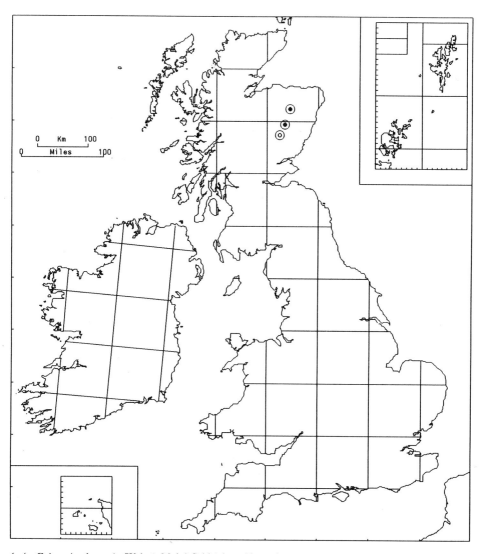

64/5. **Grimmia alpestris** (Web. & Mohr) Schleich. ex Hornsch.

On dry serpentine rock, forming small tight tufts. 580 m. GB 2+1*.
   Autoecious in Britain; capsules common, spring, summer.
   A montane species extending from Spain, Corsica and Romania north to Spitsbergen. Asia (including Turkey and Cyprus), N. America.

A. J. E. Smith

349

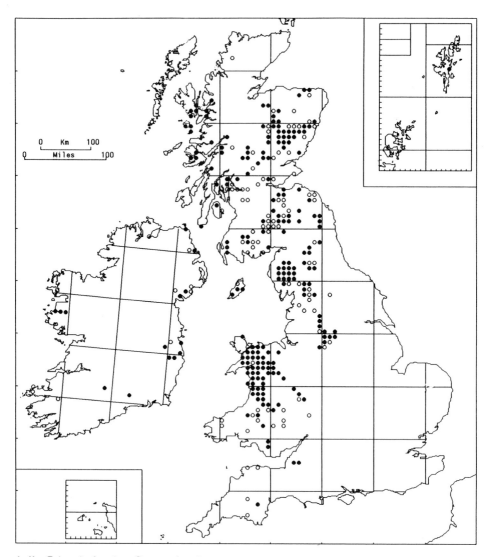

64/6a. **Grimmia donniana** Sm. var. **donniana**

A small plant growing as scattered shoots or small, sometimes coalescing tufts in crevices of hard acidic and especially slatey rock. It occurs on montane rock-outcrops and cliffs in exposed situations, in boulder-scree, and at lower altitudes on walls and waste from quarries and mines. It is possible that *G. donniana* is restricted to hard rocks because of its slow growth-rate. It is frequently found on rocks with a high concentration of heavy metals, although it is not necessarily an indicator of such (Brown & Bates, 1972). 0–950 m (Cairngorms). GB 198+91*, IR 14+9*.

Autoecious; capsules common, spring to autumn.

Boreal-montane, discontinuously circumpolar with a Southern Hemisphere disjunction. Europe north to Spitsbergen. Madeira, Tenerife, Caucasus, Himalaya, Far East, N. America, Greenland; southern S. America, Antarctica.

A. J. E. Smith

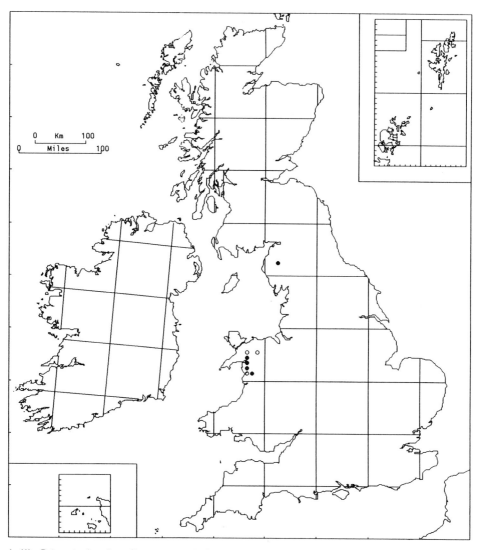

64/6b. **Grimmia donniana** Sm. var. **curvula** Spruce

It occurs in crevices of hard slatey rocks, especially on roadside walls. Lowland. GB $5+3^*$.

    Autoecious; capsules common, April.

    Rare in Europe from Spain, Italy and Romania north to Norway and Finland.

    Decreasing because of the destruction of roadside walls.

A. J. E. Smith

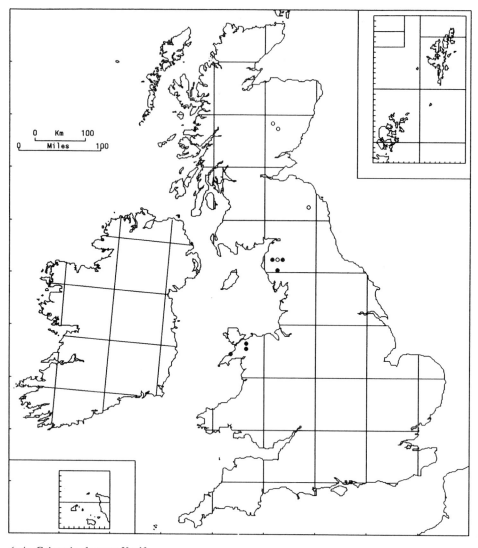

64/7. **Grimmia elongata** Kaulf.

G. elongata is a rare species forming small tufts on damp or dry acidic rocks at high altitudes. 500–1080 m (Snowdon). GB 6+4*.

Dioecious; capsules unknown in Britain.

Montane and subarctic Europe from Italy and Yugoslavia to about 67°N. Asia including Turkey, Caucasus, Mongolia, Japan and the Siberian Arctic; W. Canada, N. Ellesmere Island.

A. J. E. Smith

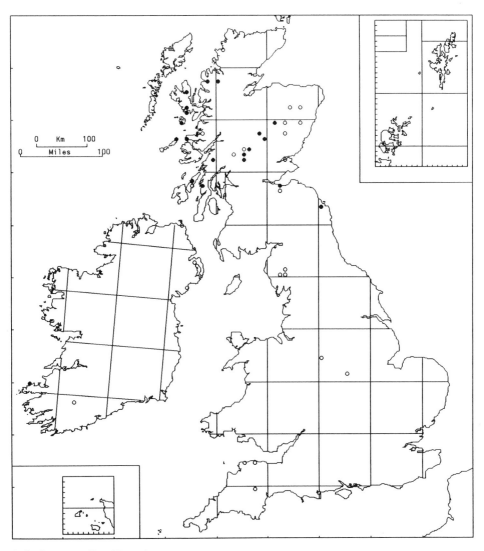

64/8. Grimmia affinis Hornsch.

Usually in small quantity, forming tufts or cushions on dry exposed acidic to ultrabasic rock. It occurs on boulders, on cliffs, in scree, and on rocks by lakes or, rarely, by the sea. 0–550 m (Kirkton Glen, Balquhidder). GB 19+21*, IR 1+4*.

Autoecious; capsules frequent, winter.

Montane and Arctic Europe from Spain and Italy north to Spitsbergen. Tenerife, Morocco, Kenya, W. and C. Asia, Himalaya, Sri Lanka, N. and C. America, southern S. America, Greenland.

G. affinis, like several other *Grimmia* species, is apparently declining. Because of nomenclatural and taxonomic confusion with *G. ovalis*, it was over-recorded in the past and some mapped pre-1950 records may be erroneous. A markedly continental species, which is commonest on dry hills and mountains, as in C. Asia and the semi-arid parts of the U.S.A.

A. J. E. SMITH

353

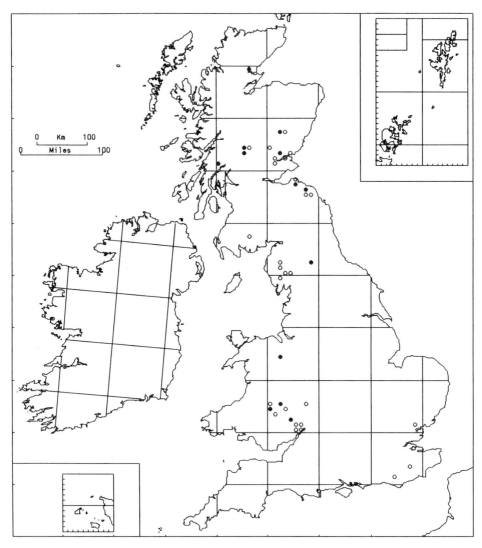

64/9. **Grimmia ovalis** (Hedw.) Lindb. (*G. commutata* Hüb.)

It forms tufts or cushions on dry exposed basic to neutral rocks, especially basalt and dolerite. It also occurs on rocks by water and, in the past, on roofing slates. Usually at low altitudes. GB 13+28*.

Dioecious; capsules rare, spring.

Europe north to N. Fennoscandia. Madeira, Tenerife, N., C. and S. Africa, Asia including Cyprus, Turkey, Caucasus, Siberia, Mongolia, Himalaya and Japan; N. and S. America.

Like several other *Grimmia* species, *G. ovalis* is decreasing. A population consisting entirely of male plants has colonized concrete bunkers in the Netherlands, apparently dispersed by small plantlets which are produced in the upper leaf axils (Greven, 1990).

A. J. E. SMITH

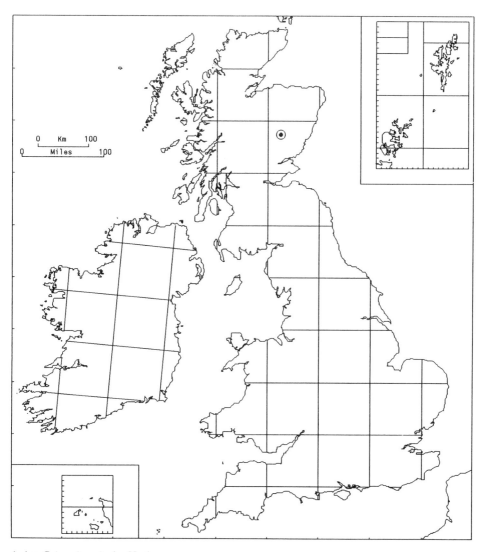

64/10. **Grimmia unicolor** Hook.

Currently known from a single locality on wet sloping schist rock in Glen Clova. There is an old record, dated 1836, from Glen Fee, but this has not been refound. 470 m. GB 1.

Dioecious; capsules common, immature (still with calyptras) in September.

Circumboreal. Montane parts of Europe from the Pyrenees, Corsica and Italy to about 68°N. Asia including Caucasus, Siberia, Mongolia and Japan; N. America.

First discovered in Glen Clova in 1823, but not seen after 1883 until it was rediscovered in quantity in 1964.

A. J. E. SMITH

355

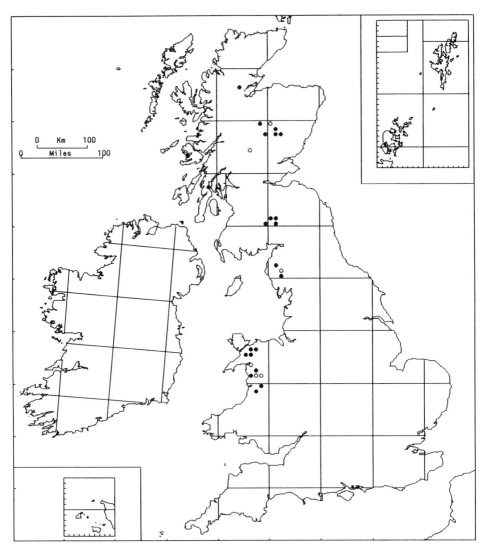

64/11. **Grimmia atrata** Mielichh.

One of the 'copper mosses' forming dark green tufts on the surface, and in crevices, of moist, sheltered or exposed, acidic heavy-metal-bearing rocks. It occurs on cliffs, on rock outcrops, and on rocks and boulders in scree and by lakes. 200 m (Cwm Rheidol) to 900 m (Carnedd Llewelyn). GB 22+6*.

Dioecious; capsules occasional, autumn.

Very rare in Europe, from Spain, Italy and Romania north to Norway and Arctic Sweden. Japan, Canada (Labrador).

A. J. E. SMITH

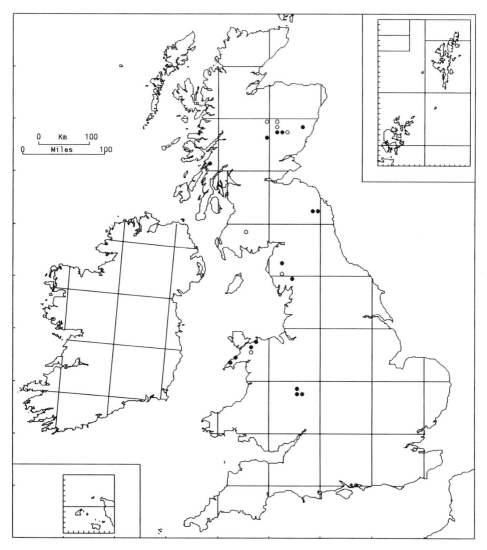

64/12. **Grimmia incurva** Schwaegr.

On dry, exposed acidic rock, especially on hill- and mountain-tops, but also in boulder-scree. 300–950 m (Foel Grach). GB 16+7*.

Dioecious; capsules very rare, spring.

Widely distributed but rare in Europe from Italy and Yugoslavia north to Spitsbergen. Azores, Caucasus, Urals, Novaya Zemlya, Mongolia, Japan, N. America, Greenland.

A. J. E. Smith

357

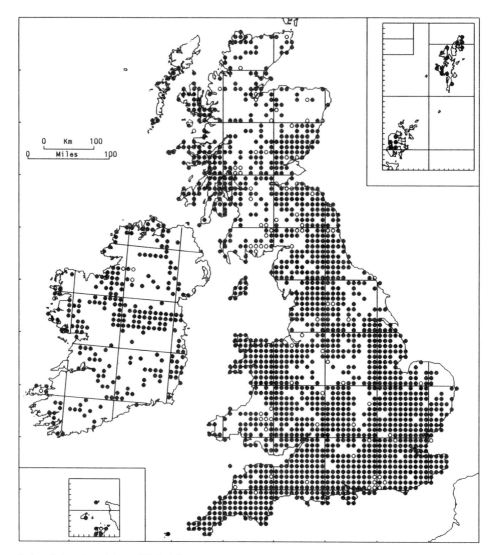

64/13. **Grimmia pulvinata** (Hedw.) Sm.

A calcicolous species forming hoary cushions on horizontal to vertical surfaces of rocks and walls, very rarely on trees. It is particularly characteristic of man-made habitats such as mortared and limestone wall-tops, bridges, roofs, tombstones, old concrete and old buildings; but it also occurs widely on rocks near the sea and on limestone outcrops and boulders. Associated species often include *Orthotrichum anomalum*, *Schistidium apocarpum* and *Tortula muralis*. *G. pulvinata* is evidently able to withstand a considerable degree of atmospheric pollution, being common in urban areas. 0–635 m (Gragareth). GB 1649+84*, IR 246+8*.

Autoecious; capsules abundant, late spring.

Europe to about 64°N. Macaronesia, Africa, Asia, N. and C. America, Australia, New Zealand.

The common plant is var. *pulvinata*. Var. *africana* (Hedw.) Hook. f. occurs in similar habitats. It has very rarely been recorded since 1950, possibly because no one has bothered to look for it. It is not mapped separately.

A. J. E. SMITH

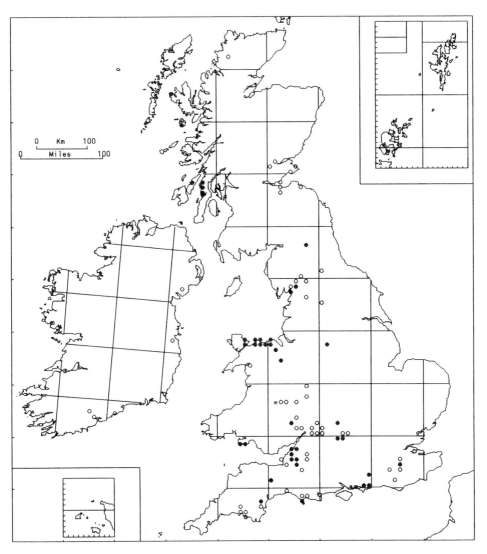

64/14. **Grimmia orbicularis** Bruch ex Wils.

G. orbicularis usually occurs on exposed dry rocks, especially Carboniferous limestone, and on walls. It may be locally common or abundant and is frequently associated with *G. pulvinata*. The species is sometimes associated with rocks with a high heavy-metal content. It is apparently susceptible to atmospheric pollution as it is decreasing except in W. Britain. 0–365 m (Creigiau Eglwyseg). GB 40+51*, IR 5*.

Autoecious; capsules abundant, spring.

In Europe from the Mediterranean region and the Crimea north to the Netherlands, Germany and Poland. Canaries, N. Africa, Cyprus, Turkey, Caucasus, Middle East, C. Asia, U.S.A. (Utah, Arizona, New Mexico), Mexico.

A. J. E. SMITH

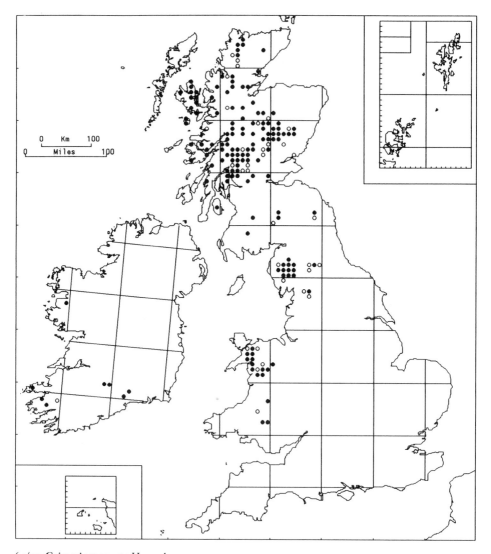

64/15. **Grimmia torquata** Hornsch.

On sheltered to exposed, periodically damp rock-ledges on cliffs and in scree, where slightly basic, rarely on basic coastal cliffs. 0–1050 m (Ben Lawers). GB 135+33*, IR 8+2*.

Dioecious; capsules unknown in the British Isles. Gemmae are produced on the backs of the older leaves.

Montane and Arctic Europe from the Pyrenees, Corsica, Sardinia and Italy north to Spitsbergen. Madeira, Tenerife, Russian Far East (Anadyr Region), Aleutians, Hawaii, N. America, Greenland.

A. J. E. SMITH

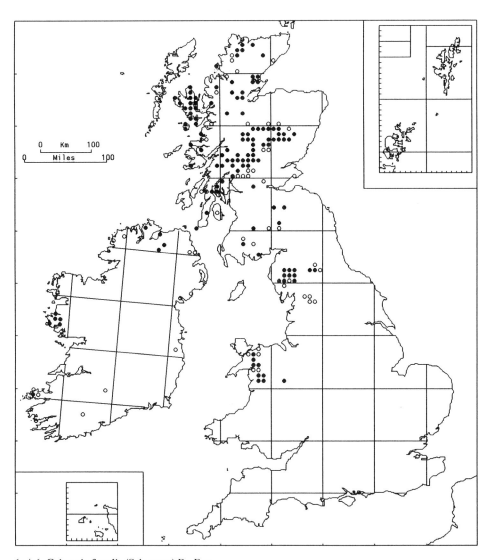

64/16. Grimmia funalis (Schwaegr.) Br. Eur.

G. funalis forms readily disintegrating tufts of string-like shoots on ledges and in crevices of dry to periodically damp, moderately to strongly basic, igneous or schistose, submontane and montane rock-ledges and scree, rarely on sea-cliffs. Mostly above 300 m, but descending to sea-level in W. Scotland (e.g. Skye) and reaching 1205 m on Ben Lawers. GB 116+43*, IR 10+12*.

Dioecious; capsules occasional, late summer. *G. funalis* sometimes produces flagelliform branches which may be propagules.

Europe from Corsica, Sardinia and Crete to Iceland and N. Norway (67°N). Canaries (Hierro, Tenerife), Algeria, Turkey, Caucasus, Himalaya, Altai Mts, Japan, Canada (Labrador), Greenland.

A. J. E. SMITH

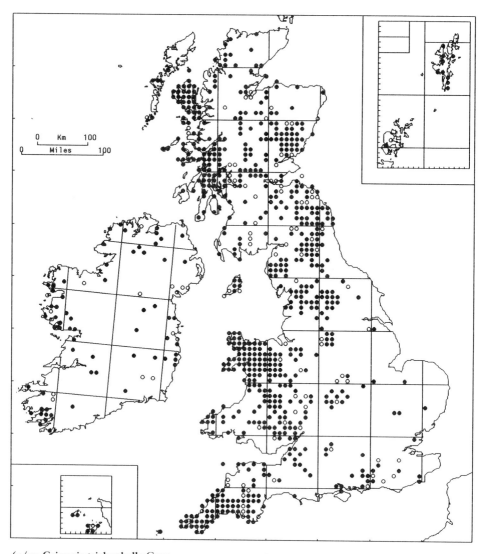

64/17. **Grimmia trichophylla** Grev.

On dry, exposed acidic rocks of various types ranging from granite to sandstone, rarely on basic substrata such as limestone. Habitats include cliffs, rock outcrops, sarsen-stones, tombstones, stones in moorland, in fields, on mountain-sides, in walls, by streams and near the coast, and, rarely, bricks and roofs. 0–990 m (Ben Lawers). GB 641+92*, IR 63+16*.

Dioecious; capsules rare, spring. Gemmae are occasionally present on the upper surface of the leaves.

Europe north to about 66°N in Fennoscandia. Macaronesia, Morocco, Algeria, Palestine, Turkey, Caucasus, E. Siberia, Hawaii, N. and C. America, southern S. America, Australasia.

Apart from var. *robusta* (Ferg.) A. J. E. Smith, the varieties recognized by Smith (1978) have not been recorded consistently and are not mapped separately. Var. *trichophylla* is the commonest. Var. *stirtonii* (Schimp.) Möll., var. *subsquarrosa* (Wils.) A. J. E. Smith and var. *tenuis* (Wahlenb.) Wijk & Marg. are often poorly defined.

A. J. E. SMITH

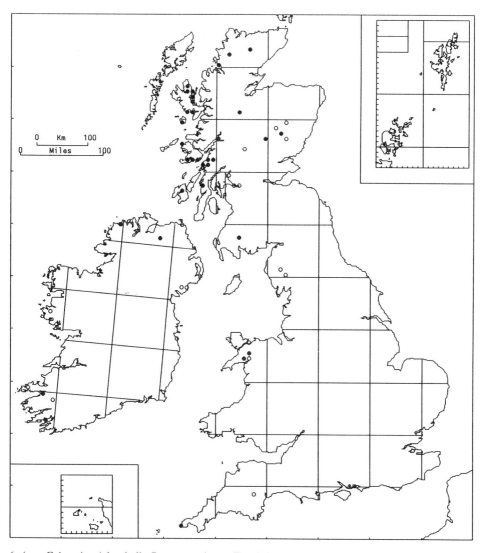

64/17e. **Grimmia trichophylla** Grev. var. **robusta** (Ferg.) A. J. E. Smith

Forming hoary tufts on basic and ultrabasic siliceous rocks (rarely limestone) in submontane and montane habitats, where it may be locally abundant. It also occurs on acidic coastal cliffs and rocks. 0–450 m (Moel Hebog), probably ascending to over 600 m in the E. Highlands of Scotland, but precise altitudinal data are not available. GB 29+13*, IR 4+7*.

Dioecious; capsules unknown.

Portugal.

A. J. E. SMITH

363

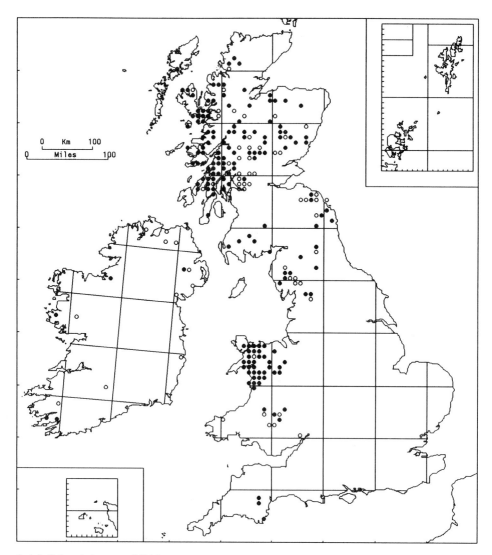

64/18. **Grimmia hartmanii** Schimp.

Sometimes locally common, forming patches on periodically moist or flooded basic to mildly acidic rocks (rarely on tree-trunks), usually near water, in woodland and wooded ravines, on flushed cliff-faces, or sometimes in exposed situations by lakes. Associated species often include *Heterocladium heteropterum*, *Hyocomium armoricum* or *Oxystegus tenuirostris*. A plant mainly of low altitudes, ascending to 915 m (Ben Lawers). GB 161+51*, IR 5+12*.

Dioecious; capsules not known in Britain or Ireland. Gemmae are frequently produced in abundance at the shoot tips.

From Mediterranean Europe to about 69°N. Madeira, N. Africa, Turkey, Caucasus, Japan, N. America.

A. J. E. SMITH

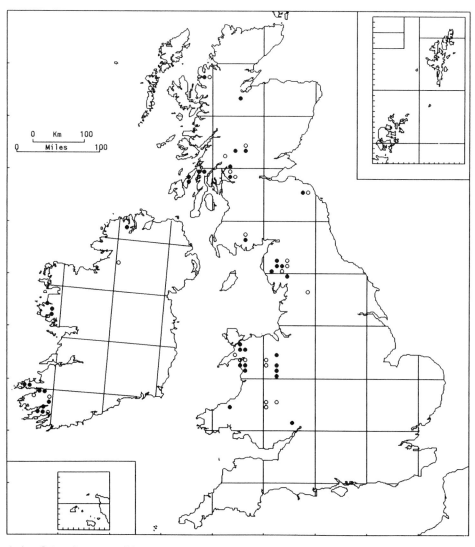

64/19. **Grimmia retracta** Stirt.

G. retracta forms tufts, usually silt- and grit-encrusted, on rocks below flood-level beside streams and rivers and by lakes. It is also found occasionally on rock outcrops away from water. It is sometimes locally abundant. 0–600 m (Snowdon). GB 29+22*, IR 12+4*.

Presumably dioecious, no data on sexuality; capsules unknown.

Spain, Portugal.

A. J. E. SMITH

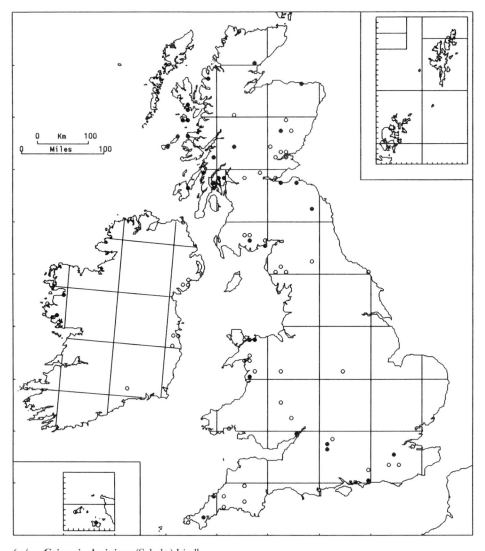

64/20. **Grimmia decipiens** (Schultz) Lindb.

G. decipiens occurs as readily disintegrating tufts on often sloping, sunny or slightly sheltered, basic or more rarely acidic siliceous rocks and walls. Lowland. GB 33+43*, IR 2+9*.

Autoecious; capsules common.

Mediterranean Europe and the Crimea to about 67°N. Canaries, Madeira, Algeria, Cyprus, Turkey.

G. decipiens occurs in similar habitats to *G. laevigata*, and likewise appears to be decreasing. For most of the period of the Mapping Scheme, *Grimmia trichophylla* var. *robusta* was regarded as a variety of *G. decipiens*; some ambiguous records have had to be discarded, and the species is probably under-recorded on the map.

A. J. E. SMITH

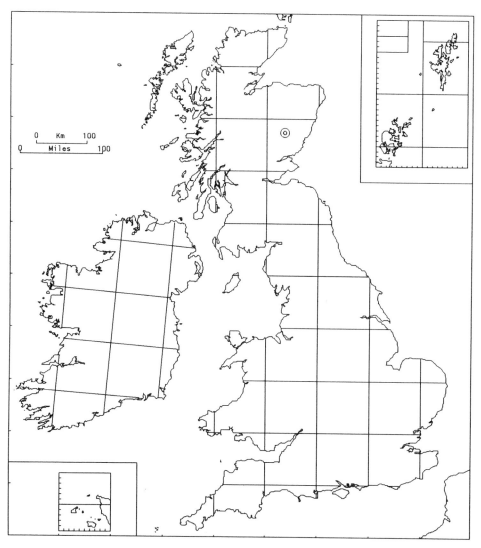

64/21. **Grimmia elatior** Bruch ex Bals. & De Not.

Formerly occurred in Glen Clova, where it was said to grow in large masses (often to a foot in length) on sloping granite rocks. Altitude not known. GB 1*.

Dioecious; capsules not known in Britain.

Montane parts of Europe from Spain, Sardinia, Sicily and Yugoslavia to about 70°N. Turkey, Caucasus, Urals, Siberia, Altai Mts, China, Japan, western N. America, Greenland.

Discovered by the Rev. J. Fergusson in 1868. Fergusson collected it again in 1870 and 1871, but it has not been seen since. In continental Europe it occurs usually on acidic rocks in submontane or montane habitats, sometimes by streams.

A. J. E. SMITH

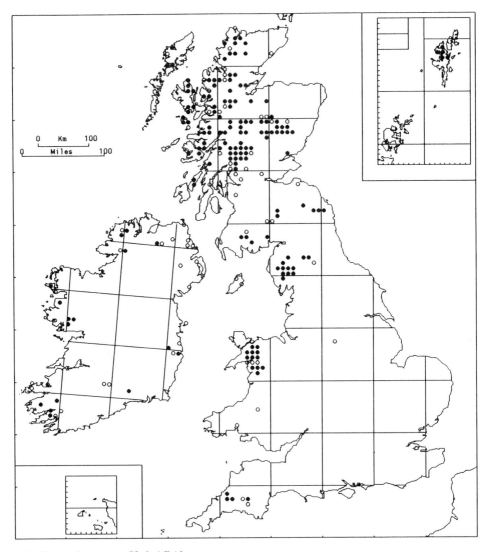

65/1. **Dryptodon patens** (Hedw.) Brid.

On moist, acidic to slightly basic rocks and boulders in and by streams, rivers and lakes, and on intermittently flushed outcropping rocks and cliffs, in sheltered or exposed sites. Sea-level (Shetland) to 1000 m (Carnedd Llewelyn). GB 140+44*, IR 16+22*.

Dioecious; capsules rare, spring.

In montane parts of Europe from Spain, Corsica, Sardinia and N. Greece to about 69°N in Iceland and Norway. W. Siberia, Altai Mts, Japan, western N. America, Michigan, Newfoundland, Greenland.

A. J. E. SMITH

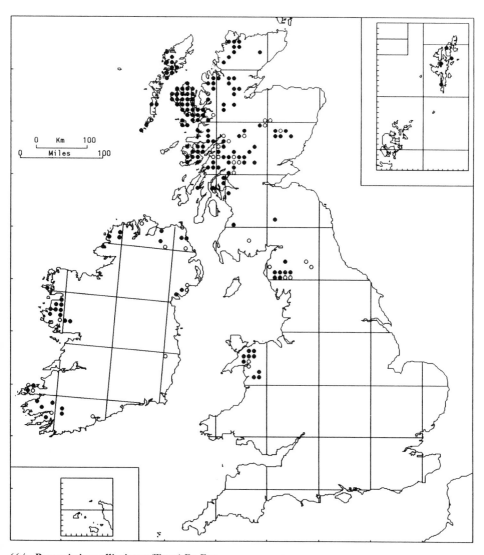

66/1. Racomitrium ellipticum (Turn.) Br. Eur.

Typically forming small dark green tufts on wet, acid or slightly basic mountain rock-faces, often in shady gullies and N.-facing corries. Mainly at medium altitudes, to at least 800 m in Wales, descending to sea-level in Knapdale and Shetland. GB 141+25*, IR 29+13*.

Dioecious; capsules are produced freely, dehiscing in early summer.

W. Europe, from Spain north to Faeroes, Iceland and Norway. Canaries.

M. C. F. PROCTOR

369

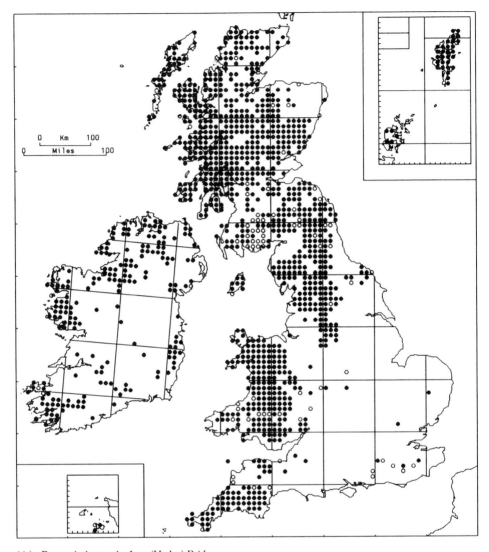

66/2. **Racomitrium aciculare** (Hedw.) Brid.

A common species forming green to blackish tufts or patches on rocks which are at least periodically wet, particularly beside streams and clear fast-flowing rivers, where it may be periodically inundated; typically in open or only lightly shaded sites. It occurs on a wide range of rock types, provided that these are hard enough for a firm attachment, but generally avoids both very acid and highly calcareous situations. Commonest at low to moderate altitudes in hilly districts. 0–830 m (Aonach Beag). GB 987+71*, IR 191+4*.

Dioecious; capsules are produced freely, dehiscing in spring.

Widespread in Europe to 71°N, but increasingly confined to the mountains southwards. Madeira, Asia Minor, N. America from California and Alabama to Alaska and Labrador.

M. C. F. Proctor

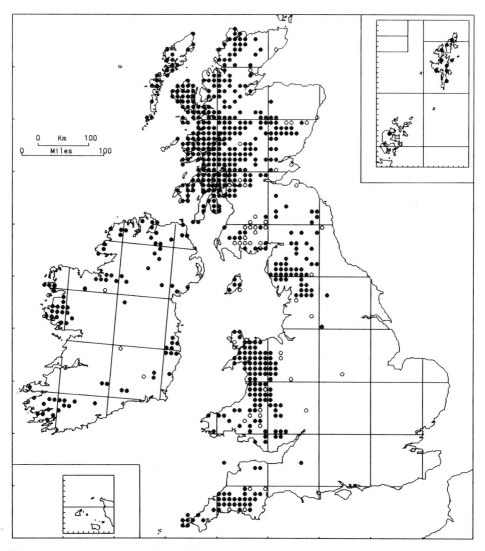

66/3. **Racomitrium aquaticum** (Schrad.) Brid.

Forming flattish olive to dark green patches on acid rocks, typically in situations kept at least intermittently wet by seepage in winter, but dry in summer; never inundated. It is often accompanied by *Andreaea rothii* and *Campylopus atrovirens*. Rarely, it occurs on soil in late-snow patches at high altitudes; on Ben Nevis it has also been noted as forming swelling tufts around high-level springs and the streams that come from them. It usually grows in open brightly-lit sites, but can tolerate moderate shade. Commonest and most conspicuous at moderate altitudes, to about 500 m, but reaching to near the highest summits in Scotland. 0–1220 m (Ben Macdui). GB 530+50*, IR 89+4*.

Dioecious; capsules produced rather regularly, dehiscing in spring.

Throughout Europe to 64°N, mainly lowland and subalpine, but increasingly confined to the mountains southwards. Macaronesia, Caucasus, Japan, N. America (Vancouver Island), Greenland, Tierra del Fuego, Kerguelen Island, New Zealand.

M. C. F. PROCTOR

371

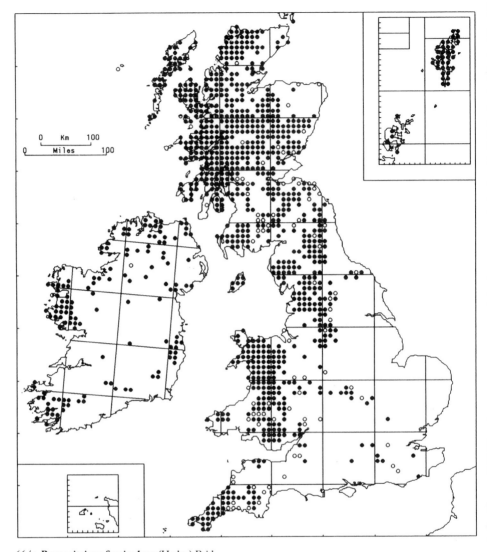

66/4. **Racomitrium fasciculare** (Hedw.) Brid.

Forms low yellow-green or brownish to olivaceous mats of richly branched shoots on dry or moist, acid rock-surfaces, including walls, roofs and isolated boulders on hillsides. Common from near sea-level to the highest summits. 0–1330 m (Ben Nevis). GB 863+88*, IR 134+2*.

Dioecious; capsules frequent, dehiscing in late spring.

Circumboreal. N. Europe and mountains of C. Europe, south to Portugal and Bulgaria. Southern S. America, New Zealand.

M. C. F. PROCTOR

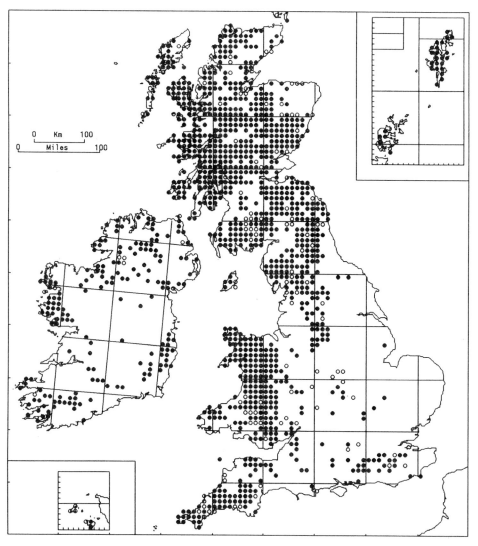

66/5. **Racomitrium heterostichum** (Hedw.) Brid. *sensu lato* (including *R. affine* (Schleich.) Lindb. and *R. microcarpon sensu* A. J. E. Smith)

The aggregate species occupies a wide range of habitats on wet or dry, acid rock-surfaces, including slate and sandstone roofs and walls; occasionally on acid soils at high altitudes. From sea-level to the highest summits. 0–1344 m (Ben Nevis). GB 996+110*, IR 172+9*.

Dioecious; frequency of capsule-production varies a good deal between the constituent taxa, and from one habitat to another. Capsules dehisce in spring.

R. heterostichum sensu lato has a wide bipolar distribution.

The aggregate species includes at least five recognizable taxa within the British Isles. Their taxonomy has been elucidated by Frisvoll (1988) and Blockeel (1991), but there is still much uncertainty as to their distribution.

M. C. F. Proctor

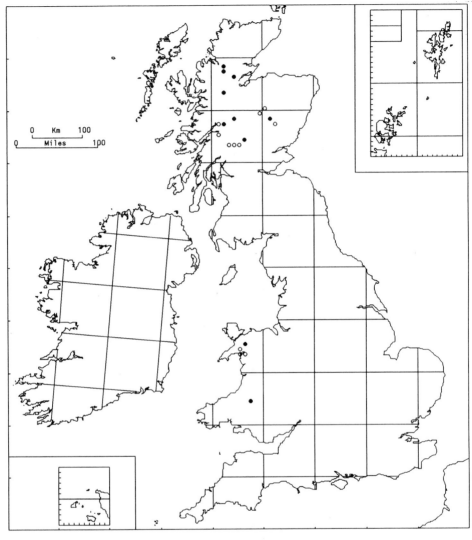

66/5A. **Racomitrium macounii** Kindb. ssp. **alpinum** (Lawt.) Frisvoll

On flat or steep, moist rocks, often by mountain streams, or where there is seepage of water across the rock surface. Less strongly calcifuge than most of the related species. 400 m (Rhum) to 1160 m (Cairn Gorm). GB 10+11*.

Dioecious; no data on gametangia or capsules.

Widespread in mountainous regions of Europe, north to Iceland and N. Norway. N.E. Turkey, N. Japan, western N. America, Greenland.

Racomitrium macounii is one of the segregates of the *R. heterostichum* complex recognized by Frisvoll (1988). It typically forms reddish-brown to olivaceous patches; the leaves have very short squarrose hair-points. The map is based on specimens checked by T. L. Blockeel, A. A. Frisvoll or M. C. F. P.

M. C. F. Proctor

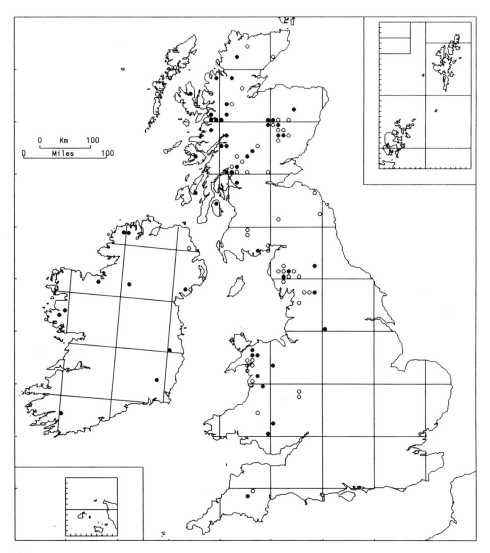

66/5B. **Racomitrium sudeticum** (Funck) Br. Eur. (*R. affine sensu* A. J. E. Smith)

On dry or moist acid rocks, generally in drier habitats than *R. macounii* ssp. *alpinum*; occasionally on soil in the mountains, especially in late-snow patches. Ascends to 1344 m on the summit cairn of Ben Nevis. GB 45+51*, IR 10+2*.

Dioecious; capsules rare.

A very widely distributed bipolar species, known from all the continents except Africa. Common in N. Europe and in mountains further south.

Although *Racomitrium sudeticum* was recognized as a subspecies of *R. heterostichum* by Dixon (1924), it is here defined more narrowly, according to the revision by Frisvoll (1988). It is usually olivaceous or greenish, forming dense cushions or patches; the hair-points are either absent or short, stout and more or less squarrose. The map is based on specimens checked by T. L. Blockeel, A. A. Frisvoll or M. C. F. P.

M. C. F. Proctor

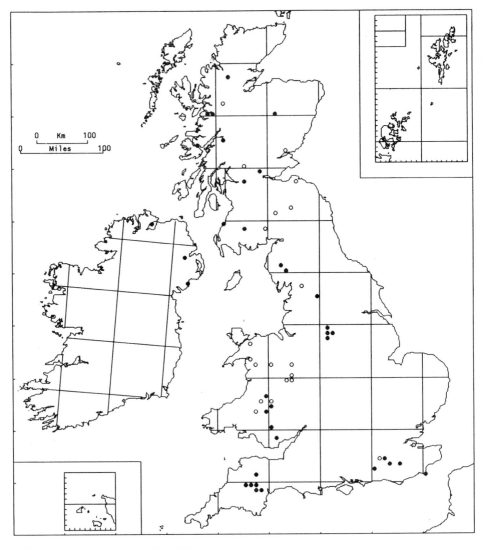

66/5C. **Racomitrium affine** (Schleich.) Lindb.

On moist or intermittently moist acid rocks, walls and roof-tiles; also recorded from turf over shingle. Mainly a plant of low elevations, upper altitudinal limit unknown. GB 33 + 21*, IR 3 + 1*.

Dioecious; capsules occasional.

S. and W. Fennoscandia, Faeroes, mountains of C. and S. Europe. Caucasus, eastern and western N. America.

Racomitrium affine is defined here in the narrow sense of Frisvoll's (1988) revision, and not according to the concept in Smith (1978). In the older literature it was often called *R. sudeticum*. It grows in loose, often wide mats or cushions; the leaves are secund, at least at the stem apices, and may or may not have erect-flexuous hair points, which may vary greatly in length. The map is based on specimens checked by T. L. Blockeel, A. A. Frisvoll or M. C. F. P.

M. C. F. PROCTOR

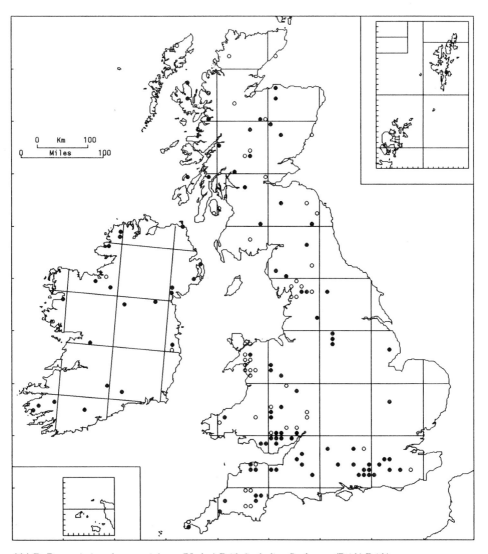

66/5D. **Racomitrium heterostichum** (Hedw.) Brid. (including *R. obtusum* (Brid.) Brid.)

On dry, acid rock-surfaces, including cliffs, quarries, rocks, boulders, stone-walls, sarsen-stones and roof-tiles; occasionally on sand and damp clinker. Mainly a plant of low to moderate elevations, up to 400 m (Ben Cleuch). GB 81+52*, IR 21+3*.

Dioecious; capsules occasional, dehiscing in spring.

W. Europe including Iceland, to nearly 70°N in W. Norway, with scattered localities in E. Europe. Macaronesia, Aleutians, western N. America.

R. heterostichum in the narrow sense of Frisvoll (1988) has proved impossible to separate consistently from *R. obtusum* in Britain (Blockeel, 1991) and the two are mapped together. It is closely related to *R. affine*, from which it is not always easy to distinguish. The map is based on specimens checked by T. L. Blockeel, A. A. Frisvoll or M. C. F. P.

M. C. F. PROCTOR

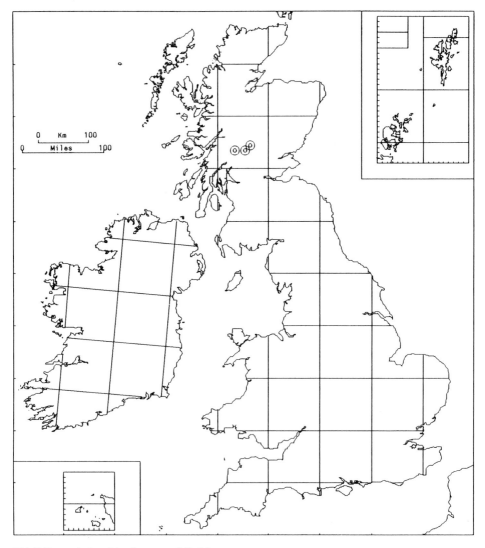

66/5E. **Racomitrium himalayanum** (Mitt.) Jaeg.

On mica-schist rocks; Ben Lawers range, Creag Mhor, Beinn Dorain. Probably at medium to high altitudes. GB 3*.

Dioecious; capsules frequent, spring.

Outside Scotland known only from the Himalaya, Tibet and China (Yunnan, Shensi).

Racomitrium himalayanum, a segregate of the *R. heterostichum* complex, was recognized as British by Frisvoll (1988). Prior to his revision, many Scottish specimens were referred to *R. microcarpon* (Hedw.) Brid., which could possibly occur in Scotland but has not yet been found there. The map is based on specimens determined by A. A. Frisvoll.

M. C. F. Proctor

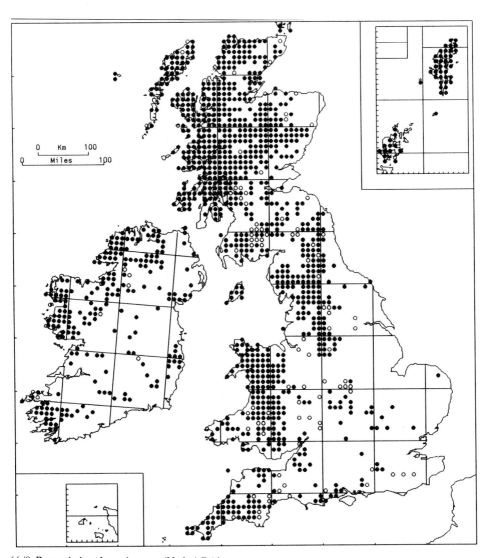

66/8. **Racomitrium lanuginosum** (Hedw.) Brid.

A common species in a wide range of unproductive sites from near sea-level to the summits of the Scottish mountains. Dominant over large tracts of stony mountain plateau in Wales, N. England and Scotland. Also conspicuous on hummocks in ombrogenous bogs, especially in the north and west. Common but local down to low altitudes on acid rocks and walls, on humus over limestone, and as a minor constituent of nutrient-poor moorlands, grasslands and damp heaths. Very local in old chalk grassland. Pollution-sensitive and probably decreasing. 0–1340 m (Ben Nevis). GB 968+91*, IR 213+8*.

Capsules occasional, spring.

Europe north to the High Arctic, becoming montane further south. A very widely distributed bipolar species occurring also on some tropical mountains.

M. C. F. PROCTOR

379

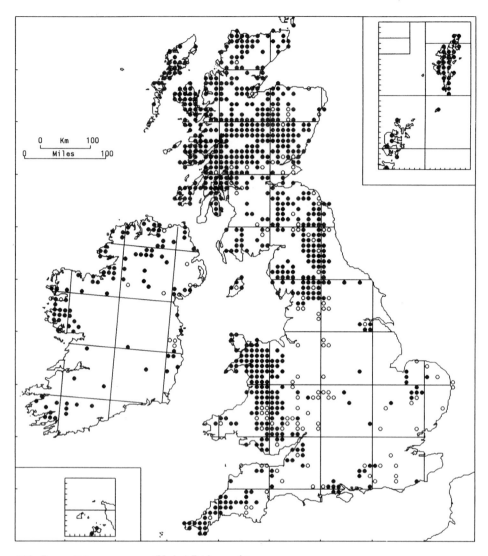

66/9. **Racomitrium canescens** (Hedw.) Brid. *sensu lato*

A complex including three related species in the British Isles. It forms extensive low patches on nutrient-poor sandy or gritty soils and fine rock-debris, in sites which range from acid to calcareous, and from dry to moist; apparent occurrences on rock are generally on a superficial layer of sandy material. 0–1340 m (Ben Nevis). GB 752+124*, IR 100+14*.

Dioecious; capsules occasional, winter.

Circumboreal.

M. C. F. Proctor

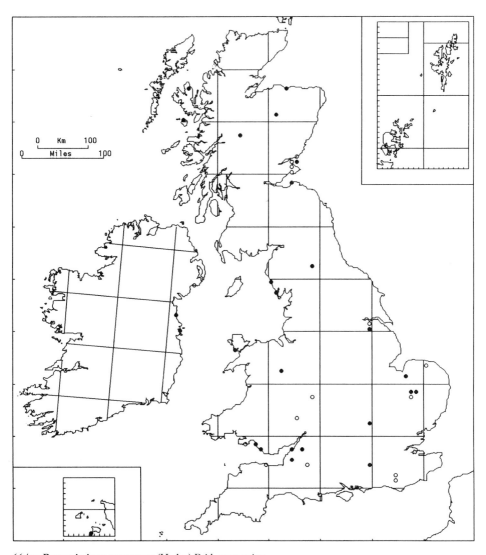

66/9a. **Racomitrium canescens** (Hedw.) Brid. *sensu stricto*

A mainly lowland plant of sandy heaths, short dune grassland and stony soil over basic rock in rather dry situations. It seems to be exclusively though rather weakly calcicolous in Britain and Ireland, and is often found on limestone or other calcareous substrata such as the sandy calcareous till of the East Anglian Breckland. In its one high-altitude locality it was found on gravelly limestone rock-ledges and slopes. o–1000 m (Coire Cheap, Ben Alder range). GB 23+14*, IR 2+1*.

Dioecious; capsules found only once in Britain, on sand dunes in Angus.

Common throughout much of C. and N. Europe. Caucasus, Urals, Siberia, N. America. Its general distribution is more continental than that of *R. elongatum* and *R. ericoides*.

The *Racomitrium canescens* complex was revised by Frisvoll (1983), and his definition of *R. canescens sensu stricto* is adopted here. It is essentially the same as that of var. *canescens* in Smith's (1978) flora; *R. elongatum* and *R. ericoides* were included in Smith's var. *ericoides*. For descriptions and illustrations of the British members of the *R. canescens* group, refer to Hill (1984).

M. C. F. Proctor & M. O. Hill

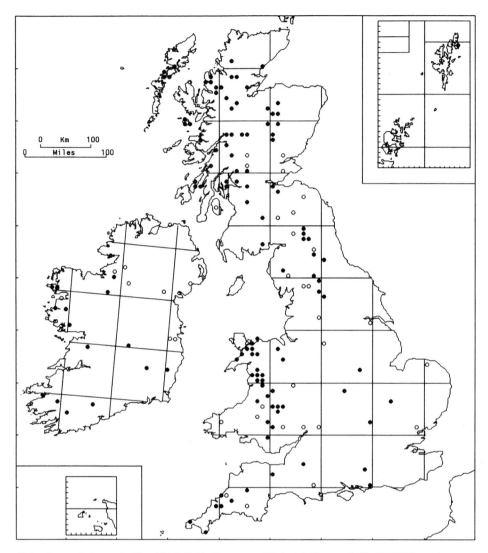

66/10. **Racomitrium ericoides** (Brid.) Brid. (*R. canescens* (Hedw.) Brid. var. *ericoides* (Brid.) Hampe)

The least xerophilous of the three British members of the *Racomitrium canescens* group. Common in a wide range of situations on dry or intermittently moist, sandy or gritty soils (occasionally on rock) from near sea-level to high altitudes in the mountains. Although commoner on non-calcareous substrata, it sometimes grows in limestone grassland or on thin soil over limestone. 0–1340 m (Ben Nevis). GB 106+30*, IR 15+11*.

Dioecious; capsules occasional.

Circumboreal, north to Svalbard, south to the Azores. Records are concentrated mainly in W. Europe. western N. America and W. Greenland. Frisvoll (1983) revised the *Racomitrium canescens* group, and defined *R. ericoides* more narrowly than previous authors. *R. ericoides* is the commonest member of the group in the British Isles, and its true distribution must be close to that shown in the aggregate map (p. 380). For a short description and a revision of its British and Irish distribution, refer to Hill (1984).

M. C. F. PROCTOR & M. O. HILL

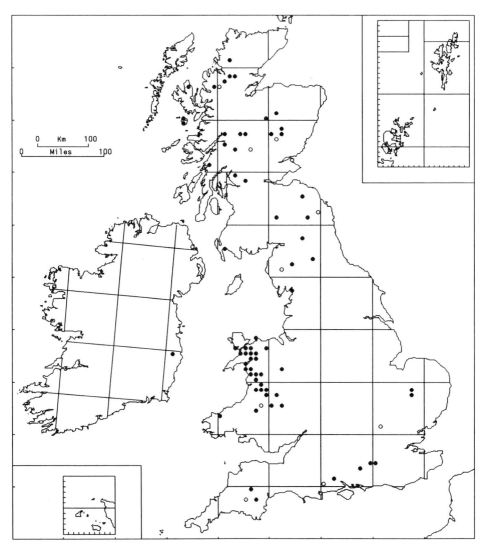

66/11. **Racomitrium elongatum** Frisvoll

In similar habitats to *R. ericoides* with which it often grows; however, it tends to favour drier, warmer and more exposed sites than that species, and perhaps more frequently grows on limestone. Widely distributed and probably common from near sea-level to high altitudes in the Scottish mountains. 0–1175 m (Ben Lawers). GB 66+10*, IR 1+1*.

Dioecious; capsules not yet found in Britain or Ireland.

Europe, from Portugal to the Caucasus and Black Sea region, north to Faeroes, Iceland and N. Norway. N. America (mainly western Canada and U.S.A.), Greenland. A markedly more southern plant than *R. ericoides*.

R. elongatum is a segregate of the *Racomitrium canescens* group, described by Frisvoll (1983). The map is based on a partial revision of British herbarium material by Hill (1984) together with subsequent finds. Inevitably, it is very incomplete.

M. C. F. Proctor & M. O. Hill

383

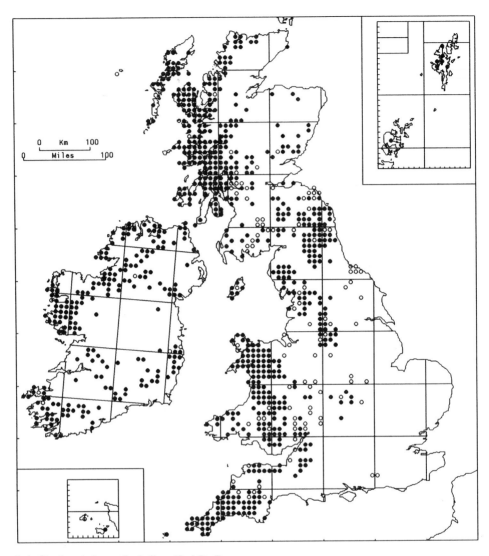

67/1. **Ptychomitrium polyphyllum** (Sw.) Br. Eur.

A locally frequent, and conspicuous plant in the north and west, occurring as dense cushions in small crevices and hollows on tops of dry, sun-exposed or slightly shaded, acid or mildly basic rocks, often where there is a thin humus-cover. Typical habitats are detached angular boulders and granite tors, stone-walls, bridges, Cornish 'hedges' and roadside rocks. *Frullania tamarisci*, *Hedwigia ciliata*, *Hypnum cupressiforme*, *Racomitrium fasciculare* and *R. heterostichum* are common associates. It also occurs occasionally in old quarries and on mine-waste, coastal rocks, and dry boulders in lightly shaded areas in wooded valleys. It is found on slate, schist, gabbro, basalt, peridotite, granite and sandstone but is absent from limestone. 0–760 m (Westmorland). GB 593+102*, IR 203+7*.

Autoecious; sporophytes abundant.

W. Europe, east to Yugoslavia and S. Sweden, north to Faeroes and S.W. Norway. Azores, Canaries, Madeira.

H. J. B. BIRKS

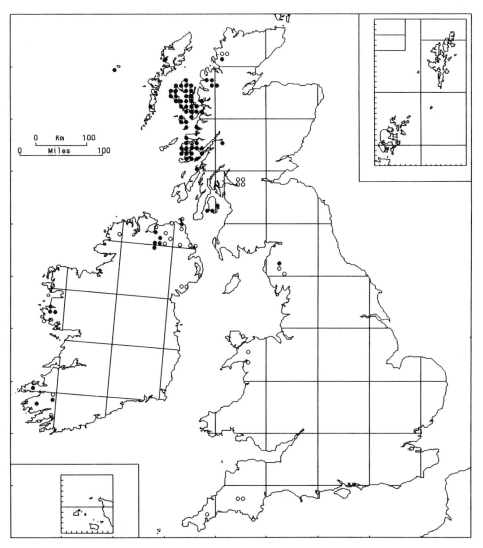

68/1. **Glyphomitrium daviesii** (With.) Brid.

Occurs most frequently on dry, often sun-exposed or periodically irrigated, S.- or W.-facing basic igneous rock-outcrops and detached angular blocks in stable block-litters. By the sea it occurs on a variety of other rock types including granite, quartzite and Lewisian gneiss, often associated with *Frullania teneriffae, Schistidium maritimum* and *Ulota phyllantha* in mildly sea-sprayed situations. In inland localities and at elevations above 200 m, it is virtually restricted to basic igneous rocks such as basalt, gabbro and peridotite, often with *Frullania tamarisci, F. teneriffae, Grimmia trichophylla, Pterogonium gracile* and *Ulota hutchinsiae.* 0–490 m (Skye). GB 61+16*, IR 12+17*.

Autoecious; sporophytes common. Chlorophyllose gemmae are nearly always present on the rhizoids (Lewinsky, 1987).

Iceland, Faeroes, Norway. Madeira, Azores.

Although largely confined today to the Tertiary volcanic districts of W. Scotland and N.E. Ireland where it is locally frequent, there are several old records from England, Wales and S. Scotland that have not been refound despite careful searching. Reasons for this decline are unclear.

H. J. B. Birks

385

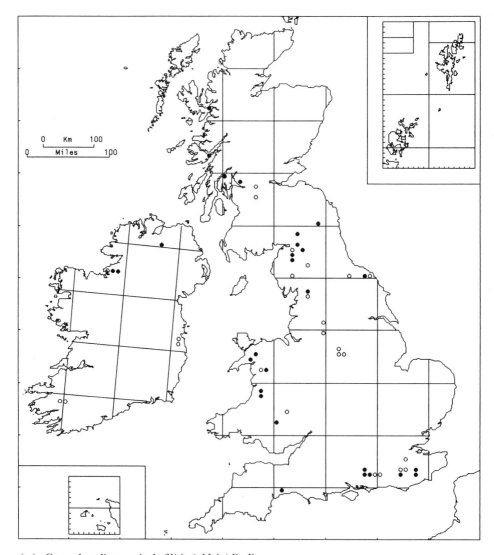

69/1. **Campylostelium saxicola** (Web. & Mohr) Br. Eur.

A species of wet or periodically flushed, acid or mildly basic sandstone, shale or slate blocks and stones in low-lying shaded ravines, gullies and wooded rivulets, often associated with *Seligeria recurvata*. It occasionally occurs in damp crevices in bedded sandstone walls of shaded wooded ravines. Also recorded in N. Wales from damp rocks by old copper-mines and in areas associated with heavy metals. Predominantly lowland, to 680 m on Snowdon. GB 25+20*, IR 3+5*.

Autoecious; sporophytes abundant.

W. and C. Europe; absent from Fennoscandia. Japan, western (Washington State) and eastern N. America. Possibly declining because of eutrophication by water pollution.

H. J. B. Birks

BIBLIOGRAPHY

Titles of periodicals are abbreviated according to the *World List of Scientific Periodicals* (Brown & Stratton, 1963–65).

Adam, P., 1976. The occurrence of bryophytes on British saltmarshes. *J. Bryol.* **9**, 265–274.

Adams, K. J., 1990. Proposals for a 5 km² mapping scheme for eastern England. *Bull. Br. bryol. Soc.* **55**, 14–17.

Anderson, L. E., Crum, H. A. & Buck, W. R., 1990. List of the mosses of North America north of Mexico. *Bryologist* **93**, 448–499.

Anon., 1939. Report on plants sent in by members. *Rep. Br. bryol. Soc.* **4**, 83–107.

Appleyard, J., 1956. *Tetraphis browniana* var. *repanda* (Funck) Hampe – new to the British Isles. *Trans. Br. bryol. Soc.* **3**, 64–65.

———, 1985. *Didymodon reedii* Robins., a moss new to the British Isles. *J. Bryol.* **13**, 319–321.

———, Hill, M. O. & Whitehouse, H. L. K., 1985. *Leptobarbula berica* (De Not.) Schimp. in Britain. *J. Bryol.* **13**, 461–470.

Arts, T., 1985. Young sporophytes of *Dicranella staphylina* Whitehouse (Musci). *Lindbergia* **11**, 55–58.

———, 1986. Drought resistant rhizoidal tubers in *Fissidens cristatus* Wils. ex Mitt. *Lindbergia* **12**, 119–120.

———, 1987a. The occurrence of rhizoidal tubers in *Atrichum tenellum* (Röhl) B. & S. and *A. crispum* (James) Sull. *Lindbergia* **13**, 72–74.

———, 1987b. *Pottia bryoides* (Dicks.) Mitt., *P. lanceolata* (Hedw.) C. Müll. and *P. truncata* (Hedw.) B. & S. with rhizoidal tubers. *Lindbergia* **13**, 130–132.

———, 1988a. Rhizoidal tubers and protonematal gemmae in *Pseudocrossidium revolutum* (Brid.) Zander var. *revolutum* and *Scopelophila cataractae* (Mitt.) Broth. *Lindbergia* **14**, 59–62.

———, 1988b. More rhizoidal tubers in European and Macaronesian *Fissidens* species. *Lindbergia* **14**, 151–154.

———, 1989. The occurrence of rhizoidal tubers in the genus *Campylopus*. *Lindbergia* **15**, 60–64.

———, 1990. Rhizoidal tubers and protonema-gemmae in *Cynodontium bruntonii*. *Lindbergia* **16**, 25–27.

——— & Frahm, J.-P., 1990. *Campylopus pyriformis* new to North America. *Bryologist* **93**, 290–294.

——— & Risse, S., 1988. The occurrence of rhizoidal tubers in *Pleuridium acuminatum*. *Lindbergia* **14**, 127–130.

Bartlett, J. K. & Vitt, D. H., 1986. A survey of species in the genus *Blindia* (Bryopsida, Seligeriaceae). *N. Z. J. Bot.* **24**, 203–246.

Bates, J. W., 1989. Growth of *Leucobryum glaucum* cushions in a Berkshire oakwood. *J. Bryol.* **15**, 785–791.

Birks, H. H. & Birks, H. J. B., 1967. *Grimmia agassizii* (Sull. & Lesq.) Jaeg. in Britain. *Trans. Br. bryol. Soc.* **5**, 215–217.

Blackstock, T. H., 1987. The male gametophores of *Leucobryum glaucum* (Hedw.) Ångstr. and *L. juniperoideum* (Brid.) C. Muell. in two Welsh woodlands. *J. Bryol.* **14**, 535–541.

Blockeel, T. L., 1981. *Barbula tomaculosa*, a new species from arable fields in Yorkshire. *J. Bryol.* **11**, 583–589.

——, 1990. The genus *Hennediella* Par.: a note on the affinities of *Tortula brevis* Whitehouse & Newton and *T. stanfordensis* Steere. *J. Bryol.* **16**, 187–192.

——, 1991. The *Racomitrium heterostichum* group in the British Isles. *Bull. Br. bryol. Soc.* **58**, 29–35.

—— & Rumsey, F. J., 1990. A new locality for *Tortula freibergii* Dix. & Loeske and notes on its taxonomy and cytology. *J. Bryol.* **16**, 179–185.

Braithwaite, R., 1887–1905. *The British Moss-Flora.* 3 vols. London.

Bremer, B., 1980. A taxonomic revision of *Schistidium* (Grimmiaceae, Bryophyta) 1. *Lindbergia* **6**, 1–16.

Brodie, H. J., 1951. The splash-cup dispersal mechanism in plants. *Can. J. Bot.* **29**, 224–234.

Brown, D. H. & Bates, J. W., 1972. Uptake of lead by two populations of *Grimmia doniana*. *J. Bryol.* **7**, 187–193.

Brown, P. & Stratton, G. B. (eds), 1963–65. *World List of Scientific Periodicals*, edn 4. 3 vols. London.

Burley, J. S., 1986. *Aspects of the Taxonomy and Biology of* Ceratodon *Brid.* Ph.D. Thesis, University of Aberdeen.

—— & Pritchard, N. M., 1990. Revision of the genus *Ceratodon* (Bryophyta). *Harvard Papers in Botany* **2**, 17–76.

Chamberlain, D. F., 1969. New combinations in *Pottia starkeana*. *Notes R. bot. Gdn Edinb.* **29**, 403–404.

Coker, P. D., 1968. Distribution maps of bryophytes in Britain. *Saelania glaucescens* (Hedw.) Broth. and *Mielichhoferia mielichhoferi*. *Trans. Br. bryol. Soc.* **5**, 594.

Corley, M. F. V., 1979. Notes on some rare Dicranaceae. *J. Bryol.* **10**, 383–386.

——, 1980. The *Fissidens viridulus* complex in the British Isles and Europe. *J. Bryol.* **11**, 191–208.

—— & Frahm, J.-P., 1982. Taxonomy and world distribution of *Campylopus pyriformis* (Schultz) Brid. *J. Bryol.* **12**, 187–190.

—— & Hill, M. O., 1981. *Distribution of Bryophytes in the British Isles: A Census Catalogue of their Occurrence in Vice-counties.* Cardiff.

—— & Perry, A. R., 1985. *Scopelophila cataractae* (Mitt.) Broth. in South Wales, new to Europe. *J. Bryol.* **13**, 323–328.

—— & Wallace, E. C., 1974. *Dicranodontium subporodictyon* Broth. in Scotland. *J. Bryol.* **8**, 185–189.

——, Crundwell, A. C., Düll, R., Hill, M. O. & Smith, A. J. E., 1981. Mosses of Europe and the Azores; an annotated list of species, with synonyms from the recent literature. *J. Bryol.* **11**, 609–689.

————, Newton, M. E. & Edwards, S. R., 1987. *Didymodon nicholsonii* Culm. with sporophytes in England. *J. Bryol.* **14**, 653–657.

Cronberg, N., 1989. Patterns of variation in morphological characters and isoenzymes in populations of *Sphagnum capillifolium* (Ehrh.) Hedw. and *S. rubellum* Wils. from two bogs in southern Sweden. *J. Bryol.* **15**, 683–696.

Crum, H. A., 1984. *Sphagnopsida, Sphagnaceae*. (North American Flora, Series II, part 11). New York.

————— & Anderson, L. E., 1981. *Mosses of Eastern North America*. 2 vols. New York.

Crundwell, A. C., 1959. *Grimmia trichodon* in Britain. *Trans. Br. bryol. Soc.* **3**, 558–562.

————, 1971. *Weissia perssonii* Kindb., a neglected west European moss. *Trans. Br. bryol. Soc.* **6**, 221–224.

————, 1972. *Leucobryum juniperoideum* (Brid.) C. Müll. in Britain. *J. Bryol.* **7**, 1–5.

————, 1976a. *Barbula mamillosa*, a new species from Scotland. *J. Bryol.* **9**, 163–166.

————, 1976b. *Ditrichum plumbicola*, a new species from lead-mine waste. *J. Bryol.* **9**, 167–169.

————, 1986. *Scopelophila cataractae* in Devonshire. *J. Bryol.* **14**, 387.

————— & Hill, M. O., 1977. *Anoectangium warburgii*, a new species of moss from the British Isles. *J. Bryol.* **9**, 435–440.

————— & Nyholm, E., 1972. *Tortula freibergii* Dix. & Loeske in Sussex, new to the British Isles. *J. Bryol.* **7**, 161–164.

————— & ————, 1973. *Seligeria diversifolia* Lindb. in Yorkshire, new to the British Isles. *J. Bryol.* **7**, 261–263.

————— & ————, 1977. *Dicranella howei* Ren. & Card. and its relationship to *D. varia* (Hedw.) Schimp. *Lindbergia* **4**, 35–38.

————— & Whitehouse, H. L. K., 1978. *Trichostomopsis umbrosa* (C. Müll.) Robins. in England. *J. Bryol.* **10**, 5–8.

Daniels, R. E., 1982. Isozyme variation in British populations of *Sphagnum pulchrum* (Braithw.) Warnst. *J. Bryol.* **12**, 65–76.

————— & Eddy, A., 1985. *Handbook of European Sphagna*. Huntingdon.

————— & ————, 1990. *Handbook of European Sphagna*, 2nd impression with additions. London.

Dirske, G.M., Rutjes, J.J., Siebel, H. & Zielman, R. (1990). *Trochobryum carniolicum* Breidler & Beck (Musci, Seligeriaceae), nouveau pour la France et pour l'Autriche. *Cryptogamie, Bryol. Lichén.* **11**, 385–389.

Dixon, H. N., 1924. *The Student's Handbook of British Mosses*, edn 3. Eastbourne.

Düll, R., 1984. Distribution of the European and Macaronesian mosses (Bryophytina). Part I. *Bryol. Beitr.* **4**, 1–113.

————, 1985. Distribution of the European and Macaronesian mosses (Bryophytina). Part II. *Bryol. Beitr.* **5**, 110–232.

During, H.J. & ter Horst, B., 1983. The diaspore bank of bryophytes and ferns in chalk grassland. *Lindbergia* **9**, 57–64.

Edwards, S. R., 1979. Taxonomic implications of cell patterns in haplolepidous moss peristomes. *In:* G. C. S. Clarke & J. G. Duckett (eds), *Bryophyte Systematics*, pp. 317–346. London.

————, 1984. Homologies and inter-relationships of moss peristomes. *In:* R. M. Schuster (ed.), *New Manual of Bryology*, pp. 658–695. Nichinan.

Ellis, L. T. & Smith, A. C., 1983. *Barbula cylindrica* (Tayl.) Schimp. with rhizoidal tubers. *J. Bryol.* **12**, 509–510.

Flatberg, K. I., 1984. A taxonomic revision of the *Sphagnum imbricatum* complex. *K. norske Vidensk. Selsk. Skr.* **1984**(3), 1–80.

———, 1985. Studies in *Sphagnum subfulvum* Sjörs, and related morphotypes. *Lindbergia* **11**, 38–54.

———, 1987. Taxonomy of *Sphagnum majus* (Russ.) C. Jens. *K. norske Vidensk. Selsk. Skr.* **1987**(2), 1–42.

———, 1988a. *Sphagnum skyense sp. nov. J. Bryol.* **15**, 101–107.

———, 1988b. *Sphagnum viridum* sp. nov., and its relation to *S. cuspidatum. K. norske Vidensk. Selsk. Skr.* **1988**(1), 1–63.

Flowers, S., 1973. *Mosses: Utah and the West.* Provo, Utah.

Frahm, J.-P., 1973. Verbreitung, Systematik und Ökologie von *Leptodontium flexifolium* (Dicks.) Hampe. *Nova Hedwigia* **24**, 413–429.

———, 1980. Synopsis of the genus *Campylopus* in North America north of Mexico. *Bryologist* **83**, 570–588.

———, 1985. The world-wide range of *Campylopus shawii* Wils. *J. Bryol.* **13**, 329–332.

Frisvoll, A. A., 1983. A taxonomic revision of the *Racomitrium canescens* group (Bryophyta, Grimmiales). *Gunneria* **41**, 1–181.

———, 1985. Lectotypifications including nomenclatural and taxonomical notes on *Ditrichum flexicaule* sensu lato. *Bryologist* **88**, 31–40.

———, 1988. A taxonomic revision of the *Racomitrium heterostichum* group (Bryophyta, Grimmiales) in N. and C. America, N. Africa, Europe and Asia. *Gunneria* **59**, 1–289.

Furness, S. B. & Hall, R. H., 1981. An explanation of the intermittent occurrence of *Physcomitrium sphaericum* (Hedw.) Brid. *J. Bryol.* **11**, 733–742.

Gardiner, J. C., 1981. A bryophyte flora of Surrey. *J. Bryol.* **11**, 747–841.

Greven, H., 1990. *Grimmia ovalis* (Hedw.) Lindb. and *G. orbicularis* Bruch ex Wils., two epilithic moss species new for The Netherlands. *Lindbergia* **16**, 19–21.

Grolle, R., 1966. *Gymnomitrion crenulatum* und Verwandte. *Trans. Br. bryol. Soc.* **5**, 86–94.

———, 1983. Hepatics of Europe including the Azores: an annotated list of species, with synonyms from the recent literature. *J. Bryol.* **12**, 403–459.

Häusler, M., 1984. Die selteneren *Tortula*-arten der Sektion *Cuneifolia* in Deutschland. *Bryol. Beitr.* **3**, 1–22.

Hesselbo, A., 1918. The bryophyta of Iceland. *In:* L. K. Rosenvinge & E. Warming (eds), *The Botany of Iceland* **1**, pp. 395–677. Copenhagen & London.

Hill, M. O., 1975. *Sphagnum subsecundum* Nees and *S. auriculatum* Schimp. in Britain. *J. Bryol.* **8**, 435–441.

———, 1979. The taxonomic position of *Oxystegus sinuosus* (Mitt.) Hilp. in relation to the genus *Barbula* Hedw. *J. Bryol.* **10**, 273–276.

———, 1980. *Seligeria brevifolia* (Lindb.) Lindb. on Snowdon, newly recorded in the British Isles. *J. Bryol.* **11**, 7–10.

———, 1982. A reassessment of *Acaulon minus* (Hook. & Tayl.) Jaeg. in Britain, with remarks on the status of *A. mediterraneum* Limpr. *J. Bryol.* **12**, 11–14.

———, 1984. *Racomitrium elongatum* Frisvoll in Britain and Ireland. *Bull. Br. bryol. Soc.* **43**, 21–25.

———, 1988a. *Sphagnum imbricatum* ssp. *austinii* (Sull.) Flatberg and ssp. *affine* (Ren. & Card.) Flatberg in Britain and Ireland. *J. Bryol.* **15**, 109–115.

———, 1988b. A bryophyte flora of North Wales. *J. Bryol.* **15**, 377–491.

————, Preston, C. D. & Smith, A. J. E., 1991. *Atlas of the Bryophytes of Britain and Ireland*, 1. *Liverworts (Hepaticae and Anthocerotae)*. Colchester.

Holmes, N. T. H., 1976. The distribution and ecology of *Grimmia agassizii* (Sull. & Lesq.) Jaeg. in Teesdale. *J. Bryol.* **9**, 275–278.

Horton, D. G., 1980. *Encalypta brevipes* and *E. brevicolla*: new records from North America, Iceland, Great Britain and Europe. *J. Bryol.* **11**, 209–212.

————, 1983. A revision of the Encalyptaceae (Musci), with particular reference to the North American taxa. Part II. *J. Hattori bot. Lab.* **54**, 353–532.

Ireland, R. R., 1982. *Moss Flora of the Maritime Provinces*. Ottawa.

————, Brassard, G. R., Schofield, W. B. & Vitt, D. H., 1987. Checklist of the mosses of Canada II. *Lindbergia* **13**, 1–62.

Jones, E. W., 1953. A bryophyte flora of Berkshire and Oxfordshire. II. Musci. *Trans. Br. bryol. Soc.* **2**, 220–277.

————, 1986. Bryophytes in Chawley brick pit, Oxford, 1948–1985. *J. Bryol.* **14**, 347–358.

Konstantinova, N. A., 1978. Ad floram hepaticarum Chibinensium notula. *Nov. Sist. nizshikh Rast.* **15**, 231–233.

————, 1985. De hepaticis curiosis in montibus Lovozerensibus (regio Murmanica) inventis. *Nov. Sist. nizshikh Rast.* **22**, 229–231.

————, 1987. De hepaticis montium Lovozericorum (regio Murmanensis) notula. *Nov. Sist. nizshikh Rast.* **24**, 218–225.

Lewinsky, J., 1987. Rhizoidal gemmae in *Glyphomitrium daviesii* (With.) Brid. Studies in the moss flora of the Faroes 6. *Lindbergia* **13**, 155–158.

Lid, J., 1929. *Sphagnum strictum* Sulliv. and *Sphagnum americanum* Warnst. in Scotland. *J. Bot., Lond.*, **67**, 170–175.

Lohammar, G., 1954. The distribution and ecology of *Fissidens julianus* in northern Europe. *Svensk bot. Tidskr.* **48**, 162–173.

Long, D. G., 1982. *Bryoerythrophyllum caledonicum*, a new moss from Scotland. *J. Bryol.* **12**, 141–157.

———— & Hill, M. O., 1982. *Tortula solmsii* (Schimp.) Limpr. in Devon and Cornwall, newly recorded in the British Isles. *J. Bryol.* **12**, 159–169.

————, Crundwell, A. C. & Townsend, C. C., 1981. New records of bryophytes from the Canary Islands. *J. Bryol.* **11**, 521–536.

Longton, R. E. & Schuster, R. M., 1983. Reproductive biology. *In:* R. M. Schuster (ed.), *New Manual of Bryology*, pp. 386–462. Nichinan.

Maass, W. S. G., 1965. *Sphagnum dusenii* and *Sphagnum balticum* in Britain. *Bryologist* **68**, 211–217.

Mahu, M., 1985. *Tetradontium brownianum* (Dicks.) Schwaegr. (Bryatae: Tetraphidaceae) nuevo para Chile. *Bryologist* **88**, 118–119.

Matthews, J. R., 1955. *Origin and Distribution of the British flora*. London.

McIntosh, T. T., 1989. Bryophyte records from the semiarid steppe of northwestern North America, including four species new to North America. *Bryologist* **92**, 356–362.

Miles, C. J. & Longton, R. E., 1990. The role of spores in reproduction in mosses. *Bot. J. Linn. Soc.* **104**, 149–173.

Mitten, W., 1885. Notes on the European and North-American species of mosses of the genus *Fissidens*. *J. Linn. Soc. bot.* **21**, 550–560.

Moen, A. & Synnott, D., 1983. *Sphagnum subfulvum* Sjörs in Ireland compared with the occurrences in Norway. *J. Bryol.* **12**, 331–336.

Müller, K., 1954. Die pflanzengeographischen Elemente in der Lebermoosflora Deutschlands. *Revue bryol. lichén.* **23**, 109–122.

Murray, B. M., 1986. *Andreaea sinuosa*, sp. nov. (Musci: Andreaeaceae), from Alaska, British Columbia and Scotland. *Bryologist* **89**, 189–194.

———, 1987. *Andreaea schofieldiana* and *A. megistospora*, species novae, and taxonomic criteria for Sect. *Nerviae* (Andreaeopsida). *Bryologist* **90**, 15–26.

———, 1988. The genus *Andreaea* in Britain and Ireland. *J. Bryol.* **15**, 17–82.

Newton, M. E., 1989. Gemma disposition in *Dichodontium pellucidum* (Hedw.) Schimp. *J. Bryol.* **15**, 806–808.

——— & Boyce, D., 1987. Gemmae in British *Leptodontium flexifolium* (With.) Hampe. *J. Bryol.* **14**, 737–740.

Nicholson, W. E., 1908. The mosses of Sussex. *Hastings E. Suss. Nat.* **1**, 79–110.

Nyholm, E., [1987]. *Illustrated Flora of Nordic Mosses*, **1**. Fissidentaceae–Seligeriaceae. Lund.

Paton, J. A., 1965. A new British moss, *Fissidens celticus* sp. nov. *Trans. Br. bryol. Soc.* **4**, 780–784.

———, 1976. *Ditrichum cornubicum*, a new moss from Cornwall. *J. Bryol.* **9**, 171–175.

Potier de la Varde, R., 1953. *Fissidens exiguus* Sull. en Grande-Bretagne. *Revue bryol. lichén.* **22**, 16.

Preston, C. D. & Whitehouse, H. L. K., 1986. The habitat of *Lythrum hyssopifolia* L. in Cambridgeshire, its only surviving English locality. *Biol. Conserv.* **35**, 41–62.

Proctor, M. C. F., 1960. Mosses and liverworts of the Malham district. *Fld Stud.* **1**(2), 61–85.

Ratcliffe, D. A., 1958. The range and habitats of *Sphagnum lindbergii* Schp. in Scotland. *Trans. Br. bryol. Soc.* **3**, 386–391.

———, 1968. An ecological account of Atlantic bryophytes in the British Isles. *New Phytol.* **67**, 365–439.

Raven, P. J., 1986. The occurrence of *Sphagnum* moss in the sublittoral of several Galloway lochs, with particular reference to Loch Fleet. *Working Papers* **13**, Palaeoecology Research Unit, Department of Geography, University College London.

Richards, P. W. & Smith, A. J. E., 1975. A progress report on *Campylopus introflexus* (Hedw.) Brid. and *C. polytrichoides* De Not. in Britain and Ireland. *J. Bryol.* 8, 293–298.

Risse, S., 1985a. *Pottia intermedia* (Turn.) Fürnr. with rhizoidal tubers. *J. Bryol.* **13**, 523–526.

———, 1985b. Rhizoidal tubers on *Ditrichum heteromallum* (Hedw.) Britt. *J. Bryol.* **13**, 527–531.

Rose, F., 1975a. Distribution maps of bryophytes in Britain and Ireland. *Atrichum angustatum* (Brid.) B. & S. *J. Bryol.* 8, 495.

———, 1975b. Distribution maps of bryophytes in Britain and Ireland. *Atrichum tenellum* (Röhl.) B. & S. *J. Bryol.* 8, 496.

Saito, K., 1975. A monograph of Japanese Pottiaceae (Musci). *J. Hattori bot. Lab.* **39**, 373–537.

Schuster, R. M., 1969. *The Hepaticae and Anthocerotae of North America East of the Hundredth Meridian*, **2**. New York.

Side, A. G., 1977. Bryophytes in arable fields in Kent. *Trans. Kent Fld Club* **6**, 63–70.

———, 1983. The occurrence of tubers on *Barbula tophacea* (Brid.) Mitt. *J. Bryol.* **12**, 620–621.

——— & Whitehouse, H. L. K., 1974. *Tortula amplexa* (Lesq.) Steere in Britain. *J. Bryol.* **8**, 15–18.

Smith, A. J. E., 1978. *The Moss Flora of Britain and Ireland.* Cambridge.

Sowter, F. A., 1972. *Octodiceras fontanum* (La Pyl.) Lindb. epiphytic on sponges. *J. Bryol.* **7**, 87.

Steere, W. C., 1978. *The Mosses of Arctic Alaska.* Vaduz.

Steven, G. & Long, D. G., 1989. An update on the status of *Buxbaumia aphylla* on bings in central Scotland. *Trans. Proc. bot. Soc. Edinb.* **45**, 389–395.

Swinscow, T. D. V., 1959. A bryophyte flora of Hertfordshire. *Trans Br. bryol. Soc.* **3**, 509–557.

Syed, H. & Crundwell, A. C., 1973. *Barbula maxima*, nom. nov., an endemic Irish species. *J. Bryol.* **7**, 527–529.

Uggla, E., 1958. Skogsbrandfält i Muddus Nationalpark. *Acta phytogeogr. suec.* **41**, 1–116.

Vitt, D. H., 1976. The genus *Seligeria* in North America. *Lindbergia* **3**, 241–275.

Warburg, E. F., 1957. Two mosses from Scotland, new to the British Isles. *Trans. Br. bryol. Soc.* **3**, 171–173.

———, 1965. *Grimmia borealis* in Britain. *Trans. Br. bryol. Soc.* **4**, 757–759.

Wheldon, J. A. & Wilson, A., 1907. *The Flora of West Lancashire.* Liverpool & Ilkley.

Whitehouse, H. L. K., 1964. Bryophyta. *In:* F. H. Perring, P. D. Sell, S. M. Walters & H. L. K. Whitehouse (eds), *A flora of Cambridgeshire*, pp. 281–328. Cambridge.

———, 1971. Some problems associated with the distribution and life-history of the moss *Tortula stanfordensis* Steere. *The Lizard* **4**(3), 17–22.

———, 1980. The production of protonemal gemmae by mosses growing in deep shade. *J. Bryol.* **11**, 133–138.

———, 1984. Survival of a moss, probably *Dicranella staphylina*, in soil stored for nearly 50 years. *J. Bryol.* **13**, 131–133.

———, 1987. Protonema-gemmae in European mosses. *Symp. biol. hung.* **35**, 227–231.

——— & Crundwell, A. C., 1991. *Gymnostomum calcareum* Nees & Hornsch. and allied plants in Europe, North Africa and the Middle East. *J. Bryol.* **16**, 561–579.

——— & During, H. J., 1986. *Leptobarbula berica* (De Not.) Schimp. in Belgium and the Netherlands. *Lindbergia* **12**, 135–138.

——— & Newton, M. E., 1988. *Tortula brevis sp. nov.* and *T. stanfordensis* Steere: morphology, cytology and geographical distribution. *J. Bryol.* **15**, 83–99.

Zander, R. H., 1978. A synopsis of *Bryoerythrophyllum* and *Morinia* (Pottiaceae) in the New World. *Bryologist* **81**, 539–560.

Zhukova, A. L., 1973. De hepaticis Insulae Rudolfii (Terra Franz-Joseph). *Nov. Sist. nizshikh Rast.* **10**, 272–277.

LIST OF LOCALITIES CITED IN THE TEXT

Localities mentioned in the text are given below with their co-ordinates. Where possible, the 10-km square or squares are given but for larger areas (e.g. New Forest, Skye) the 100-km squares are given. The accompanying map (p.395) gives the numerical equivalents of the 100-km square alphabetical codes, used below.

A
Allenheads, NY84
Allt Sowan Hill, NJ32
Aonach Beag, NN17
Arisaig, NM68
Arthur's Seat, NT27
Artle Beck, SD56
Auchterhouse Hill, NO33

B
Beinn a'Bhuird, NO09
Beinn a'Chaorrainn, NJ00
Beinn Chabhair, NN31
Beinn Dearg, NH28
Beinn Dorain, NN33
Beinn Eighe, NG95,96
Beinn Heasgarnich, NN43
Beinn nan Eachan, NN53
Beinn Odhar Bheag, NM87
Beinn Riabhach, NN17
Beinn Sgulaird, NN04
Bellavally, H12
Ben Alder, NN47
Ben Cleuch, NN90
Ben Lawers, NN64
Ben Loyal, NC54
Ben Macdui, NN99
Ben More, NM53
Ben Nevis, NN17
Ben Vorlich, Dunbartonshire, NN21
Benbulbin, G64,74
Berwyn Mountains, SJ02,03,04,13,14
Blair Atholl, NN86
Borders, NY,NT
Borrowdale, NY21
Braemar, NO19
Braes of Balquhidder, NN42
Braigh Feith Ghiubhsachain, NO07
Brampton, NY56
Brean Down, ST25
Breckland, TF,TL
Builth Wells, SO05
Burn of Longshank, NO37
Butterburn Flow, NY67

C
Cader Idris, SH71
Caenlochan Glen, NO17
Cairn Gorm, NJ00
Cairngorm Mountains, NH,NJ,NN,NO
Carn Fadryn, SH23
Carn Gorm, Glen Lyon, NN64,65
Carnedd Llewelyn, SH66
Cherry Hinton, TL45
Churn Milk Hole, Pen-y-Ghent, SD87
Clogwyn du'r Arddu, SH55
Clova, NO27,37
Cockerham Moss, SD44
Coire an Lochain, NH90
Coire an t-Sneachda, NH90
Coire Cheap, NN47
Coire-cheathaich, NN33
Common Hill, Fownhope, SO53
Cork, W67
Craig Leek, NO19
Craig y Dulyn, SH66
Craven Pennines, SD
Creag an Lochain, NN54
Creag Bhalg, Glen Quoich, NO09
Creag Meagaidh, NN48
Creag Mhor, NN74
Creigiau Eglwyseg, SJ24
Crinan Canal, NR79,88,89
Croagh Patrick, L98
Cwm Rheidol, SN77,78

D
Deiniolen, SH56
Delnabo, NJ11
Den of Airlie, NO25
Don Valley, SE,SK
Dubh Loch, NO28

E
Edinburgh, NT27
Edston, NT23
Elba, NT76

F
Fetlar, HU58,59,68,69
Foel Fawr, SN71
Foel Grach, SH66
Foula, HT93,94

G
Galtee Mountains, R81,82,92

Glas Maol, NO17
Gleann Beag, NO17
Glen Affric, NH02,12,22
Glen Callater, NO18,28
Glen Clova, NO27,36,37
Glen Coe, NN15
Glen Doll, NO27
Glen Duror, NN05
Glen Etive, NN14,15,25
Glen Fee (=Glen Phee), NO27
Glen Feshie, NN89
Glen Nevis, NN16,17
Gower Peninsula, SS48,49,58,59,68

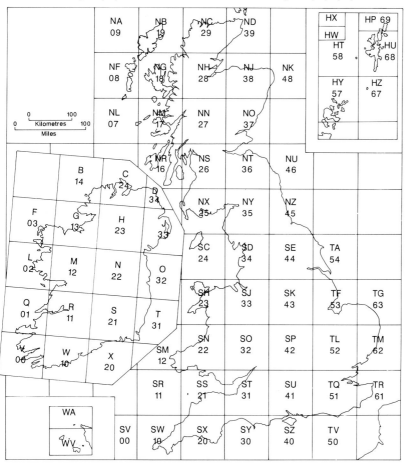

British and Irish National Grids

The map shows the numerical equivalents of the 100 km-square alphabetical codes used in the list of localities cited in the text. The Channel Islands are plotted on the UTM Grid.

INDEX TO SPECIES IN VOLUME 2